*Environmental Science and Engineering*
*Subseries: Environmental Science*

---

Series Editors: R. Allan • U. Förstner • W. Salomons

Michel De Lara · Luc Doyen

# Sustainable Management of Natural Resources

## Mathematical Models and Methods

Michel De Lara
Universite Paris-Est, CERMICS
6-8 avenue Blaise Pascal
77455 Marne la Vallee Cedex 2
France
delara@cermics.enpc.fr

Luc Doyen
Centre National de la Recherche Scientifique
CERSP, Muséum National d'Histoire Naturelle
55 rue Buffon
France 75005 Paris
lucdoyen@mnhn.fr

ISBN 978-3-540-79073-0 ISBN 978-3-540-79074-7 (eBook)

Environmental Science and Engineering ISSN: 1863-5520

Library of Congress Control Number: 2008928724



*Cover Design:* deblik, Berlin

9 8 7 6 5 4 3 2 1

springer.com

# Preface

Nowadays, environmental issues including air and water pollution, climate change, overexploitation of marine ecosystems, exhaustion of fossil resources, conservation of biodiversity are receiving major attention from the public, stakeholders and scholars from the local to the planetary scales. It is now clearly recognized that human activities yield major ecological and environmental stresses with irreversible loss of species, destruction of habitat or climate catastrophes as the most dramatic examples of their effects. In fact, these anthropogenic activities impact not only the states and dynamics of natural resources and ecosystems but also alter human health, well-being, welfare and economic wealth since these resources are support features for human life. The numerous outputs furnished by nature include direct goods such as food, drugs, energy along with indirect services such as the carbon cycle, the water cycle and pollination, to cite but a few. Hence, the various ecological changes our world is undergoing draw into question our ability to sustain economic production, wealth and the evolution of technology by taking natural systems into account.

The concept of "sustainable development" covers such concerns, although no universal consensus exists about this notion. Sustainable development emphasizes the need to organize and control the dynamics and the complex interactions between man, production activities, and natural resources in order to promote their coexistence and their common evolution. It points out the importance of studying the interfaces between society and nature, and especially the coupling between economics and ecology. It induces interdisciplinary scientific research for the assessment, the conservation and the management of natural resources.

This monograph, *Sustainable Management of Natural Resources, Mathematical Models and Methods*, exhibits and develops quantitative and formal links between issues in sustainable development, decisions and precautionary problems in the management of natural resources. The mathematical and numerical models and methods rely on dynamical systems and on control theory.

The basic concerns taken into account include management of fisheries, agriculture, biodiversity, exhaustible resources and pollution.

This book aims at reconciling economic and ecological dimensions through a common modeling framework to cope with environmental management problems from a perspective of sustainability. Particular attention is paid to multi-criteria issues and intergenerational equity.

Regarding the interdisciplinary goals, the models and methods that we present are restricted to the framework of discrete time dynamics in order to simplify the mathematical content. This approach allows for a direct entry into ecology through life-cycles, age classes and meta-population models. In economics, such a discrete time dynamic approach favors a straightforward account of the framework of decision-making under uncertainty. In the same vein, particular attention has been given to exhibiting numerous examples, together with many figures and associated computer programs (written in Scilab, a free scientific software). The main approaches presented in the book are equilibrium and stability, viability and invariance, intertemporal optimality ranging from discounted utilitarian to Rawlsian criteria. For these methods, both deterministic, stochastic and robust frameworks are examined. The case of imperfect information is also introduced at the end. The book mixes well known material and applications, with new insights, especially from viability and robust analysis.

This book targets researchers, university lecturers and students in ecology, economics and mathematics interested in interdisciplinary modeling related to sustainable development and management of natural resources. It is drawn from teachings given during several interdisciplinary French training sessions dealing with environmental economics, ecology, conservation biology and engineering. It is also the product of numerous scientific contacts made possible by the support of French scientific programs: GDR COREV (Groupement de recherche contrôle des ressources vivantes), ACI Ecologie quantitative, IFB-GICC (Institut français de la biodiversité - Gestion et impacts changement climatique), ACI MEDD (Modélisation économique du développement durable), ANR Biodiversité (Agence nationale de la recherche).

We are grateful to our institutions CNRS (Centre national de la recherche scientifique) and ENPC (École nationale des ponts et chaussées) for providing us with shelter, financial support and an intellectual environment, thus displaying the conditions for the development of our scientific work within the framework of extensive scientific freedom. Such freedom has allowed us to explore some unusual or unused roads.

The contribution of C. Lobry in the development of the French network COREV (Outils et modèles de l'automatique dans l'étude de la dynamique des écosystèmes et du contrôle des ressources renouvelables) comprising biologists and mathematicians is important. We take this opportunity to thank him and express our gratitude for so many interesting scientific discussions. At INRIA (Institut national de recherche en informatique et automatique) in Sophia-Antipolis, J.-L. Gouzé and his collaborators have been active in developing research and continue to influence our ideas on the articulation of ecology, mathematics and the framework of dynamic systems and control theory. At the Université Paris-Dauphine, we are much indebted to the very active team of mathematicians headed by J.-P. Aubin, who participated in the CEREMADE (Centre De Recherche en Mathématiques de la Décision) and CRVJC (Centre de Recherche Viabilité-Jeux-Contrôle) who significantly influenced our work on control problems and mathematical modeling and decision-making methods: D. Gabay deserves special acknowledgment regarding natural resource issues. At École nationale supérieure des mines de Paris, we are quite indebted to the team of mathematicians and automaticians at CAS (Centre automatique et systèmes) who developed a very creative environment for exploring mathematical methods devoted to real life control problems. We are particularly grateful to the influence of J. Lévine, and his legitimate preoccupation with developing methods adapted and pertinent to given applied problems. At ENPC, CERMICS (Centre d'enseignement et de recherche en mathématiques et calcul scientifique) hosts the SOWG team (Systems and Optimisation Working Group), granting freedom to explore applied paths in the mathematics of sustainable management. Our friend and colleague J.-P. Chancelier deserves a special mention for his readiness in helping us write Scilab codes and develop practical works available over the internet. The CMM (Centro de Modelamiento Matemático) in Santiago de Chile has efficiently supported the development of an activity in mathematical methods for the management of natural resources. It is a pleasure to thank our colleagues there for the pleasant conditions of work, as well as new colleagues in Peru now contributing to such development. A nice discussion with J. D. Murray was influential in devoting substantial content to uncertainty issues.

At CIRED (Centre international de recherche sur l'environnement et le développement), we are grateful to O. Godard and J.-C. Hourcade for all we learnt and understood through our contact with them regarding environmental economics and the importance of action timing and uncertainties. Our colleagues J.-C. Pereau, G. Rotillon and K. Schubert deserve special thanks for all the sound advice and challenging discussions concerning environmental economics and bio-economics to which this book owes so much.

Regarding biodiversity management, the stimulating interest and support shown for our work and modeling activities by J. Weber at IFB (Institut français de la biodiversité) has constituted a major motivation. For the modeling in fisheries management and marine biodiversity, it is a pleasure to thank F. Blanchard, M.-J. Rochet and O. Thébaud at IFREMER (Institut français de recherche pour l'exploitation de la mer) for their active investment in importing control methods in the field. We also thank J. Ferraris at IRD (Institut de recherche pour le développement). The cooperation with S. Planes (CNRS and École pratique des hautes études) has always been fruitful and pleasant. The contributions of C. Béné (World Fish Center) are major and scattered throughout several parts of this monograph.

At INRA (Institut national de recherche en agriculture), a very special thanks to M. Tichit and F. Léger for fruitful collaboration despite the complexity of agro-environmental topics. A. Rapaport deserves special mention for his long investment in control methods in the field of renewable resources management. At MNHN (Muséum national d'histoire naturelle), and especially within the Department Écologie et gestion de la biodiversité , we want to point out the support of R. Barbault and D. Couvet. Their interest in dynamic control and co-viability approaches for the management of biodiversity was very helpful. At CEMAGREF, we thank our colleague J.-P. Terreaux. At ENPC, the CEREVE (Centre d'enseignement et de recherche eau ville environnement) has been a laboratory for confronting environmental problems and mathematical methods with various researchers. Those at the Ministère de l'Équipement and at the Ministère de l'Environnement, who have allowed, encouraged and helped the development of interdisciplinary activities are too numerous to be thanked individually.

The very active and fruitful role played by young PhD and postdoc researchers such as P. Ambrosi, P. Dumas, L. Gilotte, T. Guilbaud, J.-O. Irisson and V. Martinet should be emphasized. Without the enthusiasm and work of young Master's students like F. Barnier, M. Bosseau, J. Bourgoin, I. Bouzidi, A. Daghiri, M. C. Druesne, L. Dun, C. Guerbois, C. Lebreton, A. Le Van, A. Maure, T. Mahé, P. Rabbat, M. Sbai, M.-E. Sebaoun, R. Sabatier, L. Ton That, J. Trigalo, this monograph would not have been the same. We thank them for helping us explore new tracks and developing Scilab codes.

Paris,
April 2008

*Michel De Lara*
*Luc Doyen*

# Contents

# 1

# Introduction

Over the past few decades, environmental concerns have received growing attention. Nowadays, climate change, pollution control, over-exploitation of fisheries, preservation of biodiversity and water resource management constitute important public preoccupations at the local, state and even world scales. Crises, degradation and risks affecting human health or the environment, along with the permanency of poverty, have fostered public suspicion of the evolution of technology and economic growth while encouraging doubts about the ability of public policies to handle such problems in time. The *sustainable development* concept and the *precautionary principle* both came on the scene in this context.

These concepts lead us to question the means of organizing and controlling the development and complex interactions between man, trade, production activities and natural resources. There is a need to study the interfaces between society and nature, and especially the coupling between economics and ecology. Interdisciplinary scientific studies and research into the assessment, conservation and management of natural resources are induced by such preoccupations.

The problems confronted in sustainable management share certain characteristic features: decisions must be taken throughout time and involve systems marked by complex dynamics and uncertainties. We propose mathematical approaches centered around dynamical systems and control theory to formalize and tackle such problems.

## Environmental management issues

We review the main environmental management issues before focusing on the notions of sustainable development and the precautionary principle.

## Exhaustible resources

One of the main initial environmental debates deals with the use and management of *exhaustible resource* such as coal and oil. In 1972, the Club of Rome published a famous report, "The Limits to Growth" [28], arguing that unlimited economic growth is impossible because of the exhaustibility of some resources. In response to this position, numerous economists [10, 19, 38, 39] have developed economic models to assess how the presence of an exhaustible resource might limit economic growth. These works have pointed out that *substitutability* features of natural resources are decisive in a production system economy. Moreover the question of *intergenerational equity* appears as a central point in such works.

## Renewable resources

*Renewable resources* are under extreme pressure worldwide despite efforts to design better regulation in terms of economic and/or control instruments and measures of stocks and catches.

The Food and Agricultural Organization [15] estimates for instance that, at present, 47-50% of marine fish stocks are fully exploited, 15-18% are over-exploited and 9-10% have been depleted or are recovering from depletion.

Without any regulation, it is likely that numerous stocks will be further depleted or become extinct as long as over-exploitation remains profitable for individual agents. To mitigate pressure on specific resources and prevent over-exploitation, renewable resources are regulated using quantity or price instruments. Some systems of management are thus based on quotas, limited entries or protected areas while others rely on taxing of catches or operations [6, 7, 20, 41]. The continued decline in stocks worldwide has raised serious questions about the effectiveness and sustainability of such policies for the management of renewable resources, and especially for marine resources. Among the many factors that contribute to failure in regulating renewable resources, both uncertainty and complexity play significant roles. Uncertainty includes both scientific uncertainties related to resource dynamics or assessments and the uncontrollability of catches. In this context, problems raised by non-compliance of agents or by by-catch related to multi-species management are important. The difficulties in the usual management of renewable resources have led some recent works to advocate the use of ecosystemic approaches [5, 8] as a central element of future resource management. This framework aims at capturing a major part of the complexity of the systems in a relevant way encompassing, in particular, trophic webs, habitats, spatialization and uncertainty.

## Biodiversity

More generally, the preservation, conservation and management of *biodiversity* is at stake. In the Convention on Biological Diversity (Rio de Janeiro, 1992), biodiversity is defined as "the variability among living organisms from all sources including, inter alia, terrestrial, marine and other aquatic ecosystems and the ecological complexes of which they are part; this includes diversity within species, between species and of ecosystems". Many questions arise. How can biodiversity be measured [2, 33]? How does biodiversity promote the functioning, stability, viability and productivity of ecosystems [24, 26]? What are the mechanisms responsible for perturbations ? How can the consequences of the erosion of biodiversity be evaluated at the level of society [4]? Extinction is a natural phenomenon that is part of the evolutionary cycle of species. However, little doubt now remains that the Earth's biodiversity is declining [26]. For instance, some estimates [27] indicate that endangered species encompass 11% of plants, 4.6% of vertebrates, 24% of mammals and 11% of birds worldwide. Anthropic activities and man's development is a major cause of resource depletion and weakened habitat. One main focus of biodiversity economics and management is to establish an economic basis for preservation by pointing out the advantages it procures. Consequently, there is growing interest in assessing the value and benefit of biological diversity. This is a difficult task because of the complexity of the systems under question and the non monetary values at stake. The concept of *total economic value* makes a distinction between use values (production and consumption), ecosystem services (carbon and water cycle, pollination...), existence value (intrinsic value of nature) and option values (potential future use).

Instruments for the recovery and protection of ecosystems, viable land use management and regulation of exploited ecosystems refer to conservation biology and bioeconomics. Population Viability Analysis [29] is a specific quantitative method used for conservation purposes. Within this context, protected areas or agro-environmental measures and actions are receiving growing attention to enhance biodiversity and the habitats which support it.

## Pollution

*Pollution* problems concerning water, air, land or food occur at different scales depending on whether we are looking at local or larger areas. At the global scale, *climate change* has now emerged as one, if not the most, important issue facing the international community. Over the past decade, many efforts have been directed toward evaluating policies to control the atmospheric accumulation of greenhouse gases (GHG). Particular attention has been paid to stabilizing GHG concentration [23], especially carbon dioxide ($CO_2$). However, intense debate and extensive analyses still refer to both the timing and magnitude of emission mitigation decisions and policies along with the choice between transferable permits (to emit GHG) or taxes as being relevant economic instruments for achieving such mitigation goals while maintaining economic growth. These discussions emphasize the need to take into account scientific, economic and technological uncertainties.

## Sustainable development

Since 1987, the term *sustainable development*, defined in the so-called Brundtland report *Our Common Future* [40], has been used to articulate all previous concerns. The World Commission on Environment and Development thus called for a "form of sustainable development which meets the needs of the present without compromising the ability of future generations to meet their own needs".

Many definitions of sustainable development have been introduced, as listed by [32]. Their numbers reveal the large-scale mobilization of scientific and intellectual communities around this question and the economic and political interests at stake. Although the Brundtland report has received extensive agreement – and many projects, conferences and public decisions such as the Convention on Biological Diversity (Rio de Janeiro, 1992), the United Nations Framework Convention on Climate Change (Rio de Janeiro, 1992) and the Kyoto protocol (Kyoto, 1997), the World Summit on Sustainable Development (Johannesburg 2002), nowadays refer to this general framework – the meaning of sustainability remains controversial. It is taken to mean alternatively preservation, conservation or "sustainable use" of natural resources. Such a concept questions whether humans are "a part of" or "apart from" nature. From the biological and ecological viewpoint, sustainability is generally associated with a protection perspective. In economics, it is advanced by those who favor accounting for natural resources. In particular, it examines how economic instruments like markets, taxes or quotas are appropriate to tackling so called "environmental externalities." The debate currently focuses on the substitutability between the economy and the environment or between "natural capital" and "manufactured capital" – a debate captured in terms of "weak" versus "strong" sustainability. Beyond their opposite assumptions, these different points of view refer to the apparent antagonism between preoccupations of most natural scientists – concerned with survival and viability questions – and preoccupations of economists – more motivated with efficiency and optimality. At any rate, the basic concerns of sustainability are how to reconcile environmental, social and economic requirements within the perspectivies of intra- and intergenerational equity.

## Precautionary principle

Dangers, crises, degradation and catastrophes affecting the environment or human health encourage doubt as to the ability of public policies to face such problems in time. The *precautionary principle* first appeared in such a context. For instance, the 15th Principle of the 1992 Rio Declaration on Environment and Development defines precaution by saying, "Where there are threats of serious or irreversible damage, lack of full scientific certainty shall not be used as a reason for postponing cost-effective measures to prevent environmental degradation".

Yet there is no universal precautionary principle and Sandin [34] enumerates nineteen different definitions. Graham [17] attempts to summarize the ideas and associates the principle with a "better safe than sorry" stance. He argues that the principle calls for prompt protective action rather than delay of prevention until scientific uncertainty is resolved.

Unfortunately, the precautionary principle does not clearly specify what changes one can expect in the relations between science and decision-making, or how to translate the requirements of precaution into operating standards. It is therefore vague and difficult to craft into workable policies.

What seems to be characteristic of the precaution context is that we face both *ex ante* indecision and indeterminacy. The precautionary principle is, however, the contrary of an abstention rule. This observation raises at least two main questions. Why does indecision exist *a priori*? How can such indecision be overcome? At this stage, the impact of the resolution of uncertainties on the timing of action appears as a touchstone of precaution.

## Mathematical and numerical modeling

From this brief panorama of numerous issues related to the management of natural resources, we observe that concepts such as sustainable development and precaution – initially conceived to guide the action – are not directly operational and do not mix well in any obvious manner. In such a context, qualitative and quantitative analyzes are not easy to perform on scientific grounds. This fact may be damaging both for decision-making support and production of knowledge in the environmental field. At this stage, attempts to address these issues of sustainability and natural resource management using mathematical and numerical modeling appear relevant. Such is the purpose of the present textbook. We believe that there is room for some mathematical concepts and methods to formulate *decisions*, to aid in finding solutions to environmental problems, and to mobilize the different specialized disciplines, their data, modeling approaches and methods within an interdisciplinary and integrated perspective.

### Decision-making perspective

Actions, decisions, regulations and controls often have to rely on quantitative contexts and numerical information as divers as effectiveness, precautionary indicators and reference points, costs and benefit values, amplitudes and timing of decisions. To quote but a few: at what level should $CO_2$ concentration be stabilized in the atmosphere? 450 ppm? 550 ppm? 650 ppm? What should the level of a carbon tax be? At what date should the $CO_2$ abatements start? And according to what schedule? What indicators and prices should be used for biodiversity? What viability thresholds should be considered for bird population sustainability? What harvesting quota levels for cod, hake and salmon? What size reserves will assure the conservation of elephant species in Africa and where should they be located? What land-use and degree of intensification is appropriate for agro-environmental policies in Europe? How high should compensation payments be for the biodiversity impact and damage caused by development projects? In meeting such objectives of decision-making support, two modeling orientations may be followed.

One class of models aims at capturing the large-scale complexity of the problems under concern. Such an approach may be very demanding and time consuming because such a model depends on a lot of parameters or mechanisms that may be uncertain or unknown. In this case, numerical simulations are generally the best way to display quantitative or qualitative results. They are very dependent upon the calibration and estimation of parameters and sensitivity analysis is necessary to convey robust assertions.

Another path for modeling to follow consists in constructing a low-dimensional model representing the major features and processes of the complex problem. One may speak of compact, aggregated, stylized or global models. Their mathematical study may be partly performed, which allows for very general results and a better understanding of the mechanisms under concern. It can also serve directly in decision-making by providing relevant indicators, reference points and strategies. Moreover, on this basis, an initial, simple numerical code can be developed. Using this small model and code to elaborate a more complex code with numerical simulations is certainly the second step. The results of the compact models should guide the analysis of more extended models in order to avoid sinking into a quagmire of complexity created by the numerous parameters of the model.

### Interdisciplinary perspective

Many researchers in ecology, biology, economics and environment use mathematical models to study, solve and analyze their scientific problems. These models are more or less sophisticated and complex. Integrated models are, however, required for the management of natural resources. Unfortunately, the models of each scientific area do not combine in a straightforward manner. For instance, difficulties may occur in defining common scales of time or space. Furthermore, the addition of several models extends the dimensions of the problem and makes it complicated or impossible to solve. Ecological, social and economic objectives may be contradictory. How may compromises be found? How can one build decision rules and indicators based on multiple observations and/or criteria? What should the coordination mechanism to implement heterogeneous agents exploiting natural resources be? We hope that this book favors and facilitates links between different scientific fields.

## Major mathematical material

The collection and analysis of data is of major interest for decision-making support and modeling in the concerned fields. Hence it mobilizes a huge part of the scientific research effort. Nevertheless, although quantitative information, values and data are needed and indispensable, we want to insist on the importance of mobilizing concepts and methods to formalize decisional problems.

On the basis of the previous considerations, we consider that the basic elements to combine sustainability, natural resource management and precautionary principles in some formal way are: *temporal and dynamic considerations*, *decision criteria and constraints* and *uncertainty management*. More specifically, we present *equilibrium*, *intertemporal optimality* and *viability* as concepts which may shed interesting light on sustainable decision requirements.

### Temporal and dynamic considerations

First of all, it is clear that the problems of sustainable management are intrinsically *dynamical*. Indeed, delays, accumulation effects and intertemporal externalities are important points to deal with. These dynamics are generally nonlinear (the logistic dynamics in biological modeling being a first step from linear to nonlinear growth models). By linking precaution with effects of irreversibility and flexibility, many works clearly point out the dynamical features involved in these problems. The sustainability perspective combined with intergenerational equity thus highlights the role played by the time horizon, that is to say the *temporal dimension* of the problem.

### Decisions, constraints & criteria

Secondly, by referring to regulation and prevention, the sustainability and precautionary approaches are clearly *decisional or control* problems where the timing of action is of utmost importance.

Another important feature of sustainability and precautionary actions relies on safety, viability, admissibility and feasibility along the time line in opposition to dangers, damage, crises or irreversibility. At this stage, the different modeling approaches dealing with such issues can be classified into equilibrium, cost-benefit, cost-effectiveness, invariance and effectiveness formulations.

The basic idea encompassed in the *equilibrium* approach, as in the *maximum sustainable yield* for fisheries of Gordon and Schaefer [16, 35], is to remain at a safe or satisfying state. A relevant situation is thus steady state, although *stability* allows for some dynamical processes around the equilibria.

Cost-benefit and cost-effectiveness approaches are related to *intertemporal optimal control* [6, 9] and optimal control under constraints, respectively.

In the cost-benefit case, the danger might be taken into account through a so-called monetary damage function that penalizes the intertemporal decision criteria. In contrast, the cost-effectiveness approach aims at minimizing intertemporal costs while achieving to maintain damages under safety bounds. In the optimal control framework, more exotic approaches regarding sustainability include Maximin and Chichilnisky criteria [21]. Maximin is of interest for intergenerational equity issues while Chichilnisky framework offers insights about the trade-off between future and present preferences.

The *safe minimum standards* (SMS) [31], *tolerable window approach* (TWA) [36], *population viability analysis* (PVA) [29], *viability* and invariance approaches [3, 13, 25, 30, 11, 12] indicate that tolerable margins should be maintained or reached. State constraints or targets are thus a basic issue. The so-called irreversibility constraints in the referenced works and their influence also emphasize the role played by constraints in these problems, although, in this context, irreversibility generally means decision and control constraints.

### Uncertainty management

Thirdly, the issue of *uncertainty* is also fundamental in environmental management problems [1, 22, 14]. We shall focus on two kinds of uncertainty.

On the one hand, there is *risk*, which is an event with *known probability.* To deal with risk uncertainty, policy makers have created a process called *risk assessment* which can be useful when the probability of an outcome is known from experience and statistics. In the framework of dynamic decision-making under uncertainty, the usual approach is based on the expected value of utility or cost-benefits while the general method is termed *stochastic control.*

On the other hand, there are cases presenting *ambiguity* or *uncertainty* with unknown probability or with no probability at all. Most precaution and environmental problems involve ambiguity in the sense of controversies, beliefs and irreducible scientific uncertainties. In this sense, by dealing with ambiguity, multi-prior models may appear relevant alternatives for the precaution issue. Similarly, pessimistic, worst-case, total risk-averse or guaranteed and robust control frameworks may also shed interesting light. As a first step in such directions, the present textbook proposes to introduce ambiguity through the use of "total" uncertainty and *robust control.*

## Content of the textbook

In this textbook, we advocate that concepts and methods from control theory of dynamical systems may contribute to clarifying, analyzing and providing mathematical and/or numerical tools for theoretical and applied environmental decision-making problems. Such a framework makes it possible to cover the important issues mentioned above. First, it clearly accounts for dynamical mechanisms. Second, the simple fact of exhibiting and distinguishing between states, controls, uncertainties and observations among all variables of a system is already a structuring option in the elicitation of many models. Another major interest of control theory is to focus on decision, planning and management issues. Furthermore, the different fundamental methods of control theory – that include stability, invariance and optimality – encompass the main elements of normative approaches for natural resource management, precaution and sustainability.

Regarding the interdisciplinary goal, the models and methods that we present are restricted to the framework of discrete time dynamics, in order to simplify the mathematical content. By using this approach, we avoid the introduction of too many sophisticated mathematics and notations. This should favor an easy and faster understanding of the main ideas, results and techniques. It should enable direct entry into ecology through life-cycle, age classes and meta-population models. In economics, such a discrete time dynamics approach favors a straightforward account of the framework of decision under uncertainty. In the same vein, particular attention has been given to exhibiting numerous examples, together with many figures and associated computer programs (written in Scilab, a free scientific software). Many practical works presenting management cases with Scilab computer programs can be found on the internet at the address `http://cermics.enpc.fr/~delara/BookSustain`. They may help the comprehension and serve for teaching.

We must confess that most of our examples are rather compact, global, aggregated models with few dimensions, hence taking distance with complexity in the first place. This is not because we do not aim at tackling such complex issues but our approach is rather to start up with clear models and methods before climbing higher mountains. This option helps both to "grasp" the situation from a control-theoretical point of view and also to make easier both mathematical and numerical resolution. For more complex models, we only pave the way for their study by providing examples of Scilab code in this perspective.

The emphasis in this book is not on building dynamical models, but on the formalization of decisional issues. For this reason, we shall rely on existing models without commenting them. We are aware of ongoing debate as to the validity and the empirical value of commonly used models. We send the reader to [42, 18] for useful warnings and to [37] for a mathematical point of view.

Moreover, we are aware that a lot of frustration may appear when reading this book because many important topics are not handled in depth. For instance, the integration of coordination mechanism, multi-agents and game theory is an important issue for environmental decisions and planning which is not directly developed here. These concerns represent challenging perspectives. Similarly, the use of data, estimation, calibration and identification processes constitute another important lack. Still, we had to set limits to our work. Approaches presented in the book are equilibrium and stability, viability and invariance, intertemporal optimality (going from discounted utilitarian to Rawlsian criteria). For these methods, both deterministic, stochastic and robust frameworks are exposed. The case of imperfect information is also introduced at the end. The book mixes well known material and applications with new insights, especially from viability, robust and precaution analysis.

The textbook is organized as follows. In Chap. 2, we first present some generic examples of environment and resource management detailed all along the text, then give the general form of control models under study. Chapter 3 examines the issues of equilibrium and stability. In Chap. 4, the problem of state constraints is particularly studied via viability and invariance tools, introducing the dynamic programming method. Chapter 5 is devoted to the optimal control question, still treated by dynamic programming but also by the so-called maximum principle. In Chap. 6, we introduce the natural extension of controlled dynamics to the uncertain setting, and we present different decision-making approaches including both robust and stochastic criteria. The stochastic and robust dynamic programming methods are presented for viability purposes in Chap. 7 and for optimization in Chap. 8. Chapter 9 is devoted to the case where information about the state system is partial. Proofs are relegated to Appendix A. All the numerical material may be found in the form of SCILAB codes on the internet site `http://cermics.enpc.fr/~delara/BookSustain`.

# References


[1] K. J. Arrow and A. C. Fisher. Environmental preservation, uncertainty, and irreversibity. *Quarterly Journal of Economics*, 88:312–319, 1974.

[2] R. Barbault. *Biodiversité*. Hachette, Paris, 1997.

[3] C. Béné, L. Doyen, and D. Gabay. A viability analysis for a bio-economic model. *Ecological Economics*, 36:385–396, 2001.

[4] F. S. Chapin, E. Zavaleta, and V. T. Eviner. Consequences of changing biodiversity. *Nature*, 405:234–242, 2000.

[5] V. Christensen and D. Pauly. ECOPATH II–a software for balancing steady-state models and calculating network characteristics. *Ecological Modelling*, 61:169–185, 1992.

[6] C. W. Clark. *Mathematical Bioeconomics*. Wiley, New York, second edition, 1990.

[7] J. M. Conrad. *Resource Economics*. Cambridge University Press, 1999.

[8] N. Daan, V. Christensen, and P. M. Cury. Quantitative ecosystem indicators for fisheries management. *ICES Journal of Marine Science*, 62:307–614, 2005.

[9] P. Dasgupta. *The Control of Ressources*. Basil Blackwell, Oxford, 1982.

[10] P. Dasgupta and G. Heal. The optimal depletion of exhaustible resources. *Review of Economic Studies*, 41:1–28, 1974. Symposium on the Economics of Exhaustible Resources.

[11] M. De Lara, L. Doyen, T. Guilbaud, and M.-J. Rochet. Is a management framework based on spawning-stock biomass indicators sustainable? A viability approach. *ICES J. Mar. Sci.*, 64(4):761–767, 2007.

[12] L. Doyen, M. De Lara, J. Ferraris, and D. Pelletier. Sustainability of exploited marine ecosystems through protected areas: a viability model and a coral reef case study. *Ecological Modelling*, 208(2-4):353–366, November 2007.

[13] K. Eisenack, J. Sheffran, and J. Kropp. The viability analysis of management frameworks for fisheries. *Environmental Modeling and Assessment*, 11(1):69–79, February 2006.

[14] L. G. Epstein. Decision making and temporal resolution of uncertainty. *International Economic Review*, 21:269–283, 1980.

[15] FAO. *The state of world fisheries and aquaculture.* Rome, 2000. Available on line http://www.fao.org.

[16] H. S. Gordon. The economic theory of a common property resource: the fishery. *Journal of Political Economy*, 62:124–142, 1954.

[17] J. D. Graham. Decision-analytic refinements of the precautionary principle. *Journal of Risk Research*, 4(2):127–141, 2001.

[18] C. Hall. An assessment of several of the historically most influential theoretical models used in ecology and of the data provided in their support. *Ecological Modelling*, 43(1-2):5–31, 1988.

[19] J. Hartwick. Intergenerational equity and the investing of rents from exhaustible resources. *American Economic Review*, 67:972–974, 1977.

[20] J. M. Hartwick and N. D. Olewiler. *The Economics of Natural Resource Use.* Harper and Row, New York, second edition, 1998.

[21] G. Heal. *Valuing the Future, Economic Theory and Sustainability.* Columbia University Press, New York, 1998.

[22] C. Henry. Investment decisions under uncertainty: The "irreversibility effect". *American Economic Review*, 64(6):1006–1012, 1974.

[23] IPCC. http://www.ipcc.ch/.

[24] M. Loreau, S. Naeem, and P. Inchausti. *Biodiversity and ecosystem functioning: synthesis and perspectives.* Oxford University Press, Oxford, United Kingdom, 2002.

[25] V. Martinet and L. Doyen. Sustainable management of an exhaustible resource: a viable control approach. *Resource and Energy Economics*, 29(1):p.17–39, 2007.

[26] K. S. McCann. The diversity - stability debate. *Nature*, 405:228–233, 2000.

[27] MEA. *Millennium Ecosystem Assessment.* 2005. Available on http://www.maweb.org/en/index.aspx.

[28] D. L. Meadows, J. Randers, W. Behrens, and D. H. Meadows. *The Limits to Growth.* Universe Book, New York, 1972.

[29] W. F. Morris and D. F. Doak. *Quantitative Conservation Biology: Theory and Practice of Population Viability Analysis.* Sinauer Associates, 2003.

[30] C. Mullon, P. Cury, and L. Shannon. Viability model of trophic interactions in marine ecosystems. *Natural Resource Modeling*, 17:27–58, 2004.

[31] L. J. Olson and R. Santanu. Dynamic efficiency of conservation of renewable resources under uncertainty. *Journal of Economic Theory*, 95:186–214, 2000.

[32] D. Pezzey. Economic analysis of sustainable growth and sustainable development. Technical report, Environment Department WP 15, World Bank, Washington DC, 1992.

[33] A. Purvis and A. Hector. Getting the measure of biodiversity. *Nature*, 405:212–219, 2000.

[34] P. Sandin. Dimensions of the precautionary principle. *Human and ecological risk assessment*, 5:889–907, 1999.

[35] M. B. Schaefer. Some aspects of the dynamics of populations important to the management of commercial marine fisheries. *Bulletin of the Inter-American tropical tuna commission*, 1:25–56, 1954.

[36] H. J. Schellnhuber and V. Wenzel. *Earth System Analysis, Integrating Science for Sustainability.* Springer, 1988.

[37] S. Smale. On the differential equations of species in competition. *Journal of Mathematical Biology*, 3(1):5–7, 1976.

[38] R. M. Solow. Intergenerational equity and exhaustible resources. *Review of Economic Studies*, 41:29–45, 1974. Symposium on the Economics of Exhaustible Resources.

[39] J. Stiglitz. Growth with exhaustible natural resources: Efficient and optimal growth paths. *Review of Economic Studies*, 41:123–137, 1974. Symposium on the Economics of Exhaustible Resources.

[40] WCED. *Our common Future.* Oxford University Press, 1987.

[41] J. E. Wilem. Renewable resource economists and policy: What difference have we made. *Journal of Environmental Economics and Management*, 39:306–327, 2000.

[42] P. Yodzis. Predator-prey theory and management of multispecies fisheries. *Ecological Applications*, 4(1):51–58, February 1994.

# 2

# Sequential decision models

Although the management of exhaustible and renewable resources and pollution control are issues of a different nature, their main structures are quite similar. They turn out to be decision-making problems where time plays a central role. *Control theory* of dynamic systems is well suited to tackling such situations and to building up mathematical models with analytic, algorithmic and/or numerical methods. First, such an approach clearly accounts for evolution and dynamical mechanisms. Second, it directly copes with decision-making, planning and management issues. Furthermore, control theory proposes different methods to rank and select the decisions or controls among which stability, viability or optimality appear relevant for environmental and sustainability purposes. Some major contributions in this vein are [3, 8, 9, 10, 11, 20]. As explained in the introduction, this monograph restricts all the models and methods to discrete time dynamics. In this manner, we avoid the introduction of too many sophisticated mathematics and notations. From the mathematical point of view, the specific framework of *discrete time* dynamics is not often treated by itself, contrarily to the continuous time case. Among rare references, let us mention [1]. In the framework of control theory, models then correspond to sequential decision-making problems. A sequential decision model captures a situation in which decisions are to be made at discrete stages, such as days or years. In this context, three main ingredients are generally combined: state dynamics, acceptability constraints and optimality criterion.

*State, control, dynamics.*

Each decision may influence a so-called *state* of the system: such a mechanism mainly refers to the *dynamics* or transitions, including population dynamics, capital accumulation dynamics and the carbon cycle, to quote but a few.

*Constraints*

At each stage, there may be admissibility, viability, desirability or effectiveness conditions to satisfy, corresponding to the *constraints* of the system. Such constraints may refer to non extinction conditions for populations, pollution standards, desirable consumption levels, guaranteed catches, minimal ecosystem services or basic needs. Such acceptability issues will be examined in detail in Chaps. 4 and 7.

*Criterion optimization*

An *intertemporal criterion* or performance may be *optimized* to choose among the feasible solutions. Net present value of cost-benefit or rent, discounted utility of consumption, fitness or welfare constitute the usual examples. However, "maximin" assessments stand for more exotic criteria which are also of interest for sustainability and equity purposes as will be explained in Chap. 5 and Chap. 8.

The present chapter is organized as follows. The first sections are devoted to examples and models inspired by resource and environmental management in the deterministic case, *i.e.* without uncertainty. They include models for exhaustible resources, renewable resources, biodiversity and pollution mitigation. We start with very stylized and aggregated models. More complex models are then exposed. A second part, Sect. 2.9, introduces the general mathematical framework for sequential decisions in the certain case. Some remarks, about decision strategies in Sect. 2.10 and about the curse of dimensionality in Sect. 2.11, end the chapter.

## 2.1 Exploitation of an exhaustible resource

We present a basic economic model for the evaluation and management of an exhaustible natural resource (coal, oil...). The modeling on this topic is often derived from the classic "cake eating" economy first studied by Hotelling in [21]. The usual model [21] is in continuous time with an infinite horizon but here we adapt a discrete time version with a finite horizon.

Consider an economy where the only commodity is an exhaustible natural resource. Time $t$ is an integer varying from initial time $t = t_0$ to horizon $T$ ($T < +\infty$ or $T = +\infty$). The dynamics of the resource is simply written

$$S(t+1) = S(t) - h(t) \ , \quad t = t_0, t_0 + 1, \ldots, T - 1 \tag{2.1}$$

where $S(t)$ is the stock of resource at the beginning of period $[t, t+1[$ and $h(t)$ the extraction during $[t, t+1[$, related to consumption in the economy. When $T < +\infty$, the sequence of extractions $h(t_0), h(t_0+1), \ldots, h(T-1)$ produces the sequence of stocks $S(t_0), S(t_0+1), \ldots, S(T-1), S(T)$. When the range of time $t$ is not specified, it should be understood that it runs from $t_0$ to $T-1$, or from $t_0$ to $T$, accordingly.

It is first assumed that the extraction decision $h(t)$ is irreversible in the sense that at every time $t$

$$0 \leq h(t) \ . \tag{2.2}$$

Physical constraints imply that

$$h(t) \leq S(t) \ , \tag{2.3}$$

and that

$$0 \leq S(t) \ . \tag{2.4}$$

More generally, we could consider a stronger conservation constraint for the resource as follows

$$S^\flat \leq S(t) \ , \tag{2.5}$$

where $S^\flat > 0$ stands for some minimal resource standard.

An important question is related to intergenerational equity. Can we impose some guaranteed consumption (here the extraction or consumption) level $h^\flat$

$$0 < h^\flat \leq h(t) \tag{2.6}$$

along the generations $t$? This sustainability concern can be written in terms of utility in a form close to "maximin Rawls criterion" [33]. Of course, when $T = +\infty$, such a requirement cannot be fulfilled with a finite resource $S(t_0)$.

A very common optimization problem is to maximize the sum[1] of discounted utility derived from the consumption of the resource with respect to extractions $h(t_0), h(t_0+1), \ldots, h(T-1)$, *i.e.*

$$\max_{h(t_0),\ldots,h(T-1)} \sum_{t=t_0}^{T-1} \rho^t L\big(h(t)\big)$$

where $L$ is some utility function of consumption and $\rho$ stands for a (social) *discount factor*. Generally $0 \leq \rho < 1$ as $\rho = \frac{1}{1+r_f}$ is built from the interest rate or risk-free return $r_f$, but we may also consider the case $\rho = 1$ when $T < +\infty$.

## 2.2 Assessment and management of a renewable resource

In this subsection, we start from a one-dimensional aggregated biomass dynamic model, then include harvesting *à la Schaefer* and finally introduce management criteria.

[1] The sum goes from $t = t_0$ up to $T-1$ because extractions run from $t_0$ to $T-1$ while stocks go up to $T$.

## Biological model

Most bioeconomic models addressing the problem of renewable resource exploitation (forestry, agriculture, fishery) are built upon the framework of a biological model. Such a model may account for the demographic structure (age, stages or size classes, see [5]) of the exploited stock or may attempt to deal with the trophic dimension of the exploited (eco)system. However, biologists have often found it necessary to introduce various degrees of simplification to reduce the complexity of the analysis.

In many models, the stock, measured through its biomass, is considered globally as a single unit with no consideration of the structure population. Its growth is materialized through the equation

$$B(t+1) = g\big(B(t)\big) , \tag{2.7}$$

where $B(t)$ stands for the resource biomass and $g : \mathbb{R}_+ \to \mathbb{R}_+$ is taken to satisfy $g(0) = 0$. In discrete time, examples of $g$ are given by [23, 8] and illustrated by Fig. 2.1.

1. The *linear* model
$$g(B) = RB , \tag{2.8}$$
where $r = R - 1$ is the *per capita rate of growth.*
2. The *logistic* model
$$g(B) = B + rB\left(1 - \frac{B}{K}\right) , \tag{2.9}$$
where $r \geq 0$ is the *per capita rate of growth* (for small populations), and $K$ is the *carrying capacity*[2] of the habitat. We shall also use the equivalent form
$$g(B) = (1+r)B\left(1 - \frac{rB}{(1+r)K}\right) . \tag{2.10}$$
Such a logistic model in discrete time can be easily criticized since for biomass $B$ greater than the capacity $K$ the biomass becomes negative, which of course does not make sense.
3. The *Ricker* model
$$g(B) = B\exp\left(r(1 - \frac{B}{K})\right) , \tag{2.11}$$
where again $K$ represents the carrying capacity.
4. The *Beverton-Holt* model
$$g(B) = \frac{RB}{1+bB} , \tag{2.12}$$
where the carrying capacity now corresponds to $K = \frac{R-1}{b}$.

[2] The carrying capacity $K$ is the lowest $K > 0$ which satisfies $g(K) = K$.

5. The *depensation* models

$$g(B) = B + \alpha(f(B) - B)(B - B^{\flat}) , \tag{2.13}$$

where $\alpha > 0$ and $f$ is any of the previous population dynamics, satisfying $f(B) \geq (B)$ for $B \in [0, K]$, and $B^{\flat} \in ]0, K[$ stands for some minimum viable population threshold. Indeed, $g(B) < B$ whenever $B < B^{\flat}$ and some *Allee effect* occurs in the sense that small populations decline to extinction.

The choice among the different population dynamics deeply impacts the evolution of the population, as illustrated by Fig. 2.2. The Beverton-Holt dynamics generates "stable" behaviors while logistic or Ricker may induce oscillations or chaotic paths.

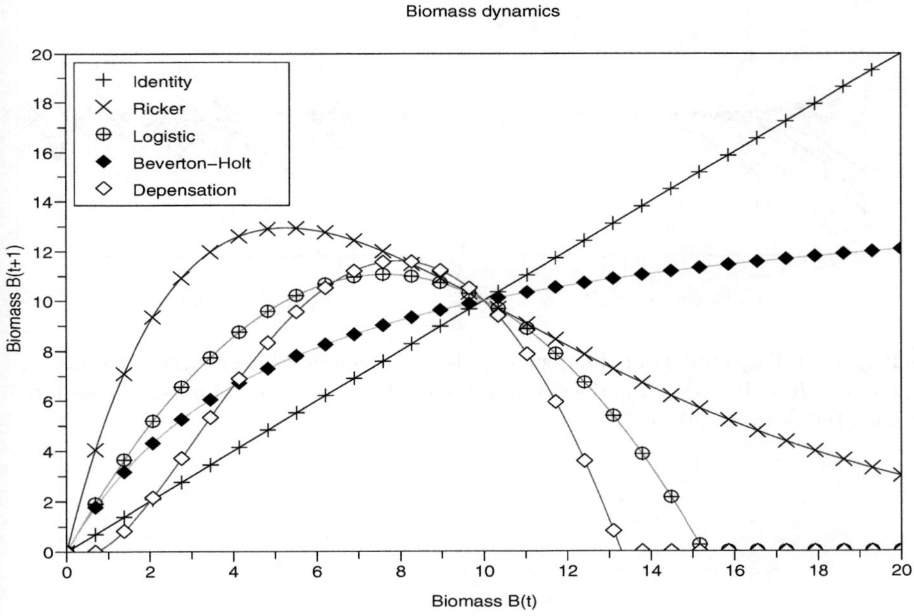


**Fig. 2.1.** Comparaison of distinct population dynamics $g$ for $r = 1.9$, $K = 10$, $B^{\flat} = 2$. Dynamics are computed with the SCILAB code 1. In $\oplus$, the logistic model; in $\times$, the Ricker dynamics; in $\diamond$, a depensation model; in $\blacklozenge$, the Beverton-Holt recruitment.

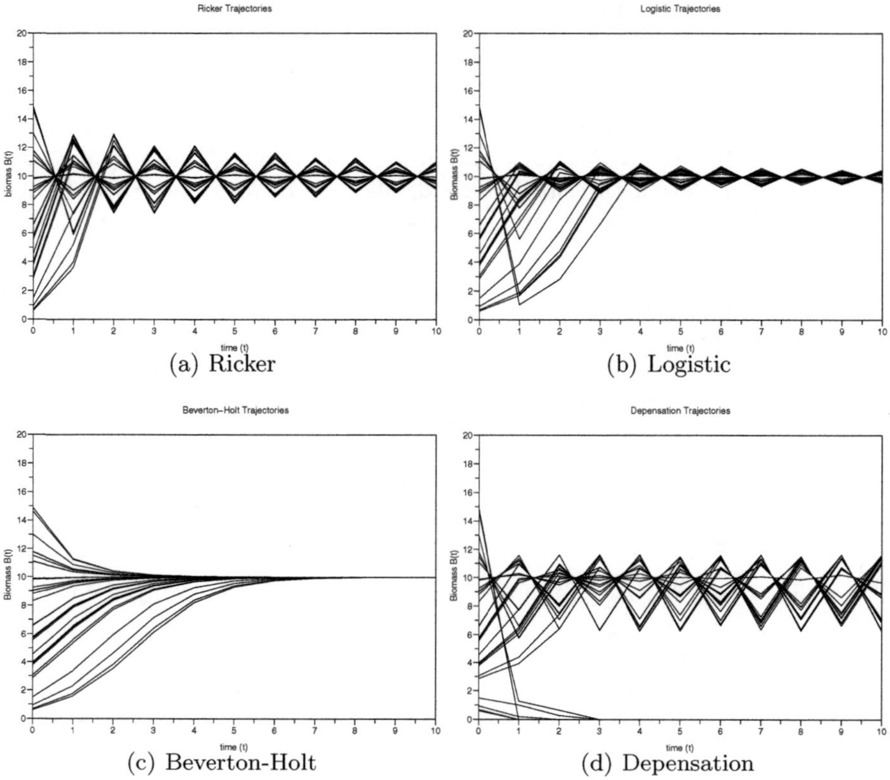


**Fig. 2.2.** Trajectories for different population dynamics with common parameters $r = 1.9$, $K = 10$, $B^\flat = 2$ and same initial conditions $B_0$. Trajectories are computed with the SCILAB code 1.

SCILAB CODE 1.

```
//
// exec dyn_pop.sce

lines(0);

r = 0.9; K= 10; R = 1+r ; MVP=K/5;
// Population dynamics parameters

k=R*K/(R-1);
function [y]=Logistic(t,B)
  y=max(0, (R*B).*(1-B/k) )
endfunction

function [y]=Ricker(t,B)
  y=B.*exp(r*(1-B/K))
  endfunction

  b=(R-1)/K ;
function [y]=Beverton(t,B)
  y=(R*B)./(1 + b*B)
endfunction

function [y]=Depensation(t,B)
//  y=max(0,B+(Beverton(t,B)-B).*(B-MVP)/MVP)
    y=max(0,B+(Beverton(t,B)-B).*(B-MVP))
  endfunction
// Dynamics

xset("window",0);xbasc(0);
B=linspace(0,2*K,1000);
plot2d(B,[B' Ricker(0,B)' Logistic(0,B)' Beverton(0,B)'...
  Depensation(0,B)']);
// drawing diamonds, crosses, etc. to identify the curves
B=linspace(0,2*K,30);
plot2d(B,[B' Ricker(0,B)' Logistic(0,B)' Beverton(0,B)'...
  Depensation(0,B)'],-[1,2,3,4,5]);
legends(["Identity";"Ricker";"Logistic";"Beverton";...
 "Depensation"],-[1,2,3,4,5],'ul');
xtitle('Biomass dynamics','Biomass B(t)','Biomass B(t+1)')
// Comparaison of the shapes of population dynamics


T=10; time=0:T;
// Time horizon

N_simu=50;
// N_simu=30;
// Number of simulations

xset("window",1:4); xbasc(1:4);
// opening windows

for i=1:N_simu
// simulation loop

B_0=rand(1)*1.5*K;
// random initial conditions

y_Ricker=ode("discrete",B_0,0,time,Ricker);
y_Logistic=ode("discrete",B_0,0,time,Logistic);
y_BH=ode("discrete",B_0,0,time,Beverton);
y_D=ode("discrete",B_0,0,time,Depensation);
// Computation of trajectories starting from B0
// along distinct population dynamics

xset("window",1);
plot2d(time,[y_Ricker'],rect=[0,0,T,2*K]);
xtitle('Ricker Trajectories','time (t)',...
'biomas B(t)')

xset("window",2);
plot2d(time,y_Logistic,rect=[0,0,T,2*K]);
 xtitle('Logistic Trajectories','time (t)',...
 'Biomass B(t)')

xset("window",3);
plot2d(time,[y_BH'],rect=[0,0,T,2*K]);
xtitle('Beverton-Holt Trajectories','time (t)',....
'Biomass B(t)')

xset("window",4);
plot2d(time,[y_D'],rect=[0,0,T,2*K]);
xtitle('Depensation Trajectories','time (t)',...
'Biomass B(t)')

end
// end simulation loop

//
```

## Harvesting

When harvesting activities are included, the model (2.7) above becomes the *Schaefer model*, originally introduced for fishing in [31],

$$B(t+1) = g\big(B(t) - h(t)\big) \ , \quad 0 \leq h(t) \leq B(t) \ , \tag{2.14}$$

where $h(t)$ is the harvesting or catch at time $t$. Notice that, in the above sequential model,

1. harvesting takes place at the beginning of the year $t$, hence the constraints $0 \leq h(t) \leq B(t)$ right above,
2. regeneration takes place at the end[3] of the year $t$.

[3] A formulation where regeneration occurs at the beginning of the year while harvesting ends would give $B(t+1) = g(B(t)) - h(t)$, with $0 \leq h(t) \leq g(B(t))$.

It is frequently assumed that the catch $h$ is proportional to both biomass and harvesting effort, namely

$$h = qeB \;, \tag{2.15}$$

where $e$ stands for the *harvesting effort* (or fishing effort, an index related for instance to the number of boats involved in the activity), and $q \geq 0$ is a *catchability coefficient.* More generally, the harvesting is related to the effort and the biomass through some relation

$$h = H(e, B) \;, \tag{2.16}$$

where the catch function $H$ is such that

- $H(0, e) = H(B, 0) = 0$
- $H$ increases in both arguments biomass $B$ and effort $e$; whenever $H$ is smooth enough, it is thus assumed that

$$\begin{cases} 0 \leq H_B(e, B) := \frac{\partial H}{\partial B}(e, B) \;, \\ 0 \leq H_e(e, B) := \frac{\partial H}{\partial B}(e, B) \;. \end{cases}$$

Ecology and economics have two distinct ways to characterize the function $H$. From the ecology point of view, such a relation $H$ relies on a functional form of predation, while from the economics viewpoint $H$ corresponds to a production function. At this stage, it is worth pointing out the case of a *Cobb-Douglas production function*

$$H(e, B) = qe^{\alpha} B^{\beta} \;, \tag{2.17}$$

where the exponents $\alpha \geq 0$ and $\beta \geq 0$ stand for the elasticities of production.

### The static Gordon-Schaefer model

A first approach consists in reasoning at equilibrium, when the a stationary exploitation induces a steady population. In this context, the well-known Schaefer model gives the so-called *sustainable yield* associated to the fishing effort by solving the implicit relation $B = g(B - h)$ giving $h$. This issue is examined in Chap. 3.

The economic model which is directly derived from the Schaefer model is the *Gordon model* [17, 8] which integrates the economic aspects of the fishing activity through the fish price $p$ and the catch costs $C(e)$ per unit of effort. The *rent*, or *profit*, is defined as the difference between benefits and cost

$$\mathcal{R}(e, B) := pH(e, B) - C(e) \;, \tag{2.18}$$

where the cost function is such that

- $C(0) = 0$;
- $C$ increases with respect to effort $e$; whenever $C$ is smooth enough, it is thus assumed that $C'(e) \geq 0$.

It is frequently assumed that the costs are linear in effort, namely:

$$C(e) = ce \quad \text{with} \quad c > 0\,.$$

Once given the cost function $C$, one can compute the effort $\overline{e}$ maximizing the rent $\mathcal{R}(e, B)$ under the constraint that $B = g(B - H(e, B))$.

Although it suffers from a large number of unrealistic assumptions, the Gordon model displays a certain degree of concordance with the empirical histories of fisheries . It is probably for this reason, along with its indisputable normative character, that it has been regularly used as the underlying framework by optimal control theory since the latter was introduced in fisheries sciences [8].

### Intertemporal profit maximization

Assuming a fixed production structure, *i.e.* stationary capital and labor, an economic model may be formulated as the intertemporal maximization of the rent with respect to the fishing effort,

$$\max_{e(t_0),\ldots,e(T-1)} \sum_{t=t_0}^{T-1} \rho^t \Big( pH\big(e(t), B(t)\big) - C\big(e(t)\big) \Big)\,,$$

where $\rho$ represents a discount factor ($0 \leq \rho \leq 1$). An important constraint is related to the limit effort $e^\sharp$ resulting from the fixed production capacity (number of boats and of fishermen):

$$0 \leq e(t) \leq e^\sharp\,.$$

Ecological viability or conservation constraint can be integrated by requiring that

$$B^\flat \leq B(t)\,,$$

where $B^\flat > 0$ is a safe minimum biomass level.

### Intertemporal utility maximization

We can also consider a social planner or a regulating agency wishing to make use, in an optimal way, of the renewable natural resource over $T$ periods. The welfare optimized by the planner is represented by the sum of updated utilities of successive harvests $h(t)$ (assumed to be related to consumption, for instance), that is

$$\max_{h(t_0),\ldots,h(T-1)} \Big( \sum_{t=t_0}^{T-1} \rho^t L\big(h(t)\big) + \rho^T L\big(B(T)\big) \Big) \tag{2.19}$$

where $\rho \in [0, 1[$ is a discount factor and $L$ is a utility function. Notice that the final term $L\big(B(T)\big)$ corresponds to an existence or inheritance value of the stock.

## 2.3 Mitigation policies for carbon dioxyde emissions

Let us consider a very stylized model of the climate-economy system. It is described by two aggregated variables, namely the atmospheric $CO_2$ *concentration level* denoted by $M(t)$ and some economic production level such as gross world product GWP denoted by $Q(t)$, measured in monetary units. The decision variable related to mitigation policy is the emission abatement rate denoted by $a(t)$. The goal of the policy makers is to minimize intertemporal discounted abatement costs while respecting a maximal sustainable $CO_2$ concentration threshold at the final time horizon: this is an example of a *cost-effectiveness problem.*

### Carbon cycle model

The description of the carbon cycle is similar to [27], namely a highly simple dynamical model

$$M(t+1) = M(t) + \alpha E_{\text{BAU}}(t)\big(1 - a(t)\big) - \delta\big(M(t) - M_{-\infty}\big) , \tag{2.20}$$

where

- $M(t)$ is the $CO_2$ atmospheric concentration, measured in ppm, parts per million (379 ppm in 2005);
- $M_{-\infty}$ is the pre-industrial atmospheric concentration (about 280 ppm);
- $E_{\text{BAU}}(t)$ is the baseline, or "business as usual" (BAU), for the $CO_2$ emissions,and is measured in GtC, Gigatonnes of carbon (about 7.2 GtC per year between 2000 and 2005);
- the abatement rate $a(t)$ corresponds to the applied reduction of $CO_2$ emissions level ($0 \leq a(t) \leq 1$);
- the parameter $\alpha$ is a conversion factor from emissions to concentration; $\alpha \approx 0.471$ ppm.GtC$^{-1}$ sums up highly complex physical mechanisms;
- the parameter $\delta$ stands for the natural rate of removal of atmospheric $CO_2$ to unspecified sinks ($\delta \approx 0.01$ year$^{-1}$).

Notice that carbon cycle dynamics can be reformulated as

$$M(t+1) - M_{-\infty} = (1-\delta)\,\big(M(t) - M_{-\infty}\big) + \alpha E_{\text{BAU}}(t)\big(1 - a(t)\big) \tag{2.21}$$

thus representing the anthropogenic perturbation of a natural system from a pre-industrial equilibrium atmospheric concentration $M_{-\infty}$. Hence, $\delta$ accounts for the inertia of a natural system, and is a most uncertain parameter[4].

---

[4] Two polar cases are worth being pointed out: when $\delta = 1$, carbon cycle inertia is nil and therefore $CO_2$ emissions induce a flow externality rather than a stock one; on the contrary, when $\delta = 0$, the stock externality reaches a maximum and $CO_2$ accumulation is irreversible.

**Emissions driven by economic production**

The baseline $E_{\mathrm{BAU}}(t)$ can be taken under the form $E_{\mathrm{BAU}}(t) = \mathfrak{E}_{\mathrm{BAU}}(Q(t))$, where the function $\mathfrak{E}_{\mathrm{BAU}}$ stands for the emissions of $CO_2$ resulting from the economic production $Q$ in a "business as usual" (BAU) scenario and accumulating in the atmosphere. The emissions depend on production $Q$ because growth is a major determinant of energy demand [24]. It can be assumed that BAU emissions increase with production $Q$, namely, when $E$ is smooth enough,

$$\frac{d\mathfrak{E}_{\mathrm{BAU}}(Q)}{dQ} > 0 \, .$$

Combined with a global economic growth assumption, a rising emissions baseline is given.

The global economics dynamic is represented by an autonomous rate of growth $g \geq 0$ for the aggregated production level $Q(t)$ related to *gross world product* GWP:

$$Q(t+1) = (1+g)Q(t) \, . \tag{2.22}$$

This dynamic means that the economy is not directly affected by abatement policies and costs. Of course, this is a restrictive assumption.

**The cost-effectiveness criteria**

A physical or environmental requirement is considered through the limitation of concentrations of $CO_2$ below a tolerable threshold $M^\sharp$ (say 450 ppm, 550 ppm, 650 ppm) at a specified date $T > 0$ (year 2050 or 2100 for instance):

$$M(T) \leq M^\sharp \, . \tag{2.23}$$

The reduction of emissions is costly. Hence, it is assumed that the abatement cost $C(a,Q)$ increases with abatement rate $a$, that is for smooth $C$:

$$\frac{\partial C(a,Q)}{\partial a} > 0 \, .$$

Furthermore, following for instance [18], we can assume that growth lowers marginal abatement costs. This means that the availability and costs of technologies for carbon switching improve with growth. Thus, if the marginal abatement cost $\frac{\partial C(a,Q)}{\partial a}$ is smooth enough, it decreases with production in the sense:

$$\frac{\partial^2 C(a,Q)}{\partial Q \partial a} < 0 \, .$$

As a result, the costs of reducing a ton of carbon decline.

The cost-effectiveness problem faced by the social planner is an optimization problem under constraints. It consists in minimizing the discounted intertemporal abatement cost $\sum_{t=t_0}^{T-1} \rho^t C\big(a(t), Q(t)\big)$ while reaching the concentration tolerable window $M(T) \leq M^\sharp$. The parameter $\rho$ stands for a discount factor. Therefore, the problem can be written as

$$\inf_{a(t_0),\dots,a(T-1)} \sum_{t=t_0}^{T-1} \rho^t C\big(a(t), Q(t)\big) , \tag{2.24}$$

under the dynamics constraints (2.20) and (2.22) and target constraint (2.23).

Some projections are displayed in Fig. 2.3 together with the ceiling target $M^\sharp = 550$ ppm. They are built from the SCILAB code 2. The "business as usual" path $a_{\text{BAU}}(t) = 0$ does not display satisfying concentrations since the ceiling target is exceeded at time $t = 2035$. The other path corresponding here to a medium stationary abatement $a(t) = 0.6$ provides a viable path.

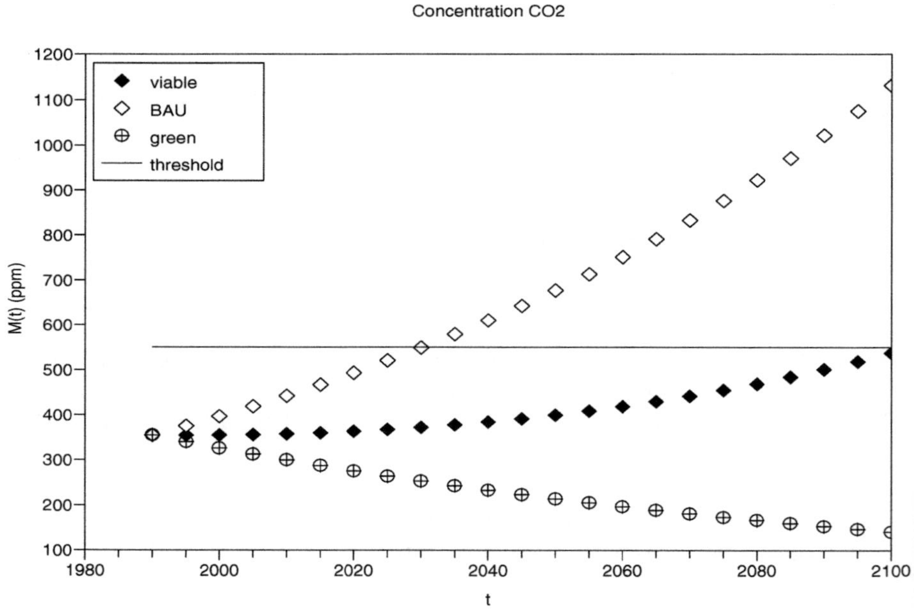


**Fig. 2.3.** Projections of $CO_2$ concentration $M(t)$ at horizon 2100 for different mitigation policies $a(t)$ together with ceiling target $M^\sharp = 550$ ppm in black. In $\diamond$, the non viable "business as usual" path $a_{\text{BAU}}(t) = 0$ and, in $\blacklozenge$, a viable medium stationary abatement $a(t) = 0.6$. The path in $\oplus$ relies on a total abatment $a(t) = 1$. Trajectories are computed with the SCILAB code 2.

SCILAB CODE 2.

```
//
clear


// PARAMETERS //

//  initial time
t_0=1990;
// Final Time
t_F=2100;
// Time step
delta_t=1;

taux_Q=0.01;
// economic growth rate
alphaa=0.64;
// marginal ratio marginal atmospheric retention
// (uncertain +- 0.15)
sigma=0.519;
absortion=1/120 ;
// concentration target (ppm)
M_sup=550;
// Initial conditions
t=t_0;
M=354; //in (ppm)
M_bau=M; M_g=M;
Q = 20.9; // in (T US$)
E = sigma * Q ;



// Distinct abatment policies
u = 1*ones(1,t_F-t_0+1);   // Strong mitigation
u = 0*ones(1,t_F-t_0+1);   // No mitigation (BAU)
u = 0.6*ones(1,t_F-t_0+1);  // medium mitigation
//u = 1*rand(1,t_F-t_0+1);  // random mitigation

// Initialisation (empty lists)
L_t=[]; L_M=[]; L_bau=[]; L_E=[];
L_Eg=[];L_Eb=[]; L_Q=[];L_g=[];
// System Dynamics
for (t=t_0:delta_t:t_F)

L_t=[L_t t];
   L_M=[L_M M];
 L_Q=[L_Q Q];
L_bau=[L_bau M_bau];
L_g=[L_g M_g];

Q=(1+taux_Q)*Q;
E = sigma * Q * (1-u(t-t_0+1));
L_E=[L_E E];
// Emissions CO2
M = M* (1-absortion) + alphaa* E;
// dynamics concentration CO2

E_bau = sigma * Q ;
L_Eb=[L_Eb E_bau];
// Emissions Business as usual (BAU)
M_bau = M_bau* (1-absortion) + alphaa* E_bau;
// dynamics BAU

E_g = 0;
L_Eg=[L_Eg E_g];
// Green: no emissions
M_g = M_g* (1-absortion) + alphaa* E_g;
// dynamics without pollution
end,

// Results printing

long=prod(size(L_t));
step=floor(long/20);
abcisse=1:step:long;
xset("window",1);xbasc(1)
plot2d(L_t(abcisse),[L_E(abcisse)' L_Eb(abcisse)' ...
    L_Eg(abcisse)'],style=-[4,5,3]) ;
legends(["viable";"BAU";"green"],-[4,5,3],'ul');
xtitle('Emissions E(t)','t','E(t) (GtC)');

xset("window",2);xbasc(2)
plot2d(L_t(abcisse),[L_M(abcisse)' L_bau(abcisse)' ...
L_g(abcisse)' ones(L_t(abcisse))'*M_sup],...
   style=-[4,5,3,-1]) ;
legends(["viable";"BAU";"green";"threshold"],...
      -[4,5,3,-1],'ul');
xtitle('Concentration CO2','t','M(t) (ppm)');

xset("window",4); xbasc(4)
plot2d(L_t(abcisse),L_Q(abcisse));
xtitle('Economie: Production Q(t)','t','Q(t) (T US$)');

//
```

## 2.4 A trophic web and sustainable use values

Consider $n$ species within a food web. An example of trophic web is given in Sect. 7.4 for a large coral reef ecosystem. To give some feelings of the numbers, 374 species were identified during a survey in the Abore reef reserve (15 000 ha) in New Caledonia, differing in mobility, taxonomy (41 families) and feeding habits. The analysis of species diets yielded 7 clusters, each cluster forming a trophic group; the model in [14] restricts them to 4 trophic groups (piscivors, macrocarnivors, herbivors and other fishes) plus coral/habitat.

Denote by $N_i(t)$ the *abundance* (number of individuals, or approximation by a continuous real) or the *density* (number of individuals per unit of surface) of species $i \in \{1, \ldots, n\}$ at the beginning of period $[t, t+1[$. The ecosystem dynamics and the interactions between the species are depicted by a Lotka-Volterra model:

$$N_i(t+1) = N_i(t)\Big(R_i + \sum_{j=1}^{n} S_{ij}N_j(t)\Big) \,. \tag{2.25}$$

- Autotrophs grow in the absence of predators (those species $i$ for which $R_i \geq 1$), while consumers die in the absence of prey (when $R_i < 1$).
- The effect of $i$ on $j$ is given by the term $S_{ij}$ so that $i$ consumes $j$ when $S_{ij} > 0$ and $i$ is the prey of $j$ if $S_{ij} < 0$. The numerical response of a consumer depends on both the number of prey captured per unit of time (functional response) and the efficiency with which captured prey are concerted into offspring. In this model, we represent prey effect $j$ on consumers $i$ by $S_{ij} = -e_{ij}S_{ji}$, where $e_{ij}$ is the conversion efficiency ($e < 1$ when the size of the consumer is larger than that of its prey).
- The strength of direct intra-specific interactions is given by $S_{ii} < 0$. Possible mechanisms behind such self-limitation include mutual interferences and competitions for non-food resources. When the index $i$ labels group of species (trophic groups for instance), it may account for intra-group interactions.

The ecosystem is also subject to human exploitation. Such an anthropogenic pressure induced by harvests and catches $h(t) = \big(h_1(t), \dots, h_n(t)\big)$ modifies the dynamics of the ecosystem as follows

$$N_i(t+1) = \big(N_i(t) - h_i(t)\big)\Big(R_i + \sum_{j=1}^{n} S_{ij}\big(N_j(t) - h_j(t)\big)\Big) \,, \tag{2.26}$$

with the constraint that the captures do not exceed the stock values

$$0 \leq h_i(t) \leq N_i(t) \,.$$

Note that many catches can be set to zero since the harvests may concentrate on certain species as top predators. We consider that catches $h(t)$ provide a *direct use value* through some utility or payoff function $L(h_1, \dots, h_n)$. The most usual case of a utility function is the separable one

$$L(h_1, \dots, h_n) = \sum_{i=1}^{n} p_i h_i = p_1 h_1 + \cdots + p_n h_n \,,$$

where $p_i$ plays the role of price for the resource $i$ as the marginal utility value of catches $h_i$. Other cases of substitutable and essential factors may impose the consideration of a utility function of the form

$$L(h_1, \dots, h_n) = \prod_{i=1}^{n} h_i^{\alpha_i} = h_1^{\alpha_1} \times \cdots \times h_n^{\alpha_n} \,.$$

An interesting problem in terms of sustainability, viability and effectiveness approaches is to guarantee some utility level $L^\flat$ at every time in the following sense:

$$L\big(h_1(t), \dots, h_n(t)\big) \geq L^\flat , \quad t = t_0, \dots, T-1 . \tag{2.27}$$

Let us remark that the direct use value constraint (2.27) induces the conservation of part of the resource involved since[5]

$$L(N(t)) \geq L(h(t)) \geq L^\flat > 0 \Longrightarrow \exists i \in \{1, \dots, n\},\ N_i(t) > 0 .$$

However, along with the direct use values, conservation requirements related to existence values may also be explicitly handled through existence constraints of the form

$$N_i(t) \geq N_i^\flat > 0 , \tag{2.28}$$

where $N_i^\flat$ stands for some *quasi-extinction threshold.*

## 2.5 A forestry management model

### An age-classified matrix model

We consider a forest whose structure in age[6] is represented in discrete time by a vector $N$ of $\mathbb{R}_+^n$

$$N(t) = \begin{pmatrix} N_n(t) \\ N_{n-1}(t) \\ \vdots \\ N_1(t) \end{pmatrix} ,$$

where $N_j(t)$ $(j = 1, \dots, n-1)$ represents the number of trees whose age, expressed in the unit of time used to define $t$, is between $j-1$ and $j$ at the beginning of yearly period $[t, t+1[$; $N_n(t)$ is the number of trees of age greater than $n-1$. We assume that the natural evolution (*i.e.* under no exploitation) of the vector $N(t)$ is described by a linear system

$$N(t+1) = A\,N(t) , \tag{2.29}$$

where the terms of the matrix $A$ are nonnegative which ensures that $N(t)$ remains positive at all times. Particular instances of matrices $A$ are of the Leslie type (see [5])

$$A = \begin{bmatrix} 1-m_n & 1-m_{n-1} & 0 & \cdots & 0 \\ 0 & 0 & 1-m_{n-2} & \ddots & 0 \\ & & & \ddots & 0 \\ 0 & \cdots & 0 & \ddots & 1-m_1 \\ \gamma_n & \gamma_{n-1} & \cdots & \cdots & \gamma_1 \end{bmatrix} \tag{2.30}$$

[5] As soon as $L(0) = 0$.

[6] Models by size classes are commonly used, because size data are more easily available than age.

where $m_j$ and $\gamma_j$ are respectively mortality and recruitment parameters belonging to $[0,1]$. The rate $m_j$ is the proportion of trees of age $j-1$ which die before reaching age $j$ while $\gamma_j$ is the proportion of new-born trees generated by trees of age $j-1$. In coordinates, (2.29) and (2.30) read

$$\begin{cases} N_n(t+1) = (1-m_n)N_n(t) + (1-m_{n-1})N_{n-1}(t) \,, \\ N_j(t+1) = (1-m_{j-1})N_{j-1}(t) \,, \quad j = 2,\dots,n-1 \,, \\ N_1(t+1) = \gamma_n N_n(t) + \cdots + \gamma_1 N_1(t) \,. \end{cases} \tag{2.31}$$

### Harvesting and replanting

Now we describe the exploitation of such a forest resource. We assume the following main hypotheses:

1. only the oldest trees may be cut (the minimum age at which it is possible to cut trees is $n-1$);
2. new trees of age 0 may be planted.

Thus, let us introduce the scalar decision variables $h(t)$, representing the trees harvested at time $t$, and $i(t)$, the new trees planted. The control is then the two dimensional vector

$$u(t) = \begin{pmatrix} h(t) \\ i(t) \end{pmatrix} .$$

Previous assumptions lead to the following controlled evolution

$$N(t+1) = A\,N(t) + B_h h(t) + B_i i(t) \,, \tag{2.32}$$

where

$$B_h = \begin{pmatrix} -1 \\ 0 \\ \vdots \\ 0 \\ 0 \end{pmatrix} \quad \text{and} \quad B_i = \begin{pmatrix} 0 \\ 0 \\ \vdots \\ 0 \\ 1 \end{pmatrix} .$$

Furthermore, since one cannot plan to harvest more than will exist at the end of the unit of time, the control variable $h(t)$ is subject to the constraint

$$0 \le h(t) \le CAN(t) \,,$$

where the row vector $C$ is equal to $(1\,0\,0\cdots\,0)$, which ensures the non negativity of the resource $N$. Thus, we have assumed implicitly that the harvesting decisions $h(t)$ are effective at the end[7] of each time interval $[t, t+1[$.

---

[7] If the harvesting decisions $h(t)$ are effective at the beginning of each unit of time $t$, we have $0 \le h(t) \le C\,N(t) = N_n(t)$.

Cutting trees is costly, but brings immediate benefits, while planting trees is costly and will bring future income. All this may be aggregated in a performance function $L(h, i)$ and the planner objective consists in maximizing the sum of discounted performance of successive cuts $h(t)$ and replanting $i(t)$, that is[8]

$$\max_{h(\cdot),i(\cdot)} \sum_{t=t_0}^{+\infty} \rho^t L\big(h(t), i(t)\big) ,$$

where again $\rho$ is a discount factor chosen in $[0, 1[$.

## 2.6 A single species age-classified model of fishing

We present an age structured abundance population model with a possibly non linear stock-recruitment relationship derived from fish stock management [28].

Time is measured in years, and the time index $t \in \mathbb{N}$ represents the beginning of year $t$ and of yearly period $[t, t+1[$. Let $A \in \mathbb{N}^*$ denote a maximum age[9], and $a \in \{1, \ldots, A\}$ an age class index, all expressed in years. The population is characterized by $N = (N_a)_{a=1,\ldots,A} \in \mathbb{R}_+^A$, the *abundances* at age: for $a = 1, \ldots, A-1$, $N_a(t)$ is the number of individuals of age between $a-1$ and $a$ at the beginning of yearly period $[t, t+1[$; $N_A(t)$ is the number of individuals of age greater than $A-1$. The evolution of the exploited population depends both on natural mortality, recruitment and human exploitation. Hence, for ages $a = 1, \ldots, A-1$, the following evolution of the abundances can be considered

$$N_{a+1}(t+1) = N_a(t) \exp\big(-(M_a + \lambda(t)F_a)\big) , \tag{2.33}$$

where

- $M_a$ is the natural *mortality rate* of individuals of age $a$;
- $F_a$ is the mortality rate of individuals of age $a$ due to harvesting between $t$ and $t+1$, taken to remain constant during yearly period $[t, t+1[$; the vector $(F_a)_{a=1,\ldots,A}$ is termed the *exploitation pattern*;
- the control $\lambda(t)$ is the *fishing effort multiplier*, taken to be applied in the middle of yearly period $[t, t+1[$.

Since $N_A(t)$ is the number of individuals of age *greater than* $A-1$, an additional term appears in the dynamical relation

$$N_A(t+1) = N_{A-1}(t) \exp\big(-(M_{A-1} + \lambda(t)F_{A-1})\big) + N_A(t)\pi \exp\big(-(M_A + \lambda(t)F_A)\big) . \tag{2.34}$$

---

[8] $h(\cdot) = (h(t_0), h(t_0+1), \ldots)$ and $i(\cdot) = (i(t_0), i(t_0+1), \ldots)$.

[9] To give some ideas, $A = 3$ for anchovy and $A = 8$ for hake are instances of maximum ages. This is partly biological, partly conventional.

The parameter $\pi \in \{0,1\}$ is related to the existence of a so-called *plus-group*: if we neglect the survivors older than age $A$ then $\pi = 0$, else $\pi = 1$ and the last age class is a plus group[10].

Recruitment involves complex biological and environmental processes that fluctuate in time and are difficult to integrate into a population model. The *recruits* $N_1(t+1)$ are taken to be a function of the *spawning stock biomass* *SSB* defined by

$$SSB(N) := \sum_{a=1}^{A} \gamma_a \upsilon_a N_a \,, \tag{2.35}$$

that sums up the contributions of individuals to reproduction, where $(\gamma_a)_{a=1,\dots,A}$ are the *proportions of mature individuals* (some may be zero) at age and $(\upsilon_a)_{a=1,\dots,A}$ are the *weights at age* (all positive). We write

$$N_1(t+1) = \varphi\Big(SSB\big(N(t)\big)\Big) \,, \tag{2.36}$$

where the function $\varphi$ describes a *stock-recruitment relationship*, of which typical examples are

- constant: $\varphi(B) = R$;
- linear: $\varphi(B) = rB$;
- Beverton-Holt: $\varphi(B) = \frac{B}{\alpha+\beta B}$;
- Ricker: $\varphi(B) = \alpha B \mathrm{e}^{-\beta B}$.

Denoting

$$N(t) = \begin{pmatrix} N_1(t) \\ N_2(t) \\ \vdots \\ N_{A-1}(t) \\ N_A(t) \end{pmatrix} \in \mathbb{R}_+^A$$

we deduce from (2.33), (2.34) and (2.36) the global dynamics of the stock controlled by catch pressure $\lambda(t)$:

$$N(t+1) = \begin{pmatrix} \varphi\Big(SSB\big(N(t)\big)\Big) \\ N_1(t)\exp\Big(-\big(M_1+\lambda(t)F_1\big)\Big) \\ N_2(t)\exp\Big(-\big(M_2+\lambda(t)F_2\big)\Big) \\ \vdots \\ N_{A-2}(t)\exp\Big(-\big(M_{A-2}+\lambda(t)F_{A-2}\big)\Big) \\ N_{A-1}(t)\exp\Big(-\big(M_{A-1}+\lambda(t)F_{A-1}\big)\Big) + \pi\exp\Big(-\big(M_A+\lambda(t)F_A\big)\Big)N_A(t) \end{pmatrix} .$$

[10] $\pi = 0$ for anchovy and $\pi = 1$ for hake, for instance.

Examples of trajectories may be found in Fig. 2.4 for the Bay of Biscay anchovy. They are built thanks to the SCILAB code 3. Figs. 2.4 (a) and (b) correspond to low constant recruitment. Figs. 2.4 (c) and (d) correspond to medium constant recruitment while Figs. 2.4 (e) and (f) show a Ricker relation. The last Figs. 2.4, (g) and (h), stand for a linear recruitment form.

The yearly exploitation is described by catch-at-age $h_a$ and yield $Y$, respectively defined for a given vector of abundance $N$ and a given control $\lambda$ by the *Baranov catch equations* [28]. The catches are the number of individuals captured over one year:

$$H_a(\lambda, N) = \frac{\lambda F_a}{\lambda F_a + M_a}\Big(1 - \exp\big(-(M_a + \lambda F_a)\big)\Big) N_a \,. \tag{2.37}$$

The production in terms of biomass at the beginning of year $t$ is

$$Y(\lambda, N) = \sum_{a=1}^{A} \upsilon_a \, H_a(\lambda, N) \,, \tag{2.38}$$

where we recall that $\upsilon_a$ is the mean weight of individuals of age $a$.

Sustainability of the resource may focus more on species conservation constraints, upon spawning stock biomass for instance, as with the *International Council for the Exploration of the Sea* (ICES) *precautionary approach*,

$$SSB\big(N(t)\big) \geq B^\flat \,.$$

Sustainability of the exploitation may stress guaranteed production

$$Y\big(\lambda(t), N(t)\big) \geq Y^\flat \,.$$

A decision problem may be to optimize discounted rent

$$\max_{\lambda(\cdot)} \sum_{t=t_0}^{+\infty} \rho^t \Big(pY\big(\lambda(t), N(t)\big) - c\lambda(t)\Big) \,,$$

where $p$ are unit prices while $c$ are unit costs.

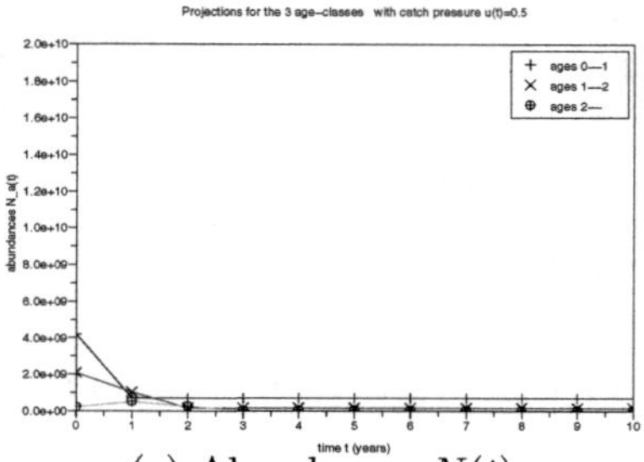


(a) Abundances $N(t)$

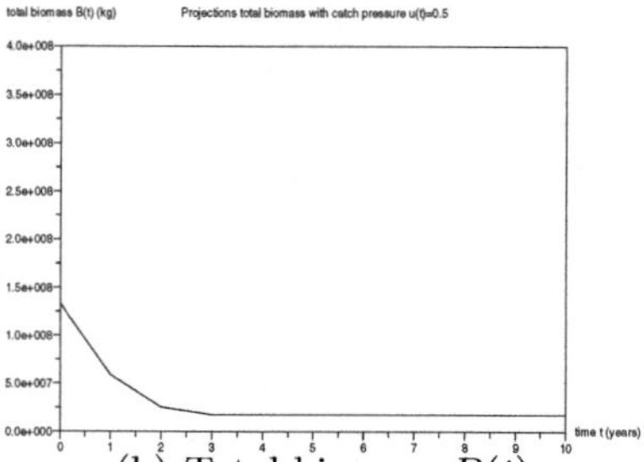


(b) Total biomass $B(t)$

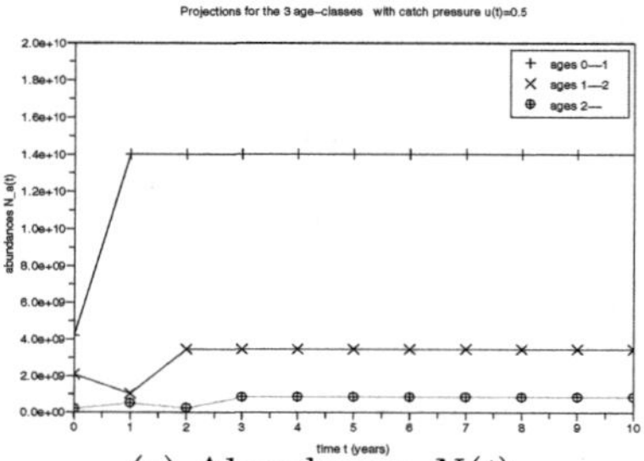


(c) Abundances $N(t)$

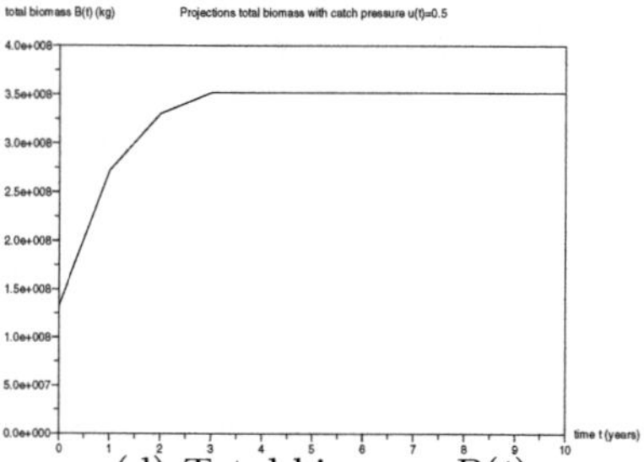


(d) Total biomass $B(t)$

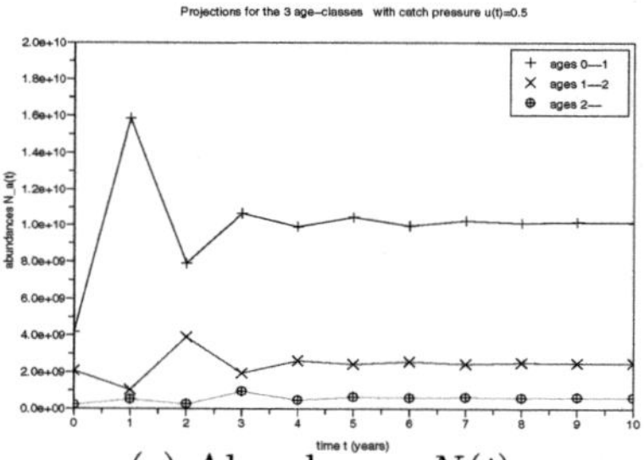


(e) Abundances $N(t)$

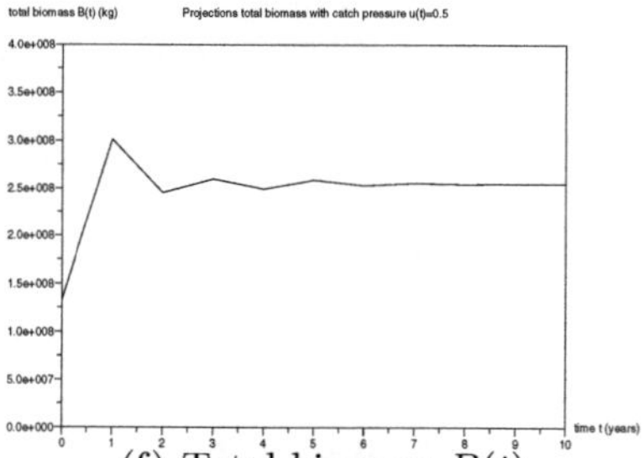


(f) Total biomass $B(t)$

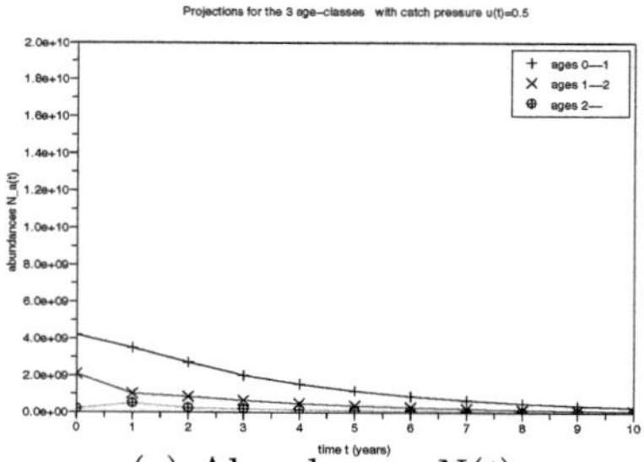


(g) Abundances $N(t)$

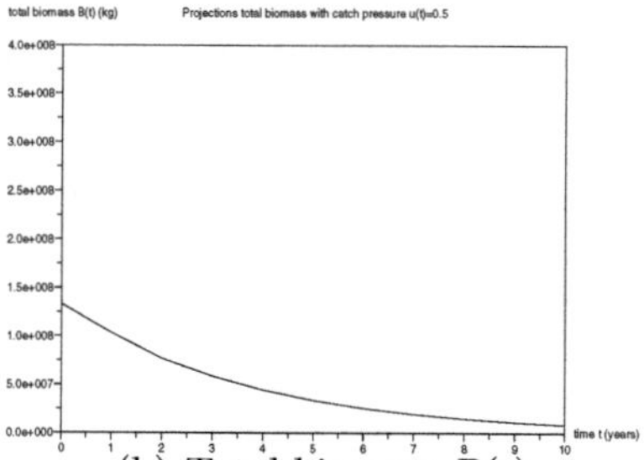


(h) Total biomass $B(t)$

**Fig. 2.4.** Bay of Biscay anchovy trajectories for different stock-recruitment relationships with same initial condition and fishing effort multiplier $\lambda(t) = 0.5$. They are built thanks to the SCILAB code 3. Figs. (a) and (b) correspond to low constant recruitment. Figs. (c) and (d) correspond to medium constant recruitment while Figs. (e) and (f) show a Ricker relation. The last Figs., (g) and (h), stand for a linear recruitment form.

SCILAB CODE 3.

```
//
// exec anchovy.sce

// ANCHOVY
// parameters for the dynamical model

sex_ratio = 0.5 ;
mature=sex_ratio*[1 1 1]; // proportion of matures at ages
weight=10^(-3)*[16, 28, 36]; // mean weights at ages (kg)
M=1.2; // natural mortality
F=[0.4 0.4 0.4]; // exploitation pattern
pi=0 ; // no plus-group
A=sum(ones(F)); // maximum age

// STOCK-RECRUITMENT RELATIONSHIPS

RR= 10^6 * [14016 7109 3964 696] ;
// R_mean R_gm R_min 2002 (ICES) R_min 2004 (ICES)

// // CONSTANT STOCK-RECRUITMENT

function y=mini_constant(x)
    y= mini(RR) ;
endfunction

function y=mean_constant(x)
    y= RR(1) ;
endfunction


// // RICKER STOCK-RECRUITMENT

a=0.79*10^6;
b=1.8*10^(-5);
// Ricker coefficients for tons

function y=Ricker(x)
  xx=10^{-3}*x // xx measured in tons
    y= a *( xx .* exp(-b* xx ) );
endfunction


// // LINEAR STOCK-RECRUITMENT

r= (500* 10^3) *21* 0.5 * 10^{-5} ;
function y=linear(x)
    y= r * x ;
endfunction


function y=SSB(N)
  // Spawning biomass
  y=  (mature.*weight) * N ;
endfunction

function Ndot=dynamics(N,lambda,phi)
  // Population dynamics
  mat=diag( exp(-M - lambda * F(1:($-1)) ) ,-1 ) +...
  diag( [zeros(F(1:($-1))) pi*exp(-M - lambda * F($)) ]);
    // sub diagonal terms // diagonal terms
  Ndot= mat*N + [phi(SSB(N)) ; zeros(N(2:$)) ] ;
endfunction

// initial values

N1999=10^6*[4195 2079 217]';
N2000=10^6*[7035 1033 381]';
N2001=10^6*[6575 1632 163]';
N2002=10^6*[1406 1535 262]';
N2003=10^6*[1192 333 255]';
N2004=10^6*[2590 254 43]';
//  N0=10^6*[1379 1506 256]';

// ICES values

B_pa = 10^6 * 33 ; // kg
F_pa=1;
B_lim=21000 * 10^3 ; // kg
 // F_lim=Inf;

// TRAJECTORIES

T=10; // horizon in years
multiplier=0.5;

stock_recruitment=list();
stock_recruitment(1)=mini_constant;
stock_recruitment(2)=mean_constant;
stock_recruitment(3)=Ricker;
stock_recruitment(4)=linear;

for i=1:4 do

  phi=stock_recruitment(i) ;
  // selecting a stock-recruitment relationship
  traj=[N1999];
  for t=0:(T-1)
traj=[traj, dynamics(traj(:,$),multiplier,phi)] ;
  end
//
  total_biomass= weight * traj ;
//
  xset("window",i) ;  xbasc(i);
  plot2d(0:T,total_biomass,rect=[0,0,T,4*10^8])
  xtitle('Projections total biomass...
 with catch pressure u(t)='...
 +string(multiplier),'time t (years)',...
 'total biomass B(t) (kg)')

  xset("window",10+i) ; xbasc(10+i);
  plot2d(0:T,traj',rect=[0,0,T,2*10^10]);
// drawing diamonds, crosses, etc. to identify the curves
  plot2d(0:T,traj',style=-[1,2,3]);
  legends(['ages 0--1';'ages 1--2';'ages 2--'],-[1,2,3],'ur')
  xtitle('Projections for the 3 age-classes...
 with catch pressure u(t)='...
 +string(multiplier),'time t (years)','abundances N_a(t)')
end

//
```

## 2.7 Economic growth with an exhaustible natural resource

Let us introduce a model referring to the management of an economy using an exhaustible natural resource as in [20]. Following [12] or [33], the classic cake eating economy first studied by Hotelling in [21] is expanded through a model of capital accumulation and consumption processes in the form of [29].

In a discrete time version, the economy with the exhaustible resource use is then described by the dynamic

$$\begin{cases} S(t+1) = S(t) - r(t) , \\ K(t+1) = (1-\delta)K(t) + Y\big(K(t), r(t)\big) - c(t) , \end{cases} \tag{2.39}$$

where $S(t)$ is the exhaustible resource stock (at the beginning of period $[t, t+1[$), $r(t)$ stands for the extraction flow per discrete unit of time, $K(t)$ represents the accumulated capital, $c(t)$ stands for the consumption and the function $Y$ represents the technology of the economy. Parameter $\delta$ is the rate of capital depreciation. The most usual example of production function is the so-called *Cobb-Douglas*

$$Y(K, r) = AK^{\alpha} r^{\beta} , \tag{2.40}$$

where the exponents $\alpha > 0$ and $\beta > 0$ represent the elasticities of production related to capital and resources respectively.

The controls of this economy are levels of consumption $c(t)$ and extraction $r(t)$ respectively. Additional constraints can be taken into account. The extraction $r(t)$ is irreversible in the sense that

$$0 \leq r(t) . \tag{2.41}$$

We take into account the scarcity of the resource by requiring

$$0 \leq S(t) .$$

More generally, we can consider a stronger conservation constraint for the resource as follows

$$S^{\flat} \leq S(t) , \tag{2.42}$$

where $S^{\flat} > 0$ stands for some guaranteed resource target, referring to a strong sustainability concern whenever it has a strictly positive value.

We also assume the investment in the reproducible capital $K$ to be irreversible in the sense that

$$0 \leq Y\big(K(t), r(t)\big) - c(t) . \tag{2.43}$$

We thus ensure the growth of capital if there is no depreciation.

We also consider that the capital is non negative:

$$0 \leq K(t) . \tag{2.44}$$

A sustainability requirement can be imposed through some guaranteed consumption level $c^{\flat}$ along the generations:

$$0 < c^{\flat} \leq c(t) . \tag{2.45}$$

The optimality problem exposed in [12] is

$$\max_{c(\cdot),r(\cdot)} \sum_{t=t_0}^{+\infty} \rho^t L(c(t)) ,$$

where $\rho \in [0,1[$ is a discount factor. Such a model questions how technology impacts the feasible or optimal extraction and consumptions paths.

## 2.8 An exploited metapopulation and protected area

Many works advocate the use of reserves as a central element of ecosystems and biodiversity sustainable management. The idea of closing areas as a fishery management instrument appeared two decades ago among marine ecologists. The first proposals focused on protected areas as laboratories, calling for modest areas in which ecologists could examine unexploited systems. By the early 1990s, the idea had evolved into a larger vision that called for significant areas to be set aside, often on the order of $20-30\%$ of the coastline. This change in scale coincided with several important papers on fisheries management. Most of these studies claimed that the world's fisheries were in a state of crisis, that conventional methods were to blame, and that a new approach to management was required. The main effects expected from the establishment of reserves are increased abundances and biomasses of spawning stocks and recruitment inside the protected area and, in surrounding areas through spillover, rebuilding of ecosystems and protection of habitat. Other potentially significant conservation and resource-enhancement benefits also include enhanced biodiversity, better habitat, increased catches and a hedge against management failures.

Several models have been developed to investigate the effectiveness of MPA in terms of stock conservation and catches. An extensive review of the literature can be found in [32, 19].

Let us now present mono-specific metapopulation modeling. Consider $n$ biomasses $N_j$ located in $n$ patches which may diffuse from one patch to the other. Without interference and harvesting, each biomass would follow separately $N_j(t+1) = g\big(N_j(t)\big)$ as in (2.7). One model with migration and catches $h_j(t)$ is, for any $j = 1,\dots,n$,

$$N_j(t+1) = g\big(N_j(t)\big) + \sum_{k=1,\ k\neq j}^{n} \tau_{k,j} N_k(t) - \sum_{k=1,\ k\neq j}^{n} \tau_{jk} N_j(t) - h_j(t) \quad (2.46)$$

where $0 \leq \tau_{j,k} < 1$ measures the migration rate from area $j$ to patch $k$ and $h_j(t)$ stands for the catches in area $j$. Notice that catches $h_j(t)$ take place at the end of period $[t, t+1[$ in this model. If protected area requirements are introduced, a major harvesting constraint becomes

$$0 \leq h_j(t) \leq q_j^\sharp , \quad (2.47)$$

where $q_j^\sharp = 0$ if the area is closed. A decision problem is to maximize the utility derived from the catches as follows

$$\max_{0 \le h_j(t) \le q_j^\sharp} \sum_{t=t_0}^{+\infty} \rho^t \sum_{j=1}^{n} L\big(h_j(t)\big) ,$$

where $\rho \in [0,1[$ is a discount factor. In this perspective, a reserve effect on catches means that

$$\exists\, q_j^\sharp = 0 \quad \text{such that} \quad \max_{0 \le h_j(t) \le q_j^\sharp} \sum_{t=t_0}^{+\infty} \rho^t \sum_{j=1}^{n} L\big(h_j(t)\big) < \max_{0 \le h_j(t)} \sum_{t=t_0}^{+\infty} \rho^t \sum_{j=1}^{n} L\big(h_j(t)\big) .$$

In this framework, the role played by the geometry of the reserve in relation to the diffusion characteristics is a challenging issue.

## 2.9 State space mathematical formulation

Although the models previously introduced for the management of exhaustible and renewable resources and pollution control are different, their main structures are quite similar. They are basically decision-making problems where time plays a central role. *Control theory* of dynamic systems is well suited to tackling such situations and building up mathematical models with analytic, algorithmic and/or numerical methods. Basically, the purpose of control theory is to find relevant decision or control rules to achieve various goals. In particular, a classical approach of control theory considers such problems through a state space formulation where decisions, commands or actions influence the evolution of a *state variable* in a causal way.

### 2.9.1 The dynamics

In the usual approach to a control problem, specifying the formulation of the *dynamical model* is the first phase of the analysis. In a discrete time context, and in specific terms making the components $i = 1, \dots, n$ explicit, this is a difference equation[11]:

$$\begin{cases} x_i(t+1) = F_i\big(t, x_1(t), \dots, x_n(t), u_1(t), \dots, u_p(t)\big), & t = t_0, t_0+1, \dots, T-1, \\ x_i(t_0) = x_{i0}. \end{cases}$$

In a compact form, a *dynamical model* or *dynamical system* is:

$$\begin{cases} x(t+1) = F\big(t, x(t), u(t)\big) , & t = t_0, t_0+1, \dots, T-1 , \\ x(t_0) = x_0 . \end{cases} \tag{2.48}$$

Here we denote by:

[11] In the continuous time context, the dynamic is represented by the differential equation $\dot{x}(t) = F\big(t, x(t), u(t)\big)$, $t \ge t_0$.

- $t$, the *time index*, belonging to the set of integers $\mathbb{N}$; $t_0$ is the *initial time* and the integer $T > t_0$ stands for the *time horizon*, be it finite ($T < +\infty$) or infinite ($T = +\infty$); thus $t$ runs from $t_0$ to $T$;
- $x(t) = \big(x_1(t), \dots, x_n(t)\big)$, the *state*, embodying a set of variables which sum up the information needed together with the control to pass from one time $t$ to the following time $t+1$; this state is an element of some *state space* denoted by $\mathbb{X}$; we shall restrict ourselves to the usual case corresponding to finite dimensional space, namely $\mathbb{X} = \mathbb{R}^n$;
- $u(t) = \big(u_1(t), \dots, u_p(t)\big)$, the *control or decision*, chosen by the decision-maker and which causes the dynamic evolution of the state $x$ according to the transition equation (2.48); this decision belongs to some *decision space* denoted by $\mathbb{U}$; we shall consider a finite dimensional space, namely $\mathbb{U} = \mathbb{R}^p$;
- $F : \mathbb{N} \times \mathbb{X} \times \mathbb{U} \to \mathbb{X}$, the *dynamics*-mapping representing the system's evolution; in many cases, the dynamic $F$ does not depend on time $t$ and is said to be *autonomous* or *stationary*;
- $x_0$, the *initial state* or *initial condition*, considered at *initial time* $t = t_0 \in \{0, 1, \dots, T-1\}$; observe that all subsequent values $x(t_0+1)$, $x(t_0+2)$, $\dots$ of the state are generated by this initial state and the sequences of controls *via* the *transition equation* (2.48); $x(t)$ is a function of the initial condition $x(t_0)$ and of past controls $u(t_0)$,$\dots$, $u(t-1)$.

The linear case corresponds to the situation where the dynamic $F$ can be written in the form $F(t, x, u) = F(t)x + G(t)u$, where $F(t)$ is a square matrix of size $n$, and $G$ is a matrix with $n$ rows and $p$ columns, giving

$$x(t+1) = F(t)x(t) + G(t)u(t) \ . \tag{2.49}$$

### 2.9.2 The trajectories

A *state trajectory*, or *state path*, is any sequence

$$x(\cdot) := \big(x(t_0), x(t_0+1), \dots, x(T)\big) \quad \text{with} \quad x(t) \in \mathbb{X} \tag{2.50}$$

and a *control trajectory*, or *control path*, is any sequence

$$u(\cdot) := \big(u(t_0), u(t_0+1), \dots, u(T-1)\big) \quad \text{with} \quad u(t) \in \mathbb{U} \ . \tag{2.51}$$

The *trajectories space* is $\mathbb{X}^{T+1-t_0} \times \mathbb{U}^{T-t_0}$, consisting of state and control trajectories

$$\big(x(\cdot), u(\cdot)\big) := \big(x(t_0), x(t_0+1), \dots, x(T-1), x(T), u(t_0), u(t_0+1), \dots, u(T-1)\big) \ . \tag{2.52}$$

Among all trajectories, of particular interest are those $\big(x(\cdot), u(\cdot)\big)$ which satisfy the dynamical equation (2.48).

Notice that when $T < +\infty$ there are $T$ controls and $T+1$ states: indeed, by (2.48), the ultimate control $u(T-1)$ generates a final state $x(T)$. When $T = +\infty$, one considers instead sequences $x(\cdot) := (x(t_0), x(t_0+1), \dots)$ and $u(\cdot) := (u(t_0), u(t_0+1), \dots)$ and the trajectories space is $\mathbb{X}^{\mathbb{N}} \times \mathbb{U}^{\mathbb{N}}$.

### 2.9.3 The feasible decisions

Each value $u(t)$ of $u(\cdot)$ specifies the control that will be chosen at time $t$: such a control is called a *decision at time* $t$. A rule that assigns a sequence of decisions is called a *policy* or a *strategy*, especially when a decision at time $t$ depends on the past states $x(t_0),\ldots, x(t)$ and controls $u(t_0),\ldots, u(t-1)$.

As displayed by previous examples, we may need to impose conditions or constraints on the system, including the states and the decisions, described as follows.

*Decision or control constraints.*

The admissible decisions are described, at each time $t$, by

$$u(t) \in \mathbb{B}\big(t, x(t)\big) , \quad t = t_0, \ldots, T-1 , \tag{2.53a}$$

where $\mathbb{B}\big(t, x(t)\big) \subset \mathbb{U}$ is some non empty subset of the control space $\mathbb{U}$. This subset is generally specified under the inequality form

$$b_1^i\big(t, x(t), u(t)\big) \geq 0 , \ldots, \; b_k^i\big(t, x(t), u(t)\big) \geq 0 ,$$

and/or equality form

$$b_1^e\big(t, x(t), u(t)\big) = 0 , \ldots, \; b_l^e\big(t, x(t), u(t)\big) = 0 ,$$

where the functions $b_1^i,\ldots, b_k^i$, $b_1^e,\ldots, b_l^e$ are real-valued. Frequently, we consider a constant set of feasible control values $\mathbb{B}$ and we ask for $u(t) \in \mathbb{B}$.

*State constraints.*

It is required that, at each time $t$, the state belongs to a non empty state domain $\mathbb{A}(t) \subset \mathbb{X}$

$$x(t) \in \mathbb{A}(t), \quad t = t_0, \ldots, T-1 , \tag{2.53b}$$

generally specified under the inequality form

$$a_1^i\big(t, x(t)\big) \geq 0 , \ldots, \; a_m^i\big(t, x(t)\big) \geq 0 ,$$

and/or equality form

$$a_1^e\big(t, x(t)\big) = 0 , \ldots, \; a_q^e\big(t, x(t)\big) = 0 .$$

As shown in the examples above, the usual case corresponds to a constant set of feasible state values $\mathbb{A}$ which does not depend on time in the sense that we require $x(t) \in \mathbb{A}$.

*Final state constraints.*

Here it is required that the final state reaches some non empty state domain $\mathbb{A}(T) \subset \mathbb{X}$, called the *target*, that is

$$x(T) \in \mathbb{A}(T) , \tag{2.53c}$$

generally specified under the inequality form

$$a_1^i\big(T, x(T)\big) \geq 0 , \ldots, a_m^i\big(T, x(T)\big) \geq 0 ,$$

and/or equality form

$$a_1^e\big(T, x(T)\big) = 0 , \ldots, a_q^e\big(T, x(T)\big) = 0 .$$

This target issue is closely related to the controllability concept studied in the control system literature [15]. When $T = +\infty$, the above statements may be understood as limits when time goes to infinity.

The so-called viability or invariance approach focuses on the role played by these different admissibility constraints. This point is examined in more detail in Chap. 4. Admissible equilibria shed a particular light on such feasibility issues. They are studied especially in Chap. 3.

### 2.9.4 The criterion and the evaluation of the decisions

Now, given an initial state $x_0$ and an initial time $t_0$, one may try to select a sequence of control variables among the feasible and tolerable ones. The usual selection of a decision sequence consists in optimizing (minimizing or maximizing) some $\pi$ criterion, representing the total cost or payoff/gain/utility/performance of the decisions over $T + 1 - t_0$ stages. Hereafter, we shall rather deal with maximization problems where the criterion is a payoff.

A *criterion* $\pi$ is a function $\pi : \mathbb{X}^{T+1-t_0} \times \mathbb{U}^{T-t_0} \to \mathbb{R}$ which assigns a real number to a state and control trajectory. Following the classification of [20] in the context of sustainability, we distinguish the following criteria.

- **Additive criterion (without inheritance).** It is the most usual criterion defined[12] in the finite horizon case by the sum

$$\pi\big(x(\cdot), u(\cdot)\big) = \sum_{t=t_0}^{T-1} L\big(t, x(t), u(t)\big) . \tag{2.54}$$

  Function $L$ is referred to as the system's *instantaneous payoff* or gain, profit, benefit, utility, etc. In economics or finance, the usual *present value*

---

[12] Whenever time is considered continuous, the intertemporal criteria is defined by the following integral $\pi\big(x(\cdot), u(\cdot)\big) = \int_{t_0}^{T} L\big(t, x(t), u(t)\big)\, dt$.

(PV) approach corresponds to the time separable case with discounting criterion in the form of [16, 22]

$$\pi\big(x(\cdot), u(\cdot)\big) = \sum_{t=t_0}^{T-1} \rho^t L\big(x(t), u(t)\big) , \tag{2.55}$$

where $\rho$ stands for a *discount factor* ($0 \leq \rho \leq 1$). The instantaneous gain $L\big(x(t), u(t)\big)$ may be a profit or a utility. This approach favors the present through the discount of future value and is sometimes qualified as "dictatorship of the present" because it neglects the future needs as soon as $\rho < 1$. In the infinite horizon case, we consider

$$\pi\big(x(\cdot), u(\cdot)\big) = \sum_{t=t_0}^{+\infty} \rho^t L\big(x(t), u(t)\big) . \tag{2.56}$$

The quadratic case corresponds to the situation where $L$ and $M$ are quadratic in the sense that $L(t, x, u) = x'R(t)x + u'Q(t)u$, where $R(t)$ and $Q(t)$ are positive matrices, giving:

$$\pi\big(x(\cdot), u(\cdot)\big) = \sum_{t=t_0}^{T-1} \big(x(t)'R(t)x(t) + u(t)'Q(t)u(t)\big) .$$

The case with inheritance, with the addition of a final payoff at final time $T$, will be seen later. The present value with inheritance will also be presented in the Chichilnisky type criterion definition.

- **The Maximin.** The *Rawlsian* or *maximin* form in the finite horizon is

$$\pi\big(x(\cdot), u(\cdot)\big) = \min_{t=t_0,\ldots,T-1} L\big(t, x(t), u(t)\big) . \tag{2.57}$$

The criterion focuses on the instantaneous value of the poorest generation, as in [30]. In economics literature, the Maximin approach has been discussed as an equity criterion [2, 30]. In particular, it has been used in environmental economics to deal with sustainability and intergenerational equity [4]. For instance, [33] examines the optimal and equitable allocation of an exhaustible resource. Other references include [13, 25, 26, 34].
In the infinite horizon, we obtain

$$\pi\big(x(\cdot), u(\cdot)\big) = \inf_{t=t_0,\ldots,+\infty} L\big(t, x(t), u(t)\big) .$$

The ultimate generation $T$ may also be taken into account by considering

$$\pi\big(x(\cdot), u(\cdot)\big) = \min\Big(\min_{t=t_0,\ldots,T-1} L\big(t, x(t), u(t)\big), M\big(T, x(T)\big)\Big) . \tag{2.58}$$

- **Green Golden type criterion.** Originally formulated in the infinite horizon, the so-called *"Green Golden"* form introduced by [7] and discussed in [20] puts emphasis on the ultimate payoff. In our discrete time framework when the horizon $T$ is finite, we label *Green Golden* a criterion of the form

$$\pi\big(x(\cdot), u(\cdot)\big) = M\big(T, x(T)\big) , \tag{2.59}$$

which puts weight only on the *final payoff* associated with the state $x(T)$ of the resource.
This approach considers only the far future and is qualified as "dictatorship of the future" because it neglects the present needs. As the state $x$ generally refers to the natural resource, such an approach may also favor the ecological or environmental dimensions justifying the term "green rule."
In the infinite horizon case, it corresponds to

$$\pi\big(x(\cdot), u(\cdot)\big) = \liminf_{T \to +\infty} M\big(T, x(T)\big) . \tag{2.60}$$

- **Chichilnisky type criterion.** Originally formulated in infinite horizon, the so-called *Chichilnisky criterion* is a convex combination of present value and Green Golden criteria [6].
In our discrete time framework, when the horizon $T$ is finite, we shall label *of Chichilnisky type* a criterion of the form

$$\pi\big(x(\cdot), u(\cdot)\big) = \theta \sum_{t=t_0}^{T-1} L\big(t, x(t), u(t)\big) + (1-\theta) M\big(T, x(T)\big) \tag{2.61}$$

where $\theta \in [0, 1]$ stands for the coefficient of present dictatorship. This criterion makes it possible to avoid both the dictatorship of the present and the dictatorship of the future.
- **Additive criterion (with inheritance).** It is defined in the finite horizon case by the sum:

$$\pi\big(x(\cdot), u(\cdot)\big) = \sum_{t=t_0}^{T-1} L\big(t, x(t), u(t)\big) + M\big(T, x(T)\big) . \tag{2.62}$$

Let us remark that such a case with scrap value can be associated with a Chichinilsky formulation for $\theta = 1/2$.

### 2.9.5 The optimization problem

The constraints (2.53a), (2.53b) and (2.53c) specified beforehand, combined with the dynamics (2.48), settle the set of all possible and feasible state and decision trajectories. Such a feasibility set, denoted by $\mathcal{T}^{ad}(t_0, x_0) \subset \mathbb{X}^{T+1-t_0} \times \mathbb{U}^{T-t_0}$, is defined by

$$\big(x(\cdot),u(\cdot)\big) \in \mathcal{T}^{ad}(t_0,x_0) \iff \big(x(\cdot),u(\cdot)\big) \text{ satisfies } (2.48)–(2.53a)–(2.53b)–(2.53c)\ .$$

In other words,

$$\mathcal{T}^{ad}(t_0,x_0) := \left\{ \big(x(\cdot),u(\cdot)\big) \;\middle|\; \begin{array}{ll} x(t_0) = x_0, & \\ x(t+1) = F\big(t,x(t),u(t)\big), & t = t_0,\dots,T-1 \\ u(t) \in \mathbb{B}\big(t,x(t)\big), & t = t_0,\dots,T-1 \\ x(t) \in \mathbb{A}(t), & t = t_0,\dots,T \end{array} \right\}. \tag{2.63}$$

We now aim at ranking the admissible paths according to a given criterion $\pi$ previously defined. Of particular interest is an optimal solution. Hence, the optimal control problem is defined as the following optimization problem:

$$\sup_{\big(x(\cdot),u(\cdot)\big)\in\mathcal{T}^{ad}(t_0,x_0)} \pi\big(x(\cdot),u(\cdot)\big)\ . \tag{2.64}$$

Abusively, we shall often formulate the hereabove optimization problem in the simplified form

$$\sup_{u(\cdot)} \pi\big(x(\cdot),u(\cdot)\big)\ . \tag{2.65}$$

Since we are interested in the existence of optimal admissible decisions, we generally assume that the supremum is achieved. Hence the supremum sup becomes a maximum max (or inf = min for minimization) and the problem (2.64) reads

$$\pi\big(x^\star(\cdot),u^\star(\cdot)\big) = \max_{\big(x(\cdot),u(\cdot)\big)\in\mathcal{T}^{ad}(t_0,x_0)} \pi\big(x(\cdot),u(\cdot)\big)\ , \tag{2.66}$$

where $u^\star(\cdot) = \big(u^\star(t_0),u^\star(t_0+1),\dots,u^\star(T-1)\big)$ denotes a feasible optimal decision trajectory and $x^\star(\cdot) = \big(x^\star(t_0),x^\star(t_0+1),\dots,x^\star(T)\big)$ an optimal state trajectory. Equivalently, the following notation is used:

$$\big(x^\star(\cdot),u^\star(\cdot)\big) \in \operatorname*{arg\,max}_{\big(x(\cdot),u(\cdot)\big)\in\mathcal{T}^{ad}(t_0,x_0)} \pi\big(x(\cdot),u(\cdot)\big)\ . \tag{2.67}$$

## 2.10 Open versus closed loop decisions

We point out two distinct approaches to compute relevant decision sequences. By relevant is meant at least admissible decisions, that is to say elements of the set $\mathcal{T}^{ad}(t_0,x_0)$ defined in (2.63), and, possibly, optimal ones, such as solutions of (2.64).

## Open loop

In this case, the key concept is to manipulate control trajectories depending only on time

$$u : t \mapsto u(t) ,$$

then compute the state by the dynamical equation (2.48):

$$x(t+1) = F\big(t, x(t), u(t)\big) .$$

For the maximization problem (2.64), when the horizon $T$ is finite, it boils down to maximizing the criterion $\pi$ with respect to $T - t_0$ variables $(u(t_0), u(t_0 + 1), \ldots, u(T-1))$. Thus, if every control $u(t)$ lies in a finite dimensional space $\mathbb{R}^p$ it follows that we are coping with an optimization problem on $\mathbb{R}^{p(T-t_0)}$.

## Closed loop

A second approach consists in searching for a control rule, called *feedback*, a mapping[13] $\mathfrak{u}$ depending on both the time $t$ and the state $x$:

$$\mathfrak{u} : (t, x) \mapsto \mathfrak{u}(t, x) \in \mathbb{U} .$$

Then the control and state trajectories are recovered by the relations

$$u(t) = \mathfrak{u}\big(t, x(t)\big) \quad \text{and} \quad x(t+1) = F\big(t, x(t), u(t)\big) .$$

Let us indicate that in the deterministic case closed and open loops are closely related because the state $x(t)$ in $u(t) = \mathfrak{u}\big(t, x(t)\big)$ is completely computed from the dynamic $F$ and the previous $x(t_0)$ and $u(t_0)$, ..., $u(t-1)$. Hence, the mapping $\mathfrak{u}(t, x)$ is not used for all values of $x$, but only for the predictable $x(t)$.

In fact, looking for feedbacks turns out to be a relevant method for implementation in real time, whenever the system is under disturbances or uncertainty $x(t+1) = F\big(t, x(t), u(t), w(t)\big)$ not directly taken into account by the modeling. Such ideas will be scrutinized in Chaps. 6, 7 and 8.

This second approach is more difficult in the sense that we now look for a function of the state and not only a sequence $u(t_0), \ldots, u(T-1)$. A solution is developed in Chaps. 4 and 5 where we introduce the *dynamic programming method.*

Furthermore, one can specify more restrictive conditions on the feedback. For instance, linear or continuous or smooth feedbacks may yield interesting additional properties for the control and decision laws.

---

[13] Note that $\mathfrak{u}$ denotes a *mapping* from $\mathbb{N} \times \mathbb{X}$ to $\mathbb{U}$, while $u$ denotes a *variable* belonging to $\mathbb{U}$.

## 2.11 Decision tree and the "curse of the dimensionality"

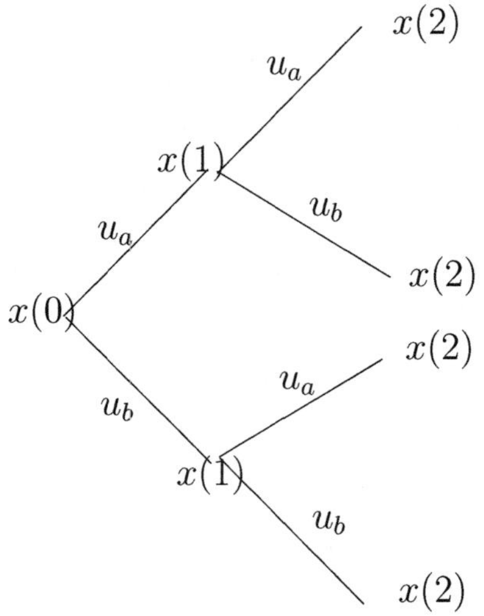


**Fig. 2.5.** Binary decision tree

Whenever the time horizon term $T$ along with the set of controls $\mathbb{U}$ are finite, the *graph theory* together with *operational research* constitute relevant frameworks to handle dynamical decision issues. Binary decisions where $u \in \mathbb{U} = \{0, 1\}$ illustrate this configuration. Indeed, given an initial condition $x(0)$ at initial time $t_0 = 0$, we can:

- represent the set of all feasible trajectories $x(\cdot)$ generated by (2.48) starting from $x(0)$ by a tree, as in Fig. 2.5, where the states $x(t)$ are associated with nodes and decisions $u(t)$ correspond to edges of the graph defined by the relation

$$x\mathcal{R}y \iff \exists t \geq 0 \,, \quad \exists u \in \mathbb{B}(t, x) \quad \text{such that} \quad y = F(t, x, u) \,;$$

- evaluate the performance and the criterion on every admissible path in order to choose the optimal one.

Although this method seems useful and easy to implement on a computer, it may yield a difficult numerical situation and a so-called curse of the dimensionality. To get a taste of it, consider a binary decision $u \in \{0, 1\}$ on horizon $T$, providing $2^T$ possible sequences $(u(0), \ldots, u(T-1)) \in \{0, 1\}^T$.

On a computer, a double-precision real requires 8 bytes $= 2^3$ bytes. So if a computer's RAM has 8 GBytes $= 8\ (1\ 024)^3$ bytes $= 2^{33}$ bytes, we can store up to $2^{30}$ double-precision reals.

One can thus imagine the difficulties implied by the comparison of $2^{52}$ criterions' final values for a horizon $T = 52$, corresponding to an annual problem with a weekly step size.

# References

[1] D. P. Bertsekas. *Dynamic Programming and Optimal Control.* Athena Scientific, Belmont, Massachusets, second edition, 2000. Volumes 1 and 2.

[2] E. Burmeister and P.J. Hammond. Maximin paths of heterogeneous capital accumulation and the instability of paradoxical steady states. *Econometrica*, 45(4):853–870, 1977.

[3] K. W. Byron, J. D. Nichols, and M. J. Conroy. *Analysis and Management of Animal Populations.* Academic Press, 2002.

[4] R. Cairns and N. V. Long. Maximin: a direct approach to sustainability. *Environment and Development Economics*, 11:275–300, 2006.

[5] H. Caswell. *Matrix Population Models.* Sinauer Associates, Sunderland, Massachussetts, second edition, 2001.

[6] G. Chichilnisky. An axiomatic approach to sustainable development. *Social Choice and Welfare*, 13(2):219–248, 1996.

[7] G. Chichilnisky, G. Heal, and A. Beltratti. The Green Golden rule. *Economics Letters*, 49:175–179, 1995.

[8] C. W. Clark. *Mathematical Bioeconomics.* Wiley, New York, second edition, 1990.

[9] J. M. Conrad. *Resource Economics.* Cambridge University Press, 1999.

[10] J. M. Conrad and C. Clark. *Natural Resource Economics.* Cambridge University Press, 1987.

[11] P. Dasgupta. *The Control of Resources.* Basil Blackwell, Oxford, 1982.

[12] P. Dasgupta and G. Heal. The optimal depletion of exhaustible resources. *Review of Economic Studies*, 41:1–28, 1974. Symposium on the Economics of Exhaustible Resources.

[13] A. Dixit, P. Hammond, and M. Hoel. On Hartwick's rule for regular maximin paths of capital accumulation and resource depletion. *Review of Economic Studies*, 47:551–556, 1980.

[14] L. Doyen, M. De Lara, J. Ferraris, and D. Pelletier. Sustainability of exploited marine ecosystems through protected areas: a viability model and

a coral reef case study. *Ecological Modelling*, 208(2-4):353–366, November 2007.

[15] B. Friedland. *Control System Design.* Mac Graw-Hill, New York, 1986.

[16] C. Gollier. *The Economics of Risk and Time.* MIT Press, Cambridge, 2001.

[17] H. S. Gordon. The economic theory of a common property resource: the fishery. *Journal of Political Economy*, 62:124–142, 1954.

[18] L. H. Goulder and K. Mathai. Optimal $CO_2$ abatement in the presence of induced technical change. *Journal of Environmental Economics and Management*, 39:1–38, 2000.

[19] R. Q. Grafton, T. Kompas, and V. Schneider. The bioeconomics of marine reserves: A selected review with policy implications. *Journal of Bioeconomics*, 7(2):161–178, 2005.

[20] G. Heal. *Valuing the Future, Economic Theory and Sustainability.* Columbia University Press, New York, 1998.

[21] H. Hotelling. The economics of exhaustible resources. *Journal of Political Economy*, 39:137–175, april 1931.

[22] T. C. Koopmans. Representation of preference orderings over time. In *Decision and Organization*, pages 79–100. North-Holland, 1972.

[23] M. Kot. *Elements of Mathematical Ecology.* Cambridge University Press, Cambridge, 2001.

[24] A. Manne, R. Mendelsohn, and R. Richels. MERGE: A model for evaluating regional and global effects of GHG reduction policies. *Energy Policy*, 23:17–34, 1995.

[25] V. Martinet and L. Doyen. Sustainable management of an exhaustible resource: a viable control approach. *Resource and Energy Economics*, 29(1):p.17–39, 2007.

[26] T. Mitra. Intertemporal equity and efficient allocation of resources. *Journal of Economic Theory*, 107:356–376, 2002.

[27] W. D. Nordhaus. *Managing the Global Commons.* MIT Press, Cambridge, 1994.

[28] T. J. Quinn and R. B. Deriso. *Quantitative Fish Dynamics.* Biological Resource Management Series. Oxford University Press, New York, 1999. 542 pp.

[29] F. Ramsey. A mathematical theory of saving. *The Economic Journal*, 38:543–559, 1928.

[30] J. Rawls. *A theory of Justice.* Clarendon, Oxford, 1971.

[31] M. B. Schaefer. Some aspects of the dynamics of populations important to the management of commercial marine fisheries. *Bulletin of the Inter-American tropical tuna commission*, 1:25–56, 1954.

[32] M. D. Smith and J. E. Wilen. Economic impacts of marine reserves: the importance of spatial behavior. *Journal of Environmental Economics and Management*, 46:183–206, 2003.

[33] R. M. Solow. Intergenerational equity and exhaustible resources. *Review of Economic Studies*, 41:29–45, 1974. Symposium on the Economics of Exhaustible Resources.

[34] C. Withagen and G. Asheim. Characterizing sustainability: the converse of Hartwick's rule. *Journal of Economic Dynamics and Control*, 23:159–165, 1998.

# 3

# Equilibrium and stability

A key concept for the study of dynamical systems is *stability*. Basically, a dynamical system is stable with respect to a given desired trajectory if weak perturbations cause only small variations in the trajectories with respect to the one being tracked. The most commonly tracked trajectory is that of equilibrium.

Designed for autonomous systems, *i.e.* when dynamics does not directly depend on time, an equilibrium of a dynamical system corresponds to a stationary state and is associated with fixed points of the dynamics in the discrete time case.

Equilibria highlight sustainability concerns in an interesting way as emphasized by the sustainable yield approach used for the management of renewable resources [2, 4, 5]. In particular, the notions of *maximum sustainable yield*, *private property* or *common property equilibria* capture important ideas for bioeconomic modeling.

From the mathematical viewpoint, detailed concepts and results dealing with stability can be found in [8, 9] and [6] for results on stability property and the Lyapunov approach. Here we restrict the methods to the use of linearized dynamics. The basic idea of linearization is that, in a small neighborhood of an equilibrium, the dynamics behaves, in most cases, similarly to its linear approximation involving the first order derivatives. Usual tools of linear algebra including eigenvalues can then be invoked.

The present chapter is organized as follows. A first Sect. 3.1 introduces the notion of equilibrium for controlled dynamics. Examples coping with exhaustible and renewable resources or pollution management illustrate the concept in Sect. 3.2. In Sect. 3.3, particular emphasis is given to different bioeconomic issues related to the notion of sustainable yield. A second part is devoted to the concept of stability in the context of open-loop decisions. The method of linearization is exposed in Sect. 3.4 and applied to some examples in the remaining sections.

## 3.1 Equilibrium states and decisions

In this section, we deal with *autonomous dynamics*, or *stationary dynamics*, in the sense that the dynamics does not depend directly on time $t$, namely:

$$x(t+1) = F\big(x(t), u(t)\big) \ , \quad t = t_0, t_0+1, \ldots \quad \text{with} \quad x(t_0) = x_0 \ , \qquad (3.1)$$

where $x(t) \in \mathbb{X} = \mathbb{R}^n$ represents the state of the system and $u(t) \in \mathbb{U} = \mathbb{R}^p$ stands for the decision vector as in (2.48). $x_0 \in \mathbb{X}$ is the initial condition at initial time $t_0 \in \mathbb{N}$.

Similarly, decision and state constraints are time independent:

$$\begin{cases} x(t) \in \mathbb{A} \ , \\ u(t) \in \mathbb{B}\big(x(t)\big) \ . \end{cases} \qquad (3.2)$$

Basically, an equilibrium of a dynamical system corresponds to a situation where the evolution is stopped in the sense that the state becomes steady:

$$x(t+s) = x(t) \ , \quad t = t_0, t_0+1, \ldots \ , \quad s = 0, 1 \ldots$$

In the framework of controlled dynamical systems, such an equilibrium is related to both a stationary admissible state and a stationary decision. For the discrete time case on which we focus, this means that we face problems of fixed points.

**Definition 3.1.** *The state $x_{\mathrm{E}} \in \mathbb{X}$ is an admissible equilibrium, or steady state, of the autonomous dynamical system (3.1) under constraints (3.2) if there exists a decision $u_{\mathrm{E}} \in \mathbb{U}$ satisfying*

$$F(x_{\mathrm{E}}, u_{\mathrm{E}}) = x_{\mathrm{E}} \quad \textit{with} \quad u_{\mathrm{E}} \in \mathbb{B}(x_{\mathrm{E}}) \quad \textit{and} \quad x_{\mathrm{E}} \in \mathbb{A} \ . \qquad (3.3)$$

*By extension, we shall also say that $(x_{\mathrm{E}}, u_{\mathrm{E}})$ is an admissible equilibrium.*

## 3.2 Some examples of equilibria

For some of the models presented in Chap. 2, we describe the equilibria if any exist.

### 3.2.1 Exploitation of an exhaustible resource

We recall the system introduced in Sect. 2.1 for the management of an exhaustible stock $S(t)$:

$$S(t+1) = S(t) - h(t) \ .$$

The resource stock $S_{\mathrm{E}}$ and harvesting $h_{\mathrm{E}}$ are stationary whenever

$$S_E = S_E - h_E \, .$$

As expected, the only equilibrium solution for any resource level $S_E$ is zero extraction: $h_E = 0$. This observation means that the only equilibrium configuration is without exploitation. Of course, this is not a challenging result in terms of management. The equilibrium approach is of little help. The optimality approach is more fruitful and will be examined in Sect. 5.9.

### 3.2.2 Mitigation policies for carbon dioxyde emissions

Assuming stationary emissions $E_{BAU} \geq 0$ in the carbon cycle model of Sect. 2.3, the dynamic (2.20) becomes

$$M(t+1) = M(t) + \alpha E_{BAU}\big(1 - a(t)\big) - \delta(M(t) - M_{-\infty}) \, ,$$

and any equilibrium $(M_E, a_E)$ satisfies

$$M_E = M_{-\infty} + \frac{\alpha E_{BAU}(1 - a_E)}{\delta} \quad \text{with} \quad 0 \leq a_E \leq 1 \, .$$

When we write it as

$$a_E = 1 - \frac{\delta(M_E - M_{-\infty})}{\alpha E_{BAU}} \quad \text{when} \quad 0 \leq \frac{\delta(M_E - M_{-\infty})}{\alpha E_{BAU}} \leq 1 \, ,$$

we see that the level of abatement to stabilize concentration at $M_E$ is sensitive to carbon removal rate $\delta$.

### 3.2.3 Single species equilibrium in an age-structured fish stock model

The model, already introduced in Sect. 2.6, is derived from fish stock management. We compute an equilibrium $(N_E, \lambda_E)$ such that

$$g(N_E, \lambda_E) = N_E \, ,$$

where the dynamic $g$ is given, in absence of a plus-group, by

$$g(N_1, \ldots, N_A, \lambda) = \begin{pmatrix} \varphi(SSB(N)) \\ N_1 \exp\big(-(M_1 + \lambda F_1)\big) \\ N_2 \exp\big(-(M_2 + \lambda F_2)\big) \\ \vdots \\ N_{A-2} \exp\big(-(M_{A-2} + \lambda F_{A-2})\big) \\ N_{A-1} \exp\big(-(M_{A-1} + \lambda F_{A-1})\big) \end{pmatrix}$$

with *spawning stock biomass SSB* defined by

$$SSB(N) := \sum_{a=1}^{A} \gamma_a \upsilon_a N_a \ ,$$

and $\varphi$ a stock-recruitment relationship. The computation of an equilibrium $N_{\mathrm{E}}(\lambda)$, for $\lambda \geq 0$, gives

$$N_{1,\mathrm{E}}(\lambda) = Z(\lambda) \quad \text{and} \quad N_{a,\mathrm{E}}(\lambda) = s_a(\lambda) Z(\lambda) \ , \quad a = 1, \ldots, A$$

where

$$s_a(\lambda) := \exp\Big(-\big(M_1 + \cdots + M_{a-1} + u(F_1 + \cdots + F_{a-1})\big)\Big) \tag{3.4}$$

is the proportion of equilibrium recruits which survive up to age a ($a = 2, \ldots, A$) while $s_1(\lambda) = 1$. The number $Z(\lambda)$ of recruits at equilibrium is a nonnegative fixed point of the function $z \mapsto \varphi\big(z\mathrm{SPR}(\lambda)\big)$ where SPR corresponds to the *equilibrium spawners per recruits*, namely

$$\mathrm{SPR}(\lambda) := \sum_{a=1}^{A} \gamma_a \upsilon_a s_a(\lambda) \ .$$

Being a fixed point means that $Z(\lambda)$ is the solution of the equation:

$$Z(\lambda) = \varphi\big(Z(\lambda)\mathrm{SPR}(\lambda)\big) \ .$$

We do not go into detail on existence results of such a fixed point. However, in the Beverton-Holt case where the recruitment mechanism corresponds to $\varphi(B) = \frac{RB}{1+bB}$, the solution can be specified as follows:

$$Z(\lambda) = \max\left(\frac{R\,\mathrm{SPR}(\lambda) - 1}{b\,\mathrm{SPR}(\lambda)} \ , \ 0\right) \ .$$

### 3.2.4 Economic growth with an exhaustible natural resource

Recall that the economy in Sect. 2.7 is governed by the evolution of

$$\begin{cases} S(t+1) = S(t) - r(t) \ , \\ K(t+1) = (1-\delta)K(t) + Y\big(K(t), r(t)\big) - c(t) \ . \end{cases}$$

Even if the second equation has, by itself, an equilibrium $(K_{\mathrm{E}}, c_{\mathrm{E}}, r_{\mathrm{E}}) \in \mathbb{R}_+^3$ solution of

$$0 = -\delta K_{\mathrm{E}} + Y\big(K_{\mathrm{E}}, r_{\mathrm{E}}\big) - c_{\mathrm{E}} \ ,$$

the overall system has no other equilibrium than with $r_{\mathrm{E}} = 0$, as we have seen in Subsect. 3.2.1, above right. As soon as resource $r$ is needed for production, then $Y(K, 0) = 0$ and the only global admissible equilibrium is

$$r_{\mathrm{E}} = c_{\mathrm{E}} = K_{\mathrm{E}} = 0$$

and the economy collapses. The equilibrium approach is of little help in terms of management in this case.

## 3.3 Maximum sustainable yield, private property, common property, open access equilibria

In contrast to the exhaustible resource management case where the equilibrium approach does not bring challenging results, the renewable resource case is more interesting.

The concept of *maximum sustainable yield* [2] is one of the cornerstones of the management of renewable resources. Despite considerable criticism, it remains a reference. Other equilibria deriving from different property rights on the resource are also presented.

### 3.3.1 Sustainable yield for surplus model

Designed for the management of renewable resources, the Schaefer model (2.14) introduced in Sect. 2.2 corresponds to:

$$B(t+1) = g\big(B(t) - h(t)\big) \,, \quad 0 \leq h(t) \leq B(t) \,. \tag{3.5}$$

At equilibrium, the harvesting $h_{\text{E}}$ induces a steady population $B_{\text{E}}$ whenever

$$B_{\text{E}} = g(B_{\text{E}} - h_{\text{E}}) \quad \text{and} \quad 0 \leq h_{\text{E}} \leq B_{\text{E}} \,.$$

One obtains the so-called *sustainable yield* $\sigma(B_{\text{E}})$ by an implicit equation whenever, for given $B_{\text{E}}$, there exists a unique $h_{\text{E}}$ satisfying the above equation, giving:

$$\sigma(B_{\text{E}}) := h_{\text{E}} \iff B_{\text{E}} = g(B_{\text{E}} - h_{\text{E}}) \quad \text{and} \quad 0 \leq h_{\text{E}} \leq B_{\text{E}} \,. \tag{3.6}$$

The relation may also be written as

$$\underbrace{h_{\text{E}}}_{\text{surplus}} = \underbrace{g(B_{\text{E}} - h_{\text{E}})}_{\text{regeneration}} - \underbrace{(B_{\text{E}} - h_{\text{E}})}_{\text{biomass after capture}} \geq 0 \,,$$

meaning that "a surplus production exists that can be harvested in perpetuity without altering the stock level" [2, p. 1]. Indeed, $h_{\text{E}}$ can be harvested forever, while the biomass is maintained indefinitely at level $B_{\text{E}}$.

Notice that if the dynamic $g$ and the sustainable yield function $B \mapsto \sigma(B)$ are differentiable, we can deduce from $B_{\text{E}} = g\big(B_{\text{E}} - \sigma(B_{\text{E}})\big)$ in (3.6) the marginal relation

$$\sigma'(B) = \frac{g'\big(B - \sigma(B)\big) - 1}{g'\big(B - \sigma(B)\big)} \,. \tag{3.7}$$

Thus, the sustainable yield $\sigma$ is the solution of a differential equation.

As in [5], it is worth distinguishing particular cases of such sustainable yields that play an important role in the economics literature for the management of renewable resources.

### 3.3.2 Maximum sustainable equilibrium

The *maximum sustainable equilibrium* (MSE) is the solution $(B_{\text{MSE}}, h_{\text{MSE}})$ of

$$\sigma(B_{\text{MSE}}) = h_{\text{MSE}} = \max_{B \geq 0,\ h=\sigma(B)} h = \max_{B \geq 0} \sigma(B) \, . \tag{3.8}$$

The maximum catch $h_{\text{MSE}}$ is called the *maximum sustainable yield* (MSY).

Whenever the sustainable yield function $B \mapsto \sigma(B)$ is differentiable, the first order optimality condition reads

$$\sigma'(B_{\text{MSE}}) = 0 \, .$$

Consequently, using (3.7) when the dynamic $g$ is supposed to be differentiable, the maximum sustainable biomass $(B_{\text{MSE}}, h_{\text{MSE}})$ solves

$$g'(B_{\text{MSE}} - h_{\text{MSE}}) = 1 \quad \text{and} \quad g(B_{\text{MSE}} - h_{\text{MSE}}) = B_{\text{MSE}} \, . \tag{3.9}$$

It should be emphasized that such a steady state depends only on the biological features of the stock summarized by the function $g$. More specific computations of MSE are displayed in the following paragraphs for linear, Beverton-Holt or logistic growth relations.

### 3.3.3 Private property equilibrium

Assuming a constant price $p$ per unit of harvested biomass, the total revenue resulting from harvesting $h$ is $ph$. Harvesting costs are $C(h, B)$. The economic *rent* or *profit* $\mathcal{R}(h, B)$ provided by the management of the resource is defined as the difference between benefits and harvesting costs:

$$\mathcal{R}(h, B) := ph - C(h, B) \, .$$

The so-called *private property equilibrium* (PPE) is the equilibrium solution $(B_{\text{PPE}}, h_{\text{PPE}})$ which maximizes such a rent as follows:

$$\mathcal{R}(h_{\text{PPE}}, B_{\text{PPE}}) = \max_{B \geq 0,\ h=\sigma(B)} \mathcal{R}(h, B) \, . \tag{3.10}$$

Assume that the harvesting costs $C(h, B)$ are smooth, and let $C_h(h, B)$ and $C_B(h, B)$ denote the partial derivatives. By writing first order optimality conditions and using (3.7), the PPE satisfies the following marginal conditions:

$$g'(B_{\text{PPE}} - h_{\text{PPE}}) = \frac{p - C_h(h_{\text{PPE}}, B_{\text{PPE}})}{p - C_h(h_{\text{PPE}}, B_{\text{PPE}}) - C_B(h_{\text{PPE}}, B_{\text{PPE}})} \, . \tag{3.11}$$

It can be pointed out that PPE equilibrium combines ecological and economic dimensions through the marginal cost, income and growth values.

### 3.3.4 Common property equilibrium

In the *open access* perspective, it is assumed that any positive rent is dissipated [2, p. 25]. A forthcoming Subsect. 3.3.6 shows how this concept can be derived from an equilibrium perspective on an extended dynamic with one additional state component. The so-called *common property equilibrium* (CPE) captures this situation of zero-profit as a solution $(B_{\text{CPE}}, h_{\text{CPE}}) = (B_{\text{CPE}}, \sigma(B_{\text{CPE}}))$ of

$$\mathcal{R}(h_{\text{CPE}}, B_{\text{CPE}}) = 0 \ ,$$

that is,

$$h_{\text{CPE}} = \sigma(B_{\text{CPE}}) \quad \text{and} \quad ph_{\text{CPE}} = C(h_{\text{CPE}}, B_{\text{CPE}}) \ . \tag{3.12}$$

For the sake of simplicity, we assume that the harvesting costs are proportional to effort $e = h/(qB)$ (see (2.15)), which can be written

$$C(h, B) = \frac{ch}{qB} \ , \tag{3.13}$$

with $c$ a unit cost of effort, and $q$ a *catchability coefficient*. In this case, whatever the natural dynamic $g$, the common property equilibrium $B_{\text{CPE}}$ satisfies, by (3.12),

$$B_{\text{CPE}} = \frac{c}{pq} \ .$$

Under the open access assumption, the sustainability and conservation requirements are in jeopardy through the CPE as soon as unit costs $c$ decrease or price $p$ strongly increases since $B_{\text{CPE}}$ may then drop.

Notice that the common property equilibrium depends, apart from the catchability coefficient $q$, only on economics parameters (unitary cost $c$ and price $p$), unlike the private property equilibrium which also depends on dynamic $g$.

### 3.3.5 Examples for different population dynamics

We follow here the material introduced in Sect. 2.2. The harvesting costs are assumed to be proportional to effort as in (3.13).

#### The linear model

Consider $g(B) = RB$ with $R \geq 1$. In this case, we obtain that

- the sustainable yield function is

$$\sigma(B) = \frac{R-1}{R} B \ ,$$

- the maximum sustainable equilibrium does not contain much information since

$$B_{\text{MSE}} = h_{\text{MSE}} = +\infty \ ,$$

- neither is the private property equilibrium a useful concept since

$$B_{\mathrm{PPE}} = h_{\mathrm{PPE}} = +\infty \, .$$

### The Beverton-Holt model

Assume that $g(B) = \frac{RB}{1+bB}$ with growth $R > 1$ and saturation $b > 0$ parameters.

- Any admissible equilibrium $(B_{\mathrm{E}}, h_{\mathrm{E}})$ satisfies

$$B_{\mathrm{E}} = \frac{R(B_{\mathrm{E}} - h_{\mathrm{E}})}{1 + b(B_{\mathrm{E}} - h_{\mathrm{E}})} \quad \text{and} \quad 0 \leq h_{\mathrm{E}} \leq B_{\mathrm{E}} \, ,$$

  and the sustainable yield corresponds therefore to

$$\sigma(B) = B - \frac{B}{R - bB} \quad \text{for} \quad 0 \leq B \leq K := \frac{R-1}{b} \, . \tag{3.14}$$

  This is illustrated in Fig. 3.1. The last inequality ensures an admissibility requirement $0 \leq \sigma(B) \leq B$. The notation $K$ stems from the fact that $\frac{R-1}{b}$ is the carrying capacity of the biological dynamic. In other words, the biomass $K = \frac{R-1}{b}$ is the only strictly positive equilibrium of the natural dynamic without harvesting. Of course, harvesting decision constraints may more stringently restrict the domain of validity for this last relation.

SCILAB CODE 4.

```
//
// exec sustainable_yield_tuna.sce

R_tuna=2.25; // discrete-time intrinsic growth
R=R_tuna;
K_tuna = 250000;  // carrying capacity in metric tons
K = K_tuna;
c_tuna=2500;
// unit cost of effort in dollars per standard fishing day
p_tuna=600; // market price in dollars per metric ton

// BEVERTON-HOLT DYNAMICS
R_BH = R_tuna ;
b_BH = (R_BH-1) / K ;

// SUSTAINABLE YIELD FUNCTION
function [SY]=sust_yield(B)
  SY=B-( B ./ ( R_BH - b_BH *B ) ) ;
endfunction


B_MSE = ( R_BH - sqrt(R_BH) ) / b_BH ;

// maximum sustainable equilibrium

MSY=sust_yield(B_MSE) ;
// maximum sustainable yield

xset("window",1); xbasc();
abcisse=linspace(0,K,100);
plot2d(abcisse,sust_yield(abcisse),rect=[0,0,K,1.1*MSY]);
H_MSY=linspace(0,MSY,20);
plot2d(B_MSE*ones(H_MSY),H_MSY,style=-6);
xtitle("Sustainable yield for the ...
   Beverton-Holt model (tuna)",...
   "biomass (metric tons)","catches (metric tons)");

// private property equilibrium
cost=c_tuna;
price=p_tuna;
B_PPE = ...
( R_BH - sqrt(R_BH - (b_BH * cost / price) ) ) / b_BH ;

//
```

- A first order optimality condition gives the MSE:

$$B_{\mathrm{MSE}} = \frac{R - \sqrt{R}}{b} \, . \tag{3.15}$$

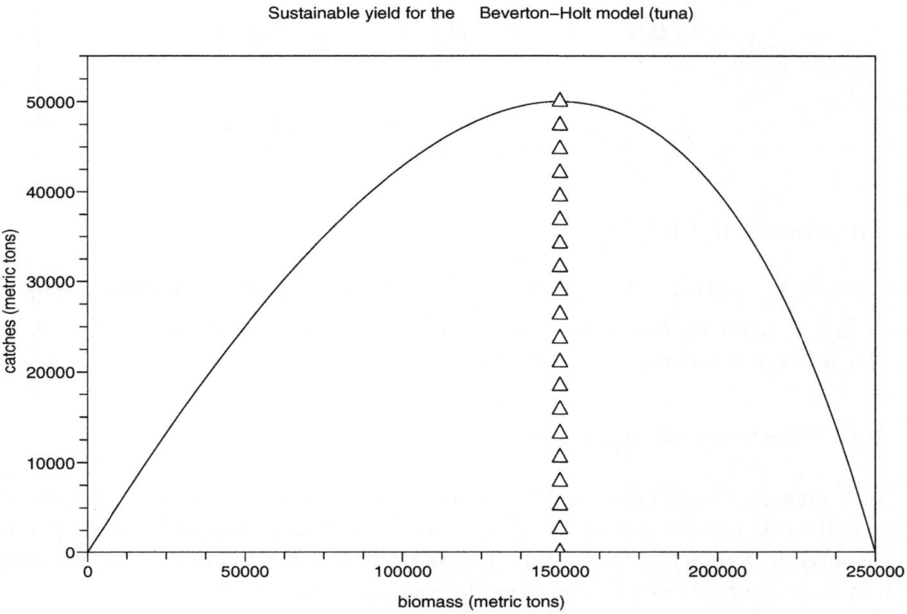


**Fig. 3.1.** Sustainable yield for Beverton-Holt model with tuna data. Biomass and catches are measured in metric tons. The maximum sustainable equilibrium is achieved at $B_{\text{MSE}} = 150\ 000$ metric tons, and provides the maximum sustainable yield $h_{\text{MSE}} = 50\ 000$ metric tons. The curve comes from SCILAB code 4.

- Whenever the common property equilibrium solution of (3.12) is admissible, *i.e.* $0 \leq \frac{c}{p} \leq K$ or $\frac{bc}{p} \leq R - 1$, the private property equilibrium is

$$B_{\text{PPE}} = \frac{R - \sqrt{R - \dfrac{bc}{p}}}{b} .$$

Remark that PPE biomass is always larger than maximum sustainable equilibrium biomass in the sense that

$$B_{\text{PPE}} > B_{\text{MSE}} . \tag{3.16}$$

In this sense, the PPE equilibrium is more conservative than the MSE.

### The logistic model

Consider $g(B) = RB(1 - \frac{B}{\kappa})$. Any equilibrium $(B_{\text{E}}, h_{\text{E}})$ satisfies

$$B_{\text{E}} = g(B_{\text{E}} - h_{\text{E}}) \iff B_{\text{E}} = R(B_{\text{E}} - h_{\text{E}}) \left(1 - \frac{(B_{\text{E}} - h_{\text{E}})}{\kappa}\right) .$$

By (3.9), the MSE $(B_{\text{MSE}}, h_{\text{MSE}})$ solves in addition

$$g'(B_{\mathrm{MSE}} - h_{\mathrm{MSE}}) = R(1 - 2\frac{B_{\mathrm{MSE}} - h_{\mathrm{MSE}}}{\kappa}) = 1 \, ,$$

giving

$$h_{\mathrm{MSE}} = \frac{(R-1)^2}{4R}\kappa \quad \text{and} \quad B_{\mathrm{MSE}} = \frac{R^2-1}{4R}\kappa \, .$$

**The Ricker model**

Now, we have $g(B) = B\exp(r(1-\frac{B}{K}))$. In this case, the sustainable yield $h_{\mathrm{E}} = \sigma(B_{\mathrm{E}})$ is given by the implicit relation $B_{\mathrm{E}} = (B_{\mathrm{E}} - h_{\mathrm{E}})\exp(r(1 - \frac{(B_{\mathrm{E}}-h_{\mathrm{E}})}{K}))$, and numerical solvers are required.

### 3.3.6 Open access dynamics

The open access situation occurs when no limitation is imposed on the harvesting effort. Consider any model of assessment and management of a renewable resource as in Sect. 2.2. We now postulate that the variations of harvesting effort are proportional to the flow of economic rent

$$\begin{cases} B(t+1) = g\big(B(t) - qe(t)B(t)\big) \, , \\ e(t+1) = e(t) + \alpha\mathcal{R}\big(e(t), B(t)\big) \, , \end{cases} \tag{3.17}$$

where parameter $\alpha > 0$ measures the sensitivity of the agents possibly exploiting the resource with respect to profits. In this formulation, the effort becomes a state of the dynamical system and there is no longer control in the dynamical system (3.17). Then, assuming that the rent is defined by $\mathcal{R}\big(e, B\big) = pqBe - ce$, any non trivial equilibrium $(B_{\mathrm{E}}, e_{\mathrm{E}})$ satisfies

$$B_{\mathrm{E}} = \frac{c}{pq} \quad \text{and} \quad e_{\mathrm{E}} = \frac{p}{c}\sigma(\frac{c}{pq}) \, , \tag{3.18}$$

where $\sigma$ stands for the sustainable yield function. Such an equilibrium $(B_{\mathrm{E}}, e_{\mathrm{E}})$ is termed Gordon's *Gordon's bionomic equilibrium.*

## 3.4 Stability of a stationary open loop equilibrium state

By stationary open loop, we mean that the control $u(t)$ is set at a stationary value $u_{\mathrm{E}}$:

$$u(t) = u_{\mathrm{E}} \, .$$

The state follows a discrete-time dynamic, for which traditional notions of stability are presented. Of particular importance is the concept of asymptotic stability which, though subtle, may generally be tested by a simple matrix analysis.

### 3.4.1 General definition

Stability of an equilibrium $(x_{\mathrm{E}}, u_{\mathrm{E}})$ is, to begin with, a *local property*. We consider the dynamics on a neighborhood of the equilibrium state $x_{\mathrm{E}}$ with the fixed equilibrium decision $u_{\mathrm{E}}$ as follows

$$x(t+1) = F(x(t), u_{\mathrm{E}}) , \quad x(t_0) = x_0 . \tag{3.19}$$

Equation (3.19) appears as a deterministic difference equation without control or decision. Being stable means that any trajectory $x(t)$ starting close enough to $x_{\mathrm{E}}$ remains in the vicinity of $x_{\mathrm{E}}$. Being asymptotically stable means that any trajectory $x(t)$ starting close enough to $x_{\mathrm{E}}$ remains in the vicinity of $x_{\mathrm{E}}$ and converges towards $x_{\mathrm{E}}$ as illustrated by Fig. 3.2.

Formal definitions dealing with the stability of the equilibrium state $x_{\mathrm{E}}$ read as follows.

**Definition 3.2.** *Consider state $x_{\mathrm{E}}$ and decision $u_{\mathrm{E}}$ satisfying (3.3). The equilibrium state $x_{\mathrm{E}}$ is said to be*

- stable *if, for any neighborhood $\mathcal{M}(x_{\mathrm{E}})$ of $x_{\mathrm{E}}$, there exists a neighborhood $\mathcal{N}(x_{\mathrm{E}})$ of $x_{\mathrm{E}}$ such that for every $x_0$ in $\mathcal{N}(x_{\mathrm{E}})$, the solution $x(t)$ of (3.19) belongs to $\mathcal{N}(x_{\mathrm{E}})$ for every $t \geq t_0$,*

$$\forall x_0 \in \mathcal{N}(x_{\mathrm{E}}) , \quad x(t) \in \mathcal{M}(x_{\mathrm{E}}) , \quad \forall t \geq t_0 ;$$

- asymptotically stable *if it is stable and if there exists a neighborhood $\mathcal{N}(x_{\mathrm{E}})$ of $x_{\mathrm{E}}$ having the following property*[1]*: for any $x_0$ in $\mathcal{N}(x_{\mathrm{E}})$, the solution $x(t)$ of (3.19) approaches $x_{\mathrm{E}}$ while $t \to +\infty$*

$$\forall x_0 \in \mathcal{N}(x_{\mathrm{E}}), \quad \lim_{t \to +\infty} x(t) = x_{\mathrm{E}} .$$

- unstable *if it is not stable.*

The *basin of attraction* of an equilibrium state $x_{\mathrm{E}}$ is the set of initial states $x_0 \in \mathbb{X}$ from which a solution $x(t)$ of (3.19) starts converging towards $x_{\mathrm{E}}$ when $t \to +\infty$. When the basin of attraction of an equilibrium state $x_{\mathrm{E}}$ is the whole state space $\mathbb{X}$, one speaks of a *globally asymptotically stable* equilibrium state.

### 3.4.2 Stability of linear systems

Whenever the dynamic (3.19) turns out to be linear, it defines a *dynamic linear system*

$$x(t+1) = Ax(t) , \tag{3.20}$$

where $A$ is a real square matrix of size $n$. In this case, stability requirements are easy to characterize using linear algebraic calculus and, in particular,

[1] This last property defines an *attractive* point.

eigenvalues[2]. Let us recall how the asymptotic behavior of the trajectories for the linear system (3.20) depends on the modulus[3] of the eigenvalues (possibly complex) of the matrix $A$. The set of all eigenvalues of the square matrix $A$ is the *spectrum* spec($A$). The sketch of the proof of the following Theorem 3.3 is recalled in Sect. A.1 in the Appendix.

**Theorem 3.3.** *The zero equilibrium state for the linear system (3.20) is asymptotically stable if the eigenvalues* $\lambda \in spec(A)$ *of the matrix* $A$ *have modulus strictly less than one:*

$$\max_{\lambda \in spec(A)} |\lambda| < 1 \iff 0 \textit{ is asymptotically stable for } (3.20) \, .$$

Thanks to this result, we obtain a very simple criterion characterizing the asymptotic stability of the equilibrium state $x_{\mathrm{E}} = 0$ of the linear system (3.20).

### 3.4.3 Linearization around the equilibrium

Unfortunately, in many cases, the dynamics around the equilibrium do not display linear features. However, whenever dynamic $F$ is continuously differentiable, the linearization of the dynamics around the equilibrium generally enables derivation of the local stability behavior of the system around this equilibrium. This result stems from the fact that, close to the equilibrium $x_{\mathrm{E}}$, dynamic $F(x, u_{\mathrm{E}})$ behaves like its first order (or linear) approximation involving the derivatives $\frac{\partial F}{\partial x}(x_{\mathrm{E}}, u_{\mathrm{E}})$ since

$$F(x, u_{\mathrm{E}}) \approx x_{\mathrm{E}} + (x - x_{\mathrm{E}}) \frac{\partial F}{\partial x}(x_{\mathrm{E}}, u_{\mathrm{E}}) \, .$$

The linearized system associated with the dynamic (3.19) is

$$\xi(t+1) \; = \; A\,\xi(t) \, , \tag{3.21}$$

where the square matrix $A = \frac{\partial F}{\partial x}(x_{\mathrm{E}}, u_{\mathrm{E}})$ is the *Jacobian matrix* of the dynamic

$$F(x, u) = \begin{pmatrix} F^1(x_1, x_2, \ldots, x_n, u) \\ \vdots \\ F^n(x_1, x_2, \ldots, x_n, u) \end{pmatrix}$$

at the equilibrium $(x_{\mathrm{E}}, u_{\mathrm{E}})$:

---

[2] Recall that an *eigenvalue* of the square matrix $A$ is a complex $\lambda$ such that there exists a non zero complex vector $v$ satisfying $Av = \lambda v$. Eigenvalues are solutions of the equation $0 = \mathrm{Det}(A - \lambda I)$, where $I$ is the identity matrix.

[3] For any complex $z = a + ib \in \mathbb{C}$ where $i^2 = -1$, $a \in \mathbb{R}$ represents the real part, $b \in \mathbb{R}$ stands for the imaginary part while the modulus is $|z| = \sqrt{a^2 + b^2}$.

$$A = \frac{\partial F}{\partial x}(x_{\rm E}, u_{\rm E}) = \begin{pmatrix} \frac{\partial F^1}{\partial x_1}(x_{\rm E}, u_{\rm E}) & \frac{\partial F^1}{\partial x_2}(x_{\rm E}, u_{\rm E}) & \cdots & \frac{\partial F^1}{\partial x_n}(x_{\rm E}, u_{\rm E}) \\ \vdots & \vdots & \ddots & \vdots \\ \frac{\partial F^n}{\partial x_1}(x_{\rm E}, u_{\rm E}) & \frac{\partial F^n}{\partial x_2}(x_{\rm E}, u_{\rm E}) & \cdots & \frac{\partial F^n}{\partial x_n}(x_{\rm E}, u_{\rm E}) \end{pmatrix} . \quad (3.22)$$

We can obtain asymptotic stability results using linearized dynamics [10].

**Theorem 3.4.** *Let $x_{\rm E}$ be an equilibrium state of the dynamical system (3.19) where $F$ is continuously differentiable in a neighborhood of $(x_{\rm E}, u_{\rm E})$. If the zero equilibrium of the linearized system (3.21)-(3.22) is asymptotically stable, then $x_{\rm E}$ is asymptotically stable. In other words*

$$\max\{|\lambda| \, , \, \lambda \in spec\,(\frac{\partial F}{\partial x}(x_{\rm E}, u_{\rm E}))\} < 1 \implies x_{\rm E} \quad \textit{asymptotically stable.}$$

*If none of the eigenvalues of the Jacobian matrix (3.22) have a modulus exactly equal to one, and if at least one eigenvalue has a modulus strictly greater than one, then $x_{\rm E}$ is unstable.*

## 3.5 What about stability for MSE, PPE and CPE?

Due to the practical importance of management based on maximum sustainable yield management, private property, common property, or open access equilibria, we provide an analysis of the stability of their equilibria.

We consider surplus models (3.5) for the management of a renewable resource. From Theorem 3.4, a sufficient condition for asymptotic stability of $(B_{\rm E}, h_{\rm E})$ relies on the marginal condition

$$|g'(B_{\rm E} - h_{\rm E})| < 1 \, .$$

We saw in (3.9) that the maximum sustainable equilibrium $(B_{\rm MSE}, h_{\rm MSE})$ satisfies $g'(B_{\rm MSE} - h_{\rm MSE}) = 1$, when $B_{\rm MSE} > 0$. The maximum sustainable equilibrium thus lies at the boundary between asymptotic stability and instability.

As far as the private property equilibrium PPE is concerned, it is asymptotically stable under some assumptions on the signs of the following partial derivatives:

- $C_h(h, B) \leq p$, meaning that an increment in catches increases the rent since $\mathcal{R}_h(h, B) = p - C_h(h, B) \geq 0$;
- $C_B(h, B) \leq 0$, meaning that the harvesting costs decrease with the biomass.

Indeed, from (3.11) we deduce that

$$0 \leq g'(B_{\rm PPE} - h_{\rm PPE}) = \frac{p - C_h(h_{\rm PPE}, B_{\rm PPE})}{p - C_h(h_{\rm PPE}, B_{\rm PPE}) - C_B(h_{\rm PPE}, B_{\rm PPE})} \leq 1 \, ,$$

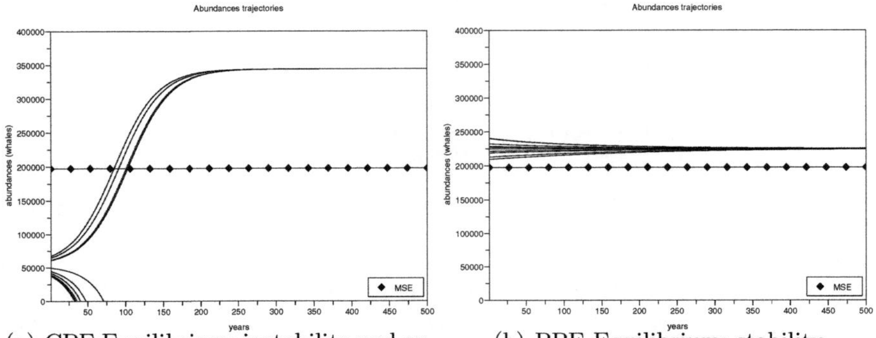


(a) CPE Equilibrium: instability and extinction from below

(b) PPE Equilibrium: stability

**Fig. 3.2.** Local stability of PPE and instability of CPE equilibria for Beverton-Holt recruitment for Antartic blue whale data from [7]: intrinsic growth $R = 1.05$, carrying capacity $K = 400\ 000$ whales, cost $c = 600\ 000$ \$ per whale-catcher year, catchability $q = 0.0\ 016$ per whale-catcher year and $p = 7\ 000$ \$ per whale. Trajectories are obtained by SCILAB code 5. The straight line is the maximum sustainable equilibrium MSE.

and the private property equilibrium PPE is asymptotically stable.

Let us now examine the stability of the equilibria for the Beverton-Holt $g(B) = \frac{RB}{1+bB}$ recruitment dynamic. Let $(B_{\text{E}}, h_{\text{E}})$ be an admissible equilibrium, namely $0 \leq B_{\text{E}} \leq \frac{R-1}{b}$ and $h_{\text{E}} = \sigma(B_{\text{E}})$ given by (3.14). We have

$$\frac{d}{dB}_{|B=B_{\text{E}}} g(B - h_{\text{E}}) = \frac{R}{\big(1 + b(B_{\text{E}} - h_{\text{E}})\big)^2} = \frac{(R - bB_{\text{E}})^2}{R} .$$

Thus, any equilibrium $(B_{\text{E}}, h_{\text{E}})$ is asymptotically stable if

$$(R - bB_{\text{E}})^2 < R .$$

Since $B_{\text{E}} \leq \frac{R-1}{b}$, we obtain that $R - bB_{\text{E}}0 \geq 1$ and we find the stability requirement:

$$\frac{R - \sqrt{R}}{b} < B_{\text{E}} .$$

Consequently, equilibria $B_{\text{E}}$ need to be small enough for a feasible harvesting to exist ($B_{\text{E}} \leq K$ by (3.14)) while large enough for asymptotic stability to hold true. Using (3.15), we thus establish the following requirements for an equilibrium to exist and to be asymptotically stable.

**Result 3.5** *For the Beverton-Holt population dynamic, an equilibrium $B_{\text{E}}$ is asymptotically stable if it lies between the maximum sustainable biomass and the carrying capacity*

$$B_{\text{MSE}} < B_{\text{E}} \leq K .$$

Fig. 3.2 illustrates the stability results for the Antartic blue whale. Both common property and private property equilibria are examined from data provided in [7]: intrinsic growth is set to $R = 1.05$ per year, carrying capacity is $K = 400\ 000$ whales, unit cost is $c = 600\ 000$ \$ per whale-catcher year, catchability is $q = 0.0\ 016$ per whale-catcher year and $p = 7\ 000$ \$ per whale. Trajectories are obtained by SCILAB code 5. Computations for equilibrium states give $B_{\mathrm{MSE}} \approx 197\ 560$ whales, $B_{\mathrm{PPE}} \approx 224\ 962$ whales and $B_{\mathrm{CPE}} \approx 53\ 571$ whales. From inequality (3.16) $B_{\mathrm{MSE}} < B_{\mathrm{PPE}}$, we deduce that the sole owner equilibrium $B_{\mathrm{PPE}}$ is asymptotically stable whereas the maximum sustainable equilibrium is not. The stability of CPE depends on costs and prices: a low fraction $\dfrac{c}{pq} = B_{\mathrm{CPE}}$ displays instability as one might then have $B_{\mathrm{CPE}} \leq B_{\mathrm{MSE}}$. Such is the case for blue whale data as depicted by Fig. 3.2 (a). Let us point out that instability of CPE also means extinction, in the sense that $B(t) \to_{t\to+\infty} 0$ for initial states $B(t_0)$ starting from below the equilibrium $B_{\mathrm{CPE}}$.

SCILAB CODE 5.

```
//
// exec stab_BH.sce ; clear

/////////////////////////////////////////////////////
// Antartic Blue whales data
/////////////////////////////////////////////////////

// Biological parameters and functions

R_B=1.05;
// Intrinsic rate of growth 5% per year
k = 400000;
// Carrying capacity (whales)
b = (R_B-1)./k;

function [y]=Beverton(N)
// Beverton-Holt population dynamics
  y=(R_B*N)./(1 + b*N)
endfunction

function [h]=Sust_yield(N)
// sustainable yield
  h=Beverton(N)-N
endfunction

function [y]=Beverton_e(t,N)
  he=Sust_yield(Ne);
  y=max(0,Beverton(N) - he);
endfunction


// Economic parameters

c_e=600000; // cost in $ per whale catcher year
p=7000; // price in $ per blue whale unit
q=0.0016; // catchability per whale catcher year



/////////////////////////////////////////////////////
// Equilibria and yields
/////////////////////////////////////////////////////

NMSE= (sqrt(R_B) - 1)/b ;
// Maximum sustainable equilibrium MSE
NPPE= ( sqrt( R_B * (1 + (b*c_e/(p*q)) ) ) - 1 ) / b ;
// Private property equilibrium PPE
NCPE=c_e/(p*q) ;
// Common property equilibrium CPE

hMSE= Sust_yield(NMSE) ;
hPPE= Sust_yield(NPPE) ;
hCPE= Sust_yield(NCPE) ;


/////////////////////////////////////////////////////
// Trajectories simulations
/////////////////////////////////////////////////////

// Only the last line counts. Change it.
Ne=NCPE     // unstable case
Ne=NPPE     // stable case

he=Beverton(Ne)-Ne

Horizon=500;  time=1:Horizon;
// Time horizon

epsil=20000;
// Perturbation level around the equilibrium

xset("window",2); xbasc(2);
plot2d2(time,ones(1,Horizon)*NMSE,rect=[1,0,Horizon,k]);
Time=linspace(1,Horizon,20);
plot2d2(Time,ones(Time)*NMSE,style=-4);
legends("MSE",-4,'lr')
// Plot of the MSE

for (i=1:10)     // trajectories loop
  N0=Ne + 2*(rand(1)-0.5)*epsil;
  // perturbation of initial state
Nt=ode("discrete",N0,0,time,Beverton_e);
// computation of the trajectory
plot2d2(time,Nt,rect=[1,0,Horizon,k]);
// plot of the trajectory
end

xtitle('Abundances trajectories','years',...
    'abundances (whales)')

//
```

## 3.6 Open access, instability and extinction

For the open access model, the Jacobian matrix of dynamics (3.17) at equilibrium $x_{\mathrm{E}} = (B_{\mathrm{E}}, e_{\mathrm{E}})$ is

$$A = \frac{dF}{dx}(x_{\mathrm{E}}) = \begin{pmatrix} (1 - qe_{\mathrm{E}})g'\big(B_{\mathrm{E}}(1 - qe_{\mathrm{E}})\big) \ ; & -qB_{\mathrm{E}}g'\big(B_{\mathrm{E}}(1 - qe_{\mathrm{E}})\big) \\ \alpha pqe_{\mathrm{E}} \quad ; & 1 \end{pmatrix},$$

where we use the property (3.18) that $B_{\mathrm{E}} = \dfrac{c}{pq}$ . In the linear case $g(B) = RB$ where $e_{\mathrm{E}} = \dfrac{R-1}{Rq}$, stability cannot be performed as illustrated by Fig. 3.3. This situation stems from the fact that the Jacobian matrix reads as follows:

$$A = \begin{pmatrix} R(1 - qe_{\mathrm{E}}) \ ; & -RqB_{\mathrm{E}} \\ \alpha pqe_{\mathrm{E}} \quad ; & 1 \end{pmatrix} = \begin{pmatrix} 1 & -\frac{Rc}{p} \\ \alpha p\frac{R-1}{R} & 1 \end{pmatrix}.$$

The two eigenvalues $(\lambda_1, \lambda_2)$ of matrix $A$ are the solution of

$$0 = \mathrm{Det}(A - \lambda I_2) = (1 - \lambda)^2 + \alpha c(R - 1) .$$

Since $R > 1$, we deduce that $\lambda_1 = 1 - i\sqrt{\alpha c(R-1)}$ and $\lambda_2 = 1 + i\sqrt{\alpha c(R-1)}$. Therefore, the modulus of each eigenvalue is

$$|\lambda_1| = |\lambda_2| = \sqrt{1 + \alpha c(R-1)}\,,$$

strictly greater than 1 whenever $\alpha > 0$. Consequently, stability cannot occur as illustrated in the Figs. 3.3 for the specific parameters $p = 1$, $q = 1$, $c = 2$, $\alpha = 10^{-2.5}$, $g(B) = (1 + r)B$, $r = 0.1$. Note that such an instability relies on oscillations that display greater and greater amplitude through time. This process ends with the collapse of the whole system, including extinction of the resource. Other illustrative numerical examples can be found in [3] for logistic dynamics.

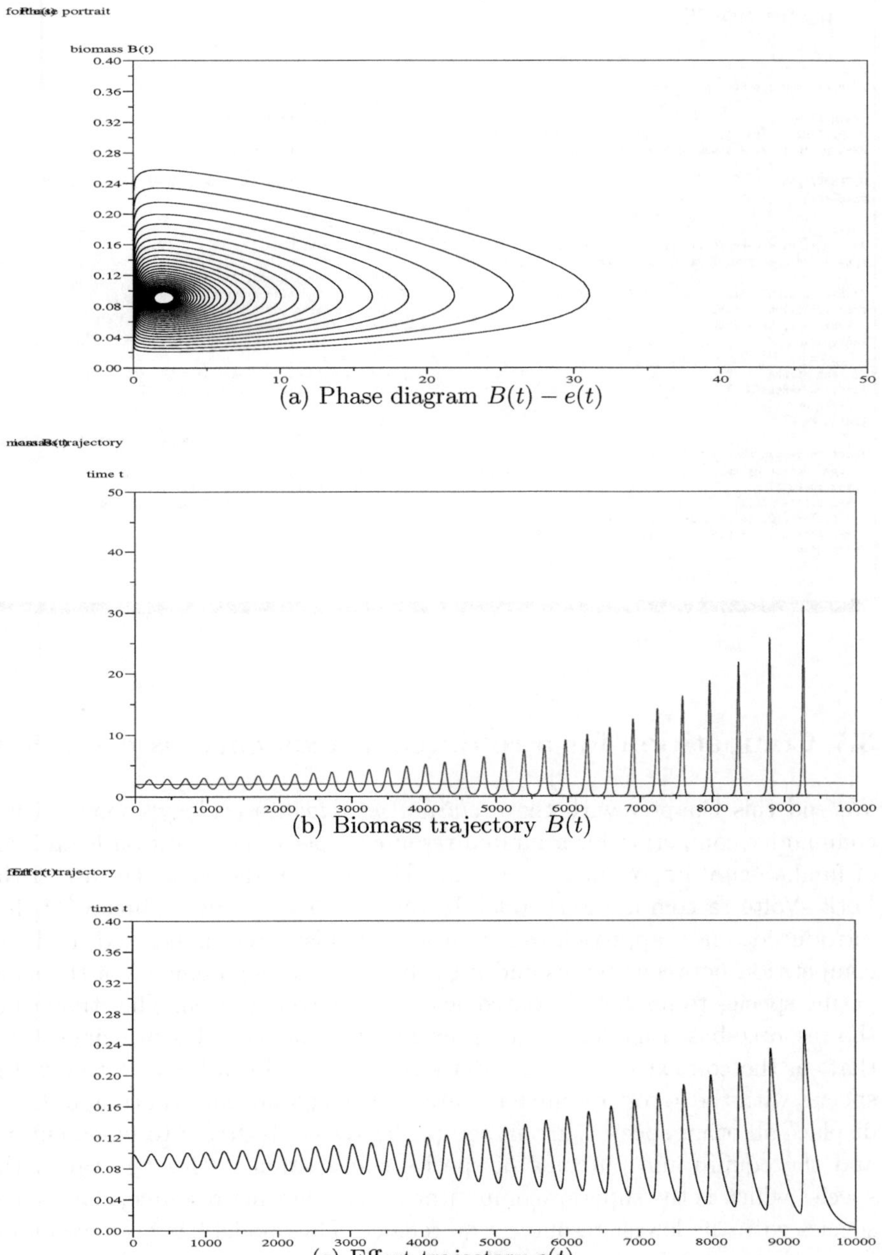


(a) Phase diagram $B(t) - e(t)$

(b) Biomass trajectory $B(t)$

(c) Effort trajectory $e(t)$

**Fig. 3.3.** Open access: a case of instability and non viability ($p = 1$, $q = 1$, $c = 2$, $\alpha = 10^{-2.5}$, $g(B) = RB$, $R = 1.1$). The trajectories were generated by SCILAB code 6.

SCILAB CODE 6.

```
//
// exec open_access.sce ; clear

p=1;q=1;c=2;
eta=10^{-2.5}; r=0.1;// Divergent case for open access
NV=%eps*10^10 // Minimum viable population

N_CPE=c/(p*q)
E_CPE=r/q

R=1+r;
A=[1 -R*c/p; eta*p*r/R 1]; // Jacobian matrix
spec(A) // Spectrum of Jacobian matrix

// Linear population growth with Allee effect
function [y]=Linear(N)
  y=N*(1+r), y=y*(y>NV)
endfunction

// rent of the catches
function R=Rent(N,e)
  R=p*q*e*N-c*e
endfunction

function y=Open_access(t,x)
// Open access dynamics
  y=zeros(2,1)
  N=x(1,1),e=x(2,1)
  y(1,1)=max(0,Linear(N-q*e*N))
  y(2,1)=max(0,(e+eta*Rent(N,e)))
endfunction

// graphics
xset("window",2); xbasc(2); ...
xtitle('Biomass trajectory','time t','biomass B(t)');
xset("window",3); xbasc(3); ...
xtitle('Effort trajectory','time t','effort e(t)')
xset("window",4); xbasc(4); ...
xtitle('Phase portrait','biomass B(t)','effort e(t)')

// trajectories
Horizon=10000;
time=0:Horizon;

x0=[N_CPE;E_CPE]+(rand(1)-0.5)/100;
// initial condition around CPE
xt=ode("discrete",x0,0,time,Open_access);
// Computation of the trajectory

xset("window",2);
plot2d(time,[ones(Horizon+1,1)*N_CPE xt(1,:)'],...
   rect=[1,0,Horizon,max(xt(1,:))*1.5]);

xset("window",3);
plot2d(time,xt(2,:)',rect=[1,0,Horizon,max(xt(2,:))*1.5]);

xset("window",4);
plot2d(xt(1,time+1), xt(2,time+1)',...
   rect=[0,0,max(xt(1,:))*1.5,max(xt(2,:))*1.5]);

//
```

## 3.7 Competition for a resource: coexistence *vs* exclusion

We end this Chapter with the *exclusion principle* in ecology coping with a community competing for a limited resource. Species competition is an issue of fundamental importance to ecology. The classical theory makes use of the Lotka-Volterra competition model. In the past few decades, Tilman [11] has introduced a new approach based on a mechanistic resource-based model of competition between species and uses the resource requirements of the competing species to predict the outcome of species competition. The strength of the resource-based model lies in an exclusion principle. This principle states that, in the context of a multi-species competition for a limiting factor, the species with the lowest resource requirement in equilibrium will competitively displace all other species. In this setup, the system is driven to mono-culture and the equilibrium outcome of species competition is the survival of the species which is the superior competitor for the limiting resource, that is, the species with the lowest resource requirement. This model has been examined from a bioeconomic perspective for the management of renewable resources. As shown in [1], an exclusion process again occurs, condemning, in this sense, biodiversity. Numerous studies are now aiming at relaxing the model to allow for the coexistence of species. The model is generally described in a continuous time framework. However, we here expose a discrete time version and study the stability of the equilibrium. The state variables are

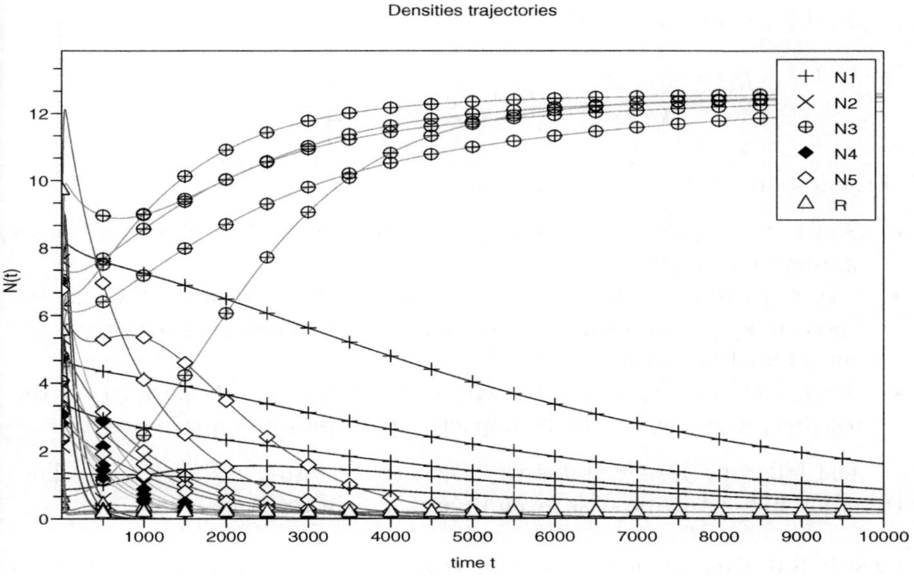


**Fig. 3.4.** An example of the exclusion principle of a resource-based model: the only stable equilibrium contains only the species (here $N_3$) with lowest resource requirement at equilibrium $R_{\rm E}$. The other species $N_1$, $N_2$, $N_4$, $N_5$ are driven to extinction.

- species $i$ density $N_i(t)$, for $i = 1, \dots, n$;
- the limited resource, represented by $R(t)$, for which the species compete.

The time interval between $t$ and $t+1$ is $\Delta_t$. We consider the dynamic

$$N_i(t+1) = N_i(t) + \Delta_t\Big(N_i(t)(f_i R(t) - d_i)\Big)\,, \quad i = 1, \dots, n\,, \tag{3.23a}$$

$$R(t+1) = R(t) + \Delta_t\Big(S(t) - aR(t) - \sum_{i=1}^{n} w_i f_i R(t) N_i(t)\Big)\,. \tag{3.23b}$$

In (3.23a), the rate of growth

$$\frac{N_i(t+1) - N_i(t)}{\Delta_t} = N_i(t)(f_i R(t) - d_i)$$

of species $i$ density is linear in $N_i(t)$ with per capita rate of growth $f_i R(t) - d_i$ where

- $f_i R(t)$ is the resource-based per capita growth of species $i$; the higher $f_i > 0$, the more species $i$ can exploit the resource $R(t)$ for its own growth;
- $d_i > 0$ is the natural death rate.

In (3.23b), the rate of growth

$$\frac{R(t+1)-R(t)}{\Delta_t} = S(t) - aR(t) - \sum_{i=1}^{n} w_i f_i R(t) N_i(t)$$

of the resource $R(t)$ is decomposed in three terms:

- $S(t)$ is the input of the resource, supposed to be known and taken as stationary in what follows;
- $-aR(t)$ corresponds to self-limitation of the resource with $a > 0$; the expression $S(t) - aR(t)$ stands for the natural evolution of the resource without interactions with species;
- $-w_i f_i R(t) N_i(t)$ represents the effect of competition of species $i$ on the resource, where $w_i > 0$ is the impact rate of species $i$ on resource $R$.

The following Result states the so-called *exclusion principle*, also illustrated by Fig. 3.4. A proof is given in Sect. A.1 in the Appendix.

**Result 3.6** *Suppose that input $S(t)$ is stationary, set to value $S_{\mathrm{E}}$. If $S_{\mathrm{E}}$ is large enough, and if the time unit $\Delta_t > 0$ is small enough*[4]*, the only asymptotically stable equilibrium of (3.23a)-(3.23b) is given by*

$$\begin{cases} R_{\mathrm{E}} = \min_{i=1,\dots,n} \dfrac{d_i}{f_i} = \dfrac{d_{i_{\mathrm{E}}}}{f_{i_{\mathrm{E}}}} , \\ N_{i,\mathrm{E}} = \begin{cases} \dfrac{S_{\mathrm{E}} - R_{\mathrm{E}} a}{R_{\mathrm{E}} w_{i_{\mathrm{E}}} f_{i_{\mathrm{E}}}} > 0 & \textit{if} \quad i = i_{\mathrm{E}} , \\ 0 & \textit{if} \quad i \neq i_{\mathrm{E}} . \end{cases} \end{cases} \tag{3.24}$$

[4] And excluding the exceptional case where $d_i/f_i = d_j/f_j$ for at least one pair $i \neq j$.

# References


[1] W. Brock and A. Xepapadeas. Optimal ecosystem management when species compete for limiting resources. *Journal of Environmental Economics and Management*, 44:189–220, 2002.

[2] C. W. Clark. *Mathematical Bioeconomics.* Wiley, New York, second edition, 1990.

[3] J. M. Conrad. *Resource Economics.* Cambridge University Press, 1999.

[4] H. S. Gordon. The economic theory of a common property resource: the fishery. *Journal of Political Economy*, 62:124–142, 1954.

[5] J. M. Hartwick and N. D. Olewiler. *The Economics of Natural Resource Use.* Harper and Row, New York, second edition, 1998.

[6] H. K. Khalil. *Nonlinear Systems.* Prentice-Hall, Englewood Cliffs, second edition, 1995.

[7] M. Kot. *Elements of Mathematical Ecology.* Cambridge University Press, Cambridge, 2001.

[8] N. Rouche and J. Mawhin. *Équations différentielles ordinaires*, volume 1. Masson, Paris, 1973.

[9] N. Rouche and J. Mawhin. *Équations différentielles ordinaires*, volume 2. Masson, Paris, 1973.

[10] S. Sastry. *Nonlinear Systems. Analysis, Stability, and Control.* Springer-Verlag, 1999.

[11] D. Tilman. *Plant Strategies and the Dynamics and Structure of Plant Communities.* Princeton Univ. Press, Princeton, 1988.

# 4

# Viable sequential decisions

Basically, viability means the ability of survival, namely the capacity for a system to maintain conditions of existence through time. By extension, viability may also refer to perennial situations of good health, safety, or effectiveness in the sense of cost-effectiveness in economics.

Of major interest are viability concerns for the sustainability sciences and especially in bioeconomics, environmental or ecological economics or conservation biology. Harvesting a resource without jeopardizing it, preventing extinction of rare species, preserving biodiversity or ecosystems, avoiding or mitigating climate change are all examples of sustainability issues where viability plays a major role. For instance, the quantitative method of conservation biology for the survival of threatened species is termed Population Viability Analysis (PVA) [19]. Safe Minimum Standards (SMS) [4] have been proposed with the goal of preserving and maintaining a renewable resource at a level that precludes irreversible extinction except in cases where social costs are prohibitive or immoderate. Similarly, the ICES precautionary framework [15] aims at conserving fish stocks and fisheries on the grounds of several indicators including spawning stock biomass or fishing effort multiplier. In this context, reference points not to be exceeded for these bioeconomic indicators stand for management objectives. In the same vein, for greenhouse gas issues, a tolerable mitigation policy is often represented by a ceiling threshold of $CO_2$ concentration (for instance 450 ppm) not to be violated [16]. Basically, in these cases, sustainability is defined as the ability to maintain the system within some satisfying normative bounds for a large or indefinite time.

Regarding these issues, a major challenge is to study the compatibility between controlled biological, physical and economical dynamics and constraints or targets accounting for economic and/or ecological objectives. Such preoccupations refer to the study of controlled dynamical systems under constraints and targets. Such a mathematical problem is termed weak or controlled invariance [6, 24] or a viability problem [1]. In applied mathematics and the systems theory, the question of constraints has mostly been neglected to concentrate rather on steady state equilibria or optimization concepts. Equilibrium and stability depicted in the previous Chap. 3 are only a partial response to viability issues. Indeed, equilibria are clearly "viable" but such an analysis does not take into account the full diversity of possible transitions of the system. Although dynamics optimization problems are usually formulated under constraints as will be seen in Chap. 5, the role played by the constraints poses difficult technical problems and is generally not tackled by itself. Furthermore, the optimization procedure reduces the diversity of feasible forms of evolution by, in general, selecting a single trajectory.

The aim of this Chapter is to provide the specific mathematical tools to deal with discrete time dynamical systems under state and control constraints and to shed new light on some applied viability problems, especially in the economic and environmental fields.

The ideas of all the mathematical statements presented in this Chapter are inspired mainly by [1] in the continuous case. Here, to restrict the mathematical content to a reasonable level, we focus on the discrete time framework as in [8, 24]. We illustrate the mathematical concepts or results through simple examples taken from the environmental and sustainability topics. We refer for instance to [17] for exhaustible resource management. The tolerable windows approach [5, 21] proposes a similar framework, mainly focusing on climate change issues. Works [2, 3, 7, 9, 10, 11, 13, 18, 20] cope with renewable resource management and especially fisheries. Agricultural and biodiversity issues are especially handled in [23, 22]. From the ecological point of view, the so-called population viability analysis (PVA) and conservation biology display concerns close to those of the viable control approach by focusing on extinction processes in an uncertain (stochastic) framework. Links between PVA and the viable control approach will be examined in Chap. 7 dealing with stochastic viability. Moreover, as emphasized in [17], the viable control approach is deeply connected with the Rawlsian or maximin approach [14] important for intergenerational equity as will be explained in Chap. 5.

The chapter is organized as follows. In Sect. 4.1, we present the viability problem on the consistency between a controlled dynamic and acceptability constraints. Resource management examples under viability constraints are given in Sect. 4.2. The main concept to tackle the viability problem is the viability kernel, introduced in Sect. 4.3 together with the *dynamic programming* method. By dynamic programming, the dynamic viability decision problem is solved sequentially: one starts at the final time horizon and then applies some backward induction mechanism at each time step. A specific Sect. 4.4 is dedicated to viability in the autonomous case. Following sections are devoted to examples. We end with invariance, or strong viability, in Sect. 4.10. This is a very demanding concept which refers to the respect of given constraints whatever the admissible options of decisions or controls taken at every time.

## 4.1 The viability problem

The viability problem relies on the consistency between a controlled dynamic and acceptability constraints applying both to states and decisions of the system.

### The dynamics

We consider again the following nonlinear dynamical system, as in Sect. 2.9,

$$x(t+1) = F\big(t, x(t), u(t)\big)\ , \quad t = t_0, \ldots, T-1 \quad \text{with} \quad x(t_0) = x_0\ , \tag{4.1}$$

where $x(t) \in \mathbb{X} = \mathbb{R}^n$ is the state ($\mathrm{CO}_2$ concentration, biomass, abundances...), $u(t) \in \mathbb{U} = \mathbb{R}^p$ is the control or decision (abatement, catch, effort...), $T \in \mathbb{N} \cup \{+\infty\}$ corresponds to the time horizon which may be finite or infinite, and $x_0 \in \mathbb{X}$ is the initial condition at initial time $t_0 \in \{0, \ldots, T-1\}$.

Viability is related to preservation, permanence and sustainability of some conditions which represent survival, safety or effectiveness of a system. We introduce both decision and state constraints, and coin them as *ex ante* or *a priori* viability conditions.

### Decision constraints

We consider the conditions

$$u(t) \in \mathbb{B}\big(t, x(t)\big)\ , \quad t = t_0, \ldots, T-1\ . \tag{4.2a}$$

The non empty control domain $\mathbb{B}\big(t, x(t)\big) \subset \mathbb{U}$ is the set of admissible and *a priori* acceptable decisions. The most usual case occurs when the set of admissible controls is constant *i.e.* $\mathbb{B}(t, x) = \mathbb{B}$ for every state $x$ and time $t$. Generally, these constraints are associated with equality and inequality (vectorial) requirements of the form

$$b_i\big(t, x(t), u(t)\big) \leq 0\ , \quad b_e\big(t, x(t), u(t)\big) = 0\ .$$

### State constraints

The safety, the admissibility or the effectiveness of the state for the system at time $t$ is represented by non empty state constraint domains $\mathbb{A}(t) \subset \mathbb{X}$ in the sense that we require

$$x(t) \in \mathbb{A}(t)\ , \quad t = t_0, \ldots, T-1\ . \tag{4.2b}$$

The usual example of such a tolerable window $\mathbb{A}(t)$ concerns equality and inequality (vectorial) constraints of the type

$$a_i\big(t, x(t)\big) \leq 0\ , \quad a_e\big(t, x(t)\big) = 0\ .$$

### Target constraints

Target problems are specific state requirements of the form

$$x(T) \in \mathbb{A}(T) . \tag{4.2c}$$

Hereafter, target constraints (4.2c) and state constraints (4.2b) are unified through the following constraints

$$x(t) \in \mathbb{A}(t) , \quad t = t_0, \ldots, T . \tag{4.2d}$$

### The question of consistency between dynamics and constraints

Dynamics (4.1) and constraints (4.2a)–(4.2d) form a set of relations for which there may or may not exist a global solution because some requirements and processes may turn out to be contradictory along time. Viability or weak or controlled invariance issues focus on the initial states for which at least one feasible solution combining dynamics and constraints exists. Mathematically speaking, the viability problem consists in identifying conditions under which the following set of state-control trajectories (already introduced in (2.63)),

$$\mathcal{T}^{ad}(t_0, x_0) = \left\{ \big(x(\cdot), u(\cdot)\big) \;\middle|\; \begin{array}{ll} x(t_0) = x_0 \;, & \\ x(t+1) = F\big(t, x(t), u(t)\big) \;, & t = t_0, \ldots, T-1 \\ u(t) \in \mathbb{B}\big(t, x(t)\big) \;, & t = t_0, \ldots, T-1 \\ x(t) \in \mathbb{A}(t) \;, & t = t_0, \ldots, T \end{array} \right\} \tag{4.3}$$

is not empty, and then exhibiting its elements.

More specifically, in the sustainability context, viability may capture the satisfaction of both economic and environmental constraints. In this sense, it is a multi-criteria approach sometimes known as co-viability.

Moreover, as soon as the constraints do not explicitely depend on time $t$ and the horizon is infinite ($T = +\infty$), let us stress that an intergenerational equity feature is naturally integrated within this framework as the viability requirements are identical along time without future or present preferences.

## 4.2 Resource management examples under viability constraints

We provide simple examples in the management of a renewable resource, in climate change mitigation, in the management of an exhaustible resource and of forestry.

### 4.2.1 Viable management of a threatened population

Suppose that the dynamics of a renewable resource, characterized by its biomass $B(t)$ and catch level $h(t)$, is governed by

$$B(t+1) = g\big(B(t) - h(t)\big) ,$$

where $g$ is some natural resource growth function, as in Sect. 2.2. The harvesting of the population is constrained by the biomass

$$0 \le h(t) \le B(t) .$$

We look for policies which maintain the biomass level within an ecological window, namely between conservation and maximal safety values, at all times.

$$0 < B^{\flat} \le B(t) \le B^{\sharp} .$$

Think of a population which may have extinction threats for small population levels because of some Allee effect while it can be dangerous for other species or the habitat from some higher level. Elephant populations in Africa or predator fish species like Nile Perch are significant examples. For social rationale, we can also require guaranteed harvests that provide food or money for local human populations in the area under concern:

$$0 < h^{\flat} \le h(t) .$$

The question that arises is whether conservation, habitat and harvesting goals can all be met simultaneously.

### 4.2.2 Mitigation for climate change

The following model, presented in Sect. 2.3, deals with the management of the interaction between economic growth and greenhouse gas emissions as in [12]. Let us consider the following dynamics of $\mathrm{CO_2}$ concentration $M(t)$ and economic production $Q(t)$:

$$\begin{cases} M(t+1) = M(t) + \alpha \mathfrak{E}_{\mathrm{BAU}}\big(Q(t)\big)(1 - a(t)) - \delta(M(t) - M_{\infty}) , \\ Q(t+1) = (1+g)Q(t) . \end{cases}$$

Here, control $a(t) \in [0,1]$ is the abatement rate of $\mathrm{CO_2}$ emissions. The $\mathrm{CO_2}$ emission function $\mathfrak{E}_{\mathrm{BAU}}(Q) > 0$ represents a "business as usual" scenario and depends on production level $Q(t)$. Production $Q(t)$ grows at rate $g$. We require the abatement costs $C(a,Q)$ not to exceed a maximal cost threshold:

$$C\big(a(t), Q(t)\big) \le c^{\sharp} , \quad t = 0, \ldots, T .$$

The concentration has to remain below the tolerable level at the horizon $T$ in a target framework:

$$M(T) \le M^\sharp .$$

A more demanding requirement consists in constraining the concentration over the whole period:

$$M(t) \le M^\sharp , \quad t = 0, \dots, T .$$

### 4.2.3 Management of an exhaustible resource

Following [17] in a discrete time context, we describe an economy with exhaustible resource use by

$$\begin{cases} S(t+1) = S(t) - h(t) , \\ K(t+1) = K(t) + Y\big(K(t), h(t)\big) - c(t) , \end{cases}$$

as in Sect. 2.7, where $S$ is the exhaustible resource stock, $h$ stands for the extraction flow, $K$ represents the accumulated capital, $c$ stands for the consumption and the function $Y$ represents the technology of the economy. Hence, the decision or controls of this economy are levels of consumption $c$ and extraction $h$ respectively.

Now let us consider the state-control constraints. First it is assumed that extraction $h(t)$ is irreversible in the sense that

$$0 \le h(t) . \tag{4.4}$$

We also consider that the capital is nonnegative

$$0 \le K(t) . \tag{4.5}$$

We take into account scarcity of the resource by requiring

$$0 \le S(t) .$$

Letting $S^\flat > 0$ stand for some minimal resource target, we can more generally consider a stronger conservation constraint for the resource as follows:

$$S^\flat \le S(t) . \tag{4.6}$$

We may choose to impose a requirement related to some guaranteed consumption level $c^\flat > 0$ along the generations:

$$c^\flat \le c(t) . \tag{4.7}$$

This constraint refers to sustainability and intergenerational equity since it can be written in terms of utility in a form close to Rawl's criteria:

$$L(c^\flat) \le \inf_{t=t_0,\dots,T-1} L\big(c(t)\big) .$$

This equivalence comes from the fact that a utility function is strictly increasing[1] in consumption $c$.

[1] We shall say that $f$ is an *increasing function* if $x \ge x' \Rightarrow f(x) \ge f(x')$, while $f$ is a *strictly increasing function* if $x > x' \Rightarrow f(x) > f(x')$.

### 4.2.4 Forestry management

A detailed version of the following model with age classes is exposed in [20]. As described in Sect. 2.5, we consider a forest whose structure in age is represented in discrete time by a vector of surfaces

$$N(t) = \begin{pmatrix} N_n(t) \\ N_{n-1}(t) \\ \vdots \\ N_1(t) \end{pmatrix} \in \mathbb{R}^n_+ ,$$

where $N_j(t)$ ($j = 1, \ldots, n-1$) represents the surface used by trees whose age, expressed in the unit of time used to define $t$, is between $j-1$ and $j$ at the beginning of yearly period $[t, t+1[$; $N_n(t)$ is the surface of trees of age greater than $n-1$. We assume that the natural evolution, *i.e.* without exploitation, of the vector $N(t)$ is described by a linear system

$$N(t+1) = A\,N(t) , \tag{4.8}$$

where the terms of the matrix $A$ are nonnegative, which ensures that $N(t)$ remains nonnegative at any time. Particular instances of matrices $A$ are of the Leslie type

$$A = \begin{bmatrix} 1-m & 1-m & 0 & \cdots & 0 \\ 0 & 0 & 1-m & \ddots & 0 \\ & & & \ddots & 0 \\ 0 & \cdots & 0 & \ddots & 1-m \\ \gamma & \gamma & \gamma & \cdots & \gamma \end{bmatrix} , \tag{4.9}$$

where $m \in [0,1]$ is a mortality parameter and $\gamma$ plays the role of the recruitment parameter.

Now we describe the exploitation of such a forest resource. For the sake of simplicity, we assume, first, that the minimum age at which trees can be cut is $n$ and, second, that each time a tree is cut, it is immediately replaced by a tree of age 0. Thus, introducing the scalar variable decision $h(t)$ which represents the surface of trees harvested at time $t$, we obtain the following controlled evolution

$$N(t+1) = A\,N(t) + B\,h(t) ,$$

where $B$ is equal to the column[2] vector $\big(-1\;0\;\cdots\;0\;1\big)'$.

Furthermore, since one cannot plan to harvest more than exists at the beginning of the unit of time, the decision or control variable $h(t)$ is subject to the constraint

[2] The superscript $'$ stands for *transposition*.

$$0 \leq h(t) \leq N_n(t) = CN(t) ,$$

where the row vector $C$ is equal to $\begin{pmatrix} 1\, 0 \cdots 0\, 0 \end{pmatrix}$, which ensures the non-negativity of the resource. Notice that by choosing this constraint, we have implicitly assumed that the harvesting decisions $h(t)$ are effective at the beginning[3] of each period $[t, t+1[$.

To encompass the economic or social feature of the exploitation, we associate the harvesting $h(t)$ with an income, a utility or a service. This harvesting is required to exceed some minimal threshold $h^\flat > 0$ at any time:

$$h^\flat \leq h(t) .$$

## 4.3 The viability kernel

Mathematically, viability or weak invariance issues refer to the consistency between dynamics (4.1) and constraints (4.2a)–(4.2d). Among different ideas[4] to display viability, we focus on the largest viable part of the state constraints: this corresponds to the viability kernel concept that we develop hereafter.

### 4.3.1 The viability kernel

Indeed, a first idea to generate viable behaviors is to consider the set of states from which at least one trajectory starts together with a control sequence, both satisfying the dynamical system equations and the constraints throughout time.

**Definition 4.1.** *The viability kernel at time $s \in \{t_0, \ldots, T\}$ for dynamics (4.1) and constraints (4.2a)–(4.2b)–(4.2c) is the subset of the state space $\mathbb{X}$, denoted by $\mathbb{V}\mathrm{iab}(s)$, defined by:*

$$\mathbb{V}\mathrm{iab}(s) := \left\{ x \in \mathbb{X} \;\middle|\; \begin{array}{l} \textit{there exist decisions } u(\cdot) \\ \textit{and states } x(\cdot) \textit{ starting from } x \textit{ at time } s \\ \textit{satisfying for any time } t \in \{s, \ldots, T-1\} \\ \textit{dynamics (4.1) and constraints (4.2a)–(4.2b)} \\ \textit{and satisfying (4.2c) at final time } T. \end{array} \right\} . \quad (4.10)$$

Notice that, as illustrated in Fig. 4.1,

$$\mathbb{V}\mathrm{iab}(s) \subset \mathbb{A}(s)$$

[3] Harvesting decisions effective at the end of each unit of time $t$ would give $0 \leq h(t) \leq CAN(t)$.

[4] One consists in enlarging the constraints and corresponds to the *viability envelope* concept, while the idea of *forcing* is to modify the dynamics by introducing new regulations or controls.

because the state constraint (4.2b) is satisfied at time $s$. It can also be emphasized that the viability kernel at horizon $T$ is the target set (4.2c):

$$\mathbb{V}\mathrm{iab}(T) = \mathbb{A}(T) .$$

Although we are especially interested in the viability kernel $\mathbb{V}\mathrm{iab}(t_0)$ at initial time $t_0$, we have introduced all the $\mathbb{V}\mathrm{iab}(s)$ for $s \in \{t_0, \ldots, T\}$. Indeed, we now describe how $\mathbb{V}\mathrm{iab}(t_0)$ is given by a backward induction involving the other viability kernels $\mathbb{V}\mathrm{iab}(T)$, $\mathbb{V}\mathrm{iab}(T-1)$, ..., $\mathbb{V}\mathrm{iab}(t_0+1)$.

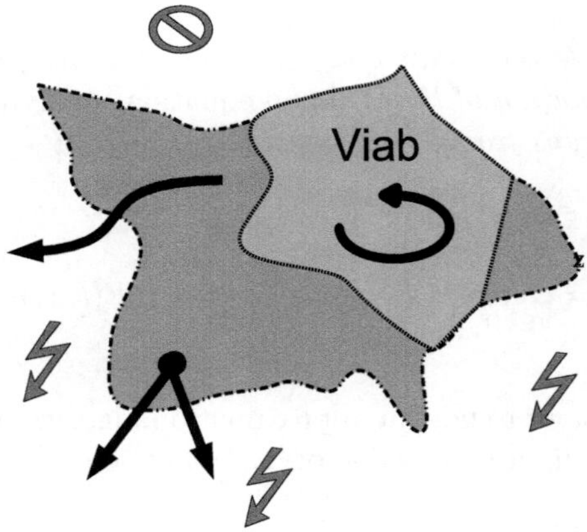


**Fig. 4.1.** The state constraint set $\mathbb{A}$ is the large set. It includes the smaller viability kernel $\mathbb{V}\mathrm{iab}$.

### 4.3.2 Maximality and Bellman properties

As in dynamics optimization problems[5], the Bellman/dynamic programming principle (DP) can be obtained within the viable control framework. Basically, the dynamic programming principle means that the dynamics decision-making problem is solved sequentially: one starts at the final time horizon and then applies some backward induction mechanism at each time. For viability problems, this principle can be described through both geometrical and functional formulations. We distinguish the infinite and finite horizons. The proofs of the two following Propositions 4.2 and 4.3 can be found in Sect. A.2 in the Appendix.

A geometric characterization in the finite horizon case is given by the following Proposition 4.2.

**Proposition 4.2.** *Assume that* $T < +\infty$. *The viability kernel* $\mathbb{V}\mathrm{iab}(t)$ *satisfies the backward induction, where* $t$ *runs from* $T-1$ *down to* $t_0$*:*

[5] See Chaps. 5 and 8.

$$\begin{cases} \mathbb{V}\mathrm{iab}(T) = \mathbb{A}(T) , \\ \mathbb{V}\mathrm{iab}(t) = \{x \in \mathbb{A}(t) \mid \exists u \in \mathbb{B}(t,x) , \quad F(t,x,u) \in \mathbb{V}\mathrm{iab}(t+1)\} . \end{cases} \tag{4.11}$$

Similarly, an equivalent functional characterization can be derived in the finite horizon case. To achieve this, we introduce the *extended*[6] *characteristic function of a set* $K \subset \mathbb{X}$:

$$\Psi_K(x) := \begin{cases} 0 & \text{if} \quad x \in K \\ +\infty & \text{otherwise.} \end{cases} \tag{4.12}$$

**Proposition 4.3.** *Assume that* $T < +\infty$. *The extended function* $V : (t,x) \mapsto \Psi_{\mathbb{V}\mathrm{iab}(t)}(x)$ *is the solution of the* Bellman equation, *or* dynamic programming equation, *where* $t$ *runs from* $T-1$ *down to* $t_0$*:*

$$\begin{cases} V(T,x) = \Psi_{\mathbb{A}(T)}(x) , \\ V(t,x) = \inf_{u \in \mathbb{B}(t,x)} \Big( \Psi_{\mathbb{A}(t)}(x) + V\big(t+1, F(t,x,u)\big)\Big) . \end{cases} \tag{4.13}$$

This Proposition may provide an approximation algorithm by replacing $\Psi_{\mathbb{A}(t)}$ with a function taking a large value on $\mathbb{A}(t)$ and zero elsewhere.

### 4.3.3 Viable controls and feedbacks

The previous characterizations of the viability kernels point out that, for every point $x$ inside the corridor $\mathbb{V}\mathrm{iab}(t)$, there exists an admissible control $u$ which yields a future state $F(t,x,u)$ remaining in $\mathbb{V}\mathrm{iab}(t+1)$ and, consequently, in $\mathbb{A}(t)$. Thus, the following *viable regulation set*

$$\mathbb{B}^{\mathrm{viab}}(t,x) := \{u \in \mathbb{B}(t,x) \mid F(t,x,u) \in \mathbb{V}\mathrm{iab}(t+1)\} \tag{4.14}$$

is not empty. Any $u \in \mathbb{B}^{\mathrm{viab}}(t,x)$ is said to be a *viable control.* A *viable feedback* is a function $\mathfrak{u} : \mathbb{N} \times \mathbb{X} \to \mathbb{U}$ such that $\mathfrak{u}(t,x) \in \mathbb{B}^{\mathrm{viab}}(t,x)$ for all $(t,x)$.

At this stage, let us note that the viability kernels elicit the "true" state constraints of the problem because the viable strategies and controls are designed to make sure that the state remains within the viability kernels $\mathbb{V}\mathrm{iab}(t)$ instead of the initial state constraints $\mathbb{A}(t)$. This situation explains why we may speak of *ex post* viability. In other words, the boundary of viability kernels stands for security barriers with respect to initial viability bounds which are larger, as depicted by Fig. 4.1.

---

6 By *extended*, we mean that a real valued function may take the values $-\infty$ and $+\infty$.

We can see at once that $\mathbb{B}^{\mathrm{viab}}(t,x)$ has no reason to be reduced to a singleton and, thus, there is no uniqueness of viable decisions. This situation advocates for multiplicity and flexibility in the viable decision process. We may however require more oriented and specified decisions among the viable ones. Reference [1] proposes some specific selection procedures such as barycentric, slow or inertial viable controls. Other optimal selections are examined in Chap. 5 dealing with optimal control.

Viability kernels and viable controls make a characterization of the feasibility problem as exposed in Subsect. 4.1 possible. The proof of the following Proposition 4.4 is a consequence of Proposition 4.2 and of the definition (4.14).

**Proposition 4.4.** *The feasibility set $\mathcal{T}^{ad}(t_0,x_0)$ in (4.3) is not empty if, and only if, the initial state belongs to the initial viability kernel:*

$$x_0 \in \mathbb{V}\mathrm{iab}(t_0) \iff \mathcal{T}^{ad}(t_0,x_0) \neq \emptyset\,. \tag{4.15}$$

*Furthermore, any trajectory in $\mathcal{T}^{ad}(t_0,x_0)$ is generated by selecting $u(t) \in \mathbb{B}^{viab}\big(t,x(t)\big)$. In other words*

$$\mathcal{T}^{ad}(t_0,x_0) = \left\{ (x(\cdot),u(\cdot)) \;\middle|\; \begin{array}{r} x(t_0) = x_0, \\ x(t+1) = F\big(t,x(t),u(t)\big),\, t = t_0,\ldots,T-1 \\ u(t) \in \mathbb{B}^{viab}\big(t,x(t)\big), \quad t = t_0,\ldots,T-1 \end{array} \right\}. \tag{4.16}$$

Notice that the state constraints (4.2d) have now disappeared, having been incorporated into the new control constraints $u(t) \in \mathbb{B}^{\mathrm{viab}}\big(t,x(t)\big)$.

### 4.3.4 Viability in the infinite horizon case

Particularly when $T = +\infty$, rather than considering the constraints (4.2a)–(4.2b), one may consider the *state/control constraints*which are respected at any time $t$:

$$(x(t),u(t)) \in \mathbb{D}(t) \subset \mathbb{X} \times \mathbb{U}\,. \tag{4.17}$$

The definition of the *viability kernel* is now [8]

$$\mathbb{V}\mathrm{iab} := \left\{ x \in \mathbb{X} \;\middle|\; \begin{array}{l} \text{there exist decisions } u(\cdot) \\ \text{and states } x(\cdot) \text{ starting from } x \text{ at time } t_0 \\ \text{satisfying for any time } t = t_0, t_0+1, \ldots \\ \text{dynamics (4.1) and constraints (4.17).} \end{array} \right\}. \tag{4.18}$$

## 4.4 Viability in the autonomous case

The so-called autonomous case of fixed (or autonomous, stationary) constraints and dynamics is when

$$
\begin{cases}
\mathbb{A}(t) = \mathbb{A} , \\
\mathbb{B}(t,x) = \mathbb{B}(x) , \\
F(t,x,u) = F(x,u) .
\end{cases}
\tag{4.19}
$$

The following Proposition 4.5 is a consequence of Definition 4.1 of viability kernels.

**Proposition 4.5.** *In the autonomous case (4.19), the viability kernels are increasing with respect to time:*

$$
\mathbb{V}\mathrm{iab}(t_0) \subset \mathbb{V}\mathrm{iab}(t_0+1) \subset \cdots \subset \mathbb{V}\mathrm{iab}(T) = \mathbb{A} . \tag{4.20}
$$

*If, in addition, the horizon is infinite ($T = +\infty$), the viability kernels are stationary and we write the common set* $\mathbb{V}\mathrm{iab}$*:*

$$
\mathbb{V}\mathrm{iab}(t_0) = \cdots = \mathbb{V}\mathrm{iab}(t) = \cdots = \mathbb{V}\mathrm{iab} \subset \mathbb{A} . \tag{4.21}
$$

### 4.4.1 Equilibria and viability

Chap. 3 pointed out the interest of equilibria in dealing with sustainability issues, especially for renewable resources, through the maximum sustainable yield concept. An equilibrium is only a partial response to viability issues. Indeed, admissible equilibria are clearly "viable" but such an analysis does not take into account the full diversity of tolerable transitions of the system. It turns out that the viability approach enlarges that of equilibrium. Hence, in the autonomous case (4.19), it is straightforward to prove that the admissible equilibria constitute a part of any viability kernel.

**Proposition 4.6.** *In the autonomous case (4.19), the admissible equilibria of Definition 3.1 belong to the viability kernel* $\mathbb{V}\mathrm{iab}(t)$ *at any time* $t$*:*

$$
\mathbb{X}_{\mathrm{E}}^{ad} := \{x_{\mathrm{E}} \in \mathbb{A} \mid \exists u_{\mathrm{E}} \in \mathbb{B}(x_{\mathrm{E}}) , \quad x_{\mathrm{E}} = F(x_{\mathrm{E}}, u_{\mathrm{E}})\} \subset \mathbb{V}\mathrm{iab}(t) .
$$

Notice that, by (4.20), this is equivalent to $\mathbb{X}_{\mathrm{E}}^{ad} \subset \mathbb{V}\mathrm{iab}(t_0)$.

### 4.4.2 Viability domains

Viability domains are convenient for the characterization of the viability kernel as shown below. Basically, a viability domain is a set that coincides with its viability kernel. It is defined as follows.

**Definition 4.7.** *In the autonomous case (4.19), a subset* $\mathbb{V} \subset \mathbb{X}$ *is said to be* weakly invariant, *or a* viability domain, *or a* viable set *if*

$$
\forall x \in \mathbb{V} , \quad \exists u \in \mathbb{B}(x) , \quad F(x,u) \in \mathbb{V} . \tag{4.22}
$$

An equivalent functional characterization is $\Psi_{\mathbb{V}}(x) = \min_{u \in \mathbb{B}(x)} \Psi_{\mathbb{V}}\big(F(x,u)\big)$.

The set of equilibria is a particular instance of a viability domain. Cycles or limit cycles constitute other illustrations of viability domains. Such a concept is central to characterizing the viability kernel as detailed below.

### 4.4.3 Maximality and Bellman properties

In the autonomous case (4.19), the infinite horizon case provides the simplest characterization since time disappears from the statements. A geometric assertion in the infinite horizon case is given by the following Theorem [1] involving viability domains (a proof may be found in [8]).

**Theorem 4.8.** *In the autonomous case (4.19), the viability kernel is the largest viability domain contained in* $\mathbb{A}$*, that is to say, the union of all viability domains in* $\mathbb{A}$*.*

Thus, any viability domain is a *lower approximation* of the viability kernel, being included in the latter. Similarly, we obtain a functional dynamic programming characterization in the infinite horizon case.

**Proposition 4.9.** *In autonomous case (4.19), the extended function* $\Psi_{\mathbb{V}\mathrm{iab}}$ *is the largest solution* $V$ *of the* Bellman equation*, or* dynamic programming *equation,*

$$\begin{cases} V(x) = \inf_{u \in \mathbb{B}(x)} V\big(F(x,u)\big)\,, \\ V(x) \geq \Psi_{\mathbb{A}}(x)\,. \end{cases} \tag{4.23}$$

### 4.4.4 Viable controls

In the autonomous case (4.19) and in the infinite horizon case, the time component vanishes and we obtain the *viable controls* as follows:

$$\mathbb{B}^{\mathrm{viab}}(x) := \{u \in \mathbb{B}(x) \mid F(x,u) \in \mathbb{V}\mathrm{iab}\}\ . \tag{4.24}$$

Hence, ensuring viability means remaining in the viability kernel, which stands for the "true constraints" and thus sheds light on *ex post* viability.

### 4.4.5 Hopeless, comfortable and dangerous configurations

In the autonomous case (4.19), three significant situations are worth being pointed out.

#### The comfortable case

Whenever the viability kernel is the whole set of constraints

$$\mathbb{A} = \mathbb{V}\mathrm{iab}(t_0) = \mathbb{V}\mathrm{iab}(t)\ ,$$

it turns out that the *a priori* state constraints (namely the set $\mathbb{A}$) are relevant. Indeed, from every state in $\mathbb{A}$ a feasible policy starts. In this case, the set $\mathbb{A}$ is a viability domain.

### The dangerous case and security barriers

Configuration of partial viability refers to the case where the viability kernel is neither empty nor viable, that is to say:

$$\emptyset \subsetneq \mathbb{V}\text{iab}(t_0) \subsetneq \mathbb{A} \ .$$

As shown by Fig. 4.1, this case is dangerous since parts of the initial constraint domain $\mathbb{A}$ are safe but others may lead to collapse or crisis by violating the constraints in finite time whatever the admissible decisions $u(\cdot)$ applied. Indeed, the set of boundary points of the viability kernel, $\mathbb{V}\text{iab}(t_0)$, which is located in $\mathbb{A}$ can be interpreted as the anticipative zone of dangers. If we go over this safety barrier, we have to face the crisis (to be outside $\mathbb{A}$) in finite time whatever the future decisions taken. In this sense, one may speak of viability margins for the viability kernel.

### The hopeless case

Whenever the viability kernel is empty, that is

$$\mathbb{V}\text{iab}(t_0) = \emptyset \ ,$$

unavoidable crisis and irreversibility is everywhere within domain $\mathbb{A}$. In this sense, we have a hopeless situation, and set $\mathbb{A}$ is said to be a repeller. Other strategies, such as enlarging the constraints or modifying the dynamics should be applied in order to solve the viability problem.

The following sections present the viability kernels together with related viable feedbacks for some of the examples exposed in Sect. 4.2.

## 4.5 Viable control of an invasive species

We consider the discrete dynamics of an invasive species characterized by its biomass $B(t)$ and harvesting effort decision $e(t)$. For the sake of simplicity, we restrict the study to a non-density-dependent and linear case

$$B(t+1) = RB(t)\big(1 - e(t)\big) \ ,$$

where the productivity of the resource is strictly larger than 1 ($R > 1$). The effort is constrained by

$$0 \leq e^\flat \leq e(t) \leq e^\sharp \leq 1 \ .$$

We assume that the policy is to constrain the ultimate biomass level within an ecological window, namely between conservation and maximal safety values at time horizon $T$, giving:

$$0 < B^\flat \leq B(T) \leq B^\sharp \ .$$

We can show that viability occurs if the initial biomass is sufficiently high, but not too high.

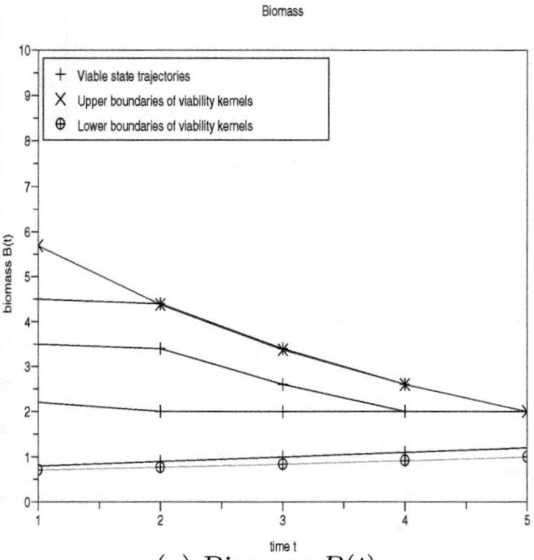


(a) Biomass $B(t)$

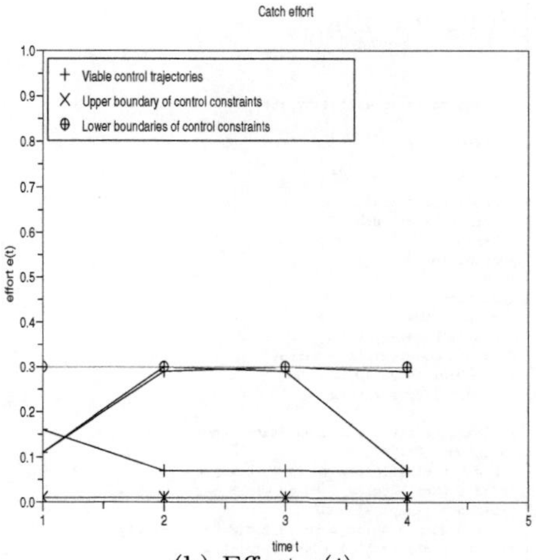


(b) Effort $e(t)$

**Fig. 4.2.** Viable resource management for time horizon $T = 5$. The tolerable conservation window is $C = [B^\flat, B^\sharp] = [1, 2]$. Growth parameter is $R = 1.1$. Effort bounds are $e^\flat = 0.1$ and $e^\sharp = 0.3$. We plot (a) viable trajectories of biomass states $B(t)$ and the boundaries of the viability kernel $\mathbb{V}\text{iab}(t)$, (b) Effort $e(t)$ with the acceptability bounds $[e^\flat, e^\sharp]$. Trajectories have been generated by SCILAB code 7

**Result 4.10** *The viability kernels are intervals*

$$\mathbb{V}\text{iab}(t) = [B^\flat(t), B^\sharp(t)] \,,$$

*whose viability biomass bounds are given by*

$$\begin{cases} B^\flat(t) = B^\flat \left(R(1-e^\flat)\right)^{t-1-T} \\ B^\sharp(t) = B^\sharp \left(R(1-e^\sharp)\right)^{t-1-T} \end{cases} .$$

*Viable feedback efforts $\mathfrak{e}(t, B)$ are those belonging to the set*

$$E^{viab}(t, B) = \left\{ e \,\middle|\, 1 - \frac{B^\sharp(t)}{RB} \le e \le 1 - \frac{B^\flat(t)}{RB} \quad \textit{and} \quad e^\flat \le e \le e^\sharp \right\} .$$

Numerical illustrations are given in Fig. 4.2. Proof is expounded in Sect. A.2 in the Appendix. When the state constraint is not only a final one, but rather a requirement $0 < B^\flat \le B(t) \le B^\sharp$ holding over all times $t = t_0, \ldots, T$, one can verify that the viability kernels are intervals whose bounds are given by backward inductions:

$$B^\flat(t) = \max\{B^\flat, \frac{B^\flat(t+1)}{R(1-e^\flat)}\} \quad \text{and} \quad B^\sharp(t) = \min\{B^\sharp, \frac{B^\sharp(t+1)}{R(1-e^\sharp)}\} .$$

Scilab code 7.

```
//
// exec invasive_viability.sce

lines(0);

R=1.1;
// resource productivity
function y=dynpop(B)
y=R*B;
endfunction

Horizon=5 ;
// final time
B_min=1; B_max=2;
// state Constraints - target
e_min=0.01; e_max=0.3;
// control constraints

// Characteristic function State (approximation)
function y=Phi_x(t,B)
  SGN = bool2s((B_min <= B) & (B<=B_max)) ;
// is 1 when B belongs to [B_min,B_max], 0 else
y=0*SGN + 1/%eps *(1-SGN) ;
// is 0 when B belongs to [B_min,B_max], +infty else
//  if (B_min <= B) & (B<=B_max) then y=0;
//     else y=exp(100);
//     end
endfunction

// Characteristic function Control (approximation)
function [y]=Phi_u(t,B,e)
  SGN = bool2s((e_min <= e) & (e<=e_max)) ;
  y=0*SGN + 1/%eps *(1-SGN) ;
// if (e_min <= e) & (e<=e_max) then y=0;
//    else y=exp(100);
// end
endfunction

// Grid state space x
x_min=0; x_max=10; delta_x=0.1 ;
grille_x=x_min: delta_x:x_max ;
S_x=size(grille_x) ;
NN=S_x(2) ;

// Grid control u
u_min=e_min; u_max=e_max; delta_u=0.01 ;
u=u_min:delta_u:u_max ;
S_u=size(u); MM=S_u(2) ;

function [z]=Projbis(x)
// projection of state on the grid
  z=round(x./delta_x).*delta_x;
  z=min(z,x_max);
  z=max(z,x_min);
endfunction

function i=Indice(x)
  i=int((x-x_min)./delta_x)+1;
endfunction

// controled dynamics on the grid
function [z]=dyna(x,e)
  xsuiv=dynpop(x*(1-e));
  P=Projbis(xsuiv);
  z=Indice(P);
endfunction

for ii=1:NN
  Etat_x(ii)=x_min+(ii-1)*delta_x;
end

for ii=1:MM
  Control_u(ii)=u_min+(ii-1)*delta_u;
end;


/////////////////////////////////////////////
// Dynamic Programming Viability
/////////////////////////////////////////////

x=grille_x;
W=zeros(Horizon,NN); // W value function
// Initialization at final time  T

for  i=1:NN
  xx=Etat_x(i);
  W(Horizon,i)=-Phi_x(Horizon,xx);
end

// Bellman backward equation
for (t=Horizon-1:-1:1)
for (i=1:NN)
xx=Etat_x(i);
VV=[];
for (j=1:MM)
uu=Control_u(j);
VV=[VV; Phi_u(t,xx,uu)-Phi_x(t,xx)+W(t+1,dyna(xx,uu))];
end,
[Vopt,jopt]=max(VV) ;
W(t,i)=Vopt;
j_opt(t,i)=jopt;
end,
end,


/////////////////////////////////////////////////
// Simulations of some viable trajectories
/////////////////////////////////////////////////

xset("window",2); xbasc(2);
xtitle("Biomass",'time t','biomass B(t)');

xset("window",3); xbasc(3);
xtitle("Catch effort",'time t','effort e(t)');

if (max(W(1,:))>=0) then
  for k=[0.8 2.2 3.5 4.5]
     i(1)=Indice(k);
 traj_x(1)=Etat_x(i(1));
 for (t=1:Horizon-1)
e_viab(t)=Control_u(j_opt(t,i(t)));
i(t+1)=dyna(traj_x(t),e_viab(t));
traj_x(t+1)=Etat_x(i(t+1));
 end
 Tempx=1:Horizon;
 Bin=B_min*(R*(1-e_min))^(Tempx-Horizon) ;
 Bax=B_max*(R*(1-e_max))^(Tempx-Horizon) ;
 xset("window",2);
 plot2d(Tempx,[traj_x Bax' Bin' ],...
[1,2,3],rect=[1,x_min,Horizon,x_max],...
nax=[0,Horizon,1,11]);
 plot2d(Tempx,[traj_x Bax' Bin' ],-[1,2,3]);
 legends(["Viable state trajectories",...
  "Upper boundaries of viability kernels",...
  "Lower boundaries of viability kernels"],...
 -[1,2,3],"ul");
 //
 Tempu=1:Horizon-1;
 ein=e_min*ones(Tempu);   eax=e_max*ones(Tempu);
 xset("window",3);
 plot2d(Tempu,[e_viab ein' eax'],...
 [1,2,3],rect=[1,0,Horizon,1],...
nax=[0,Horizon,1,11]);
 plot2d(Tempu,[e_viab ein' eax'],-[1,2,3]);
 legends(["Viable control trajectories",...
  "Upper boundary of control constraints",...
  "Lower boundaries of control constraints"],...
 -[1,2,3],"ul");
  end
  else "empty kernel"
end

//
```

## 4.6 Viable greenhouse gas mitigation

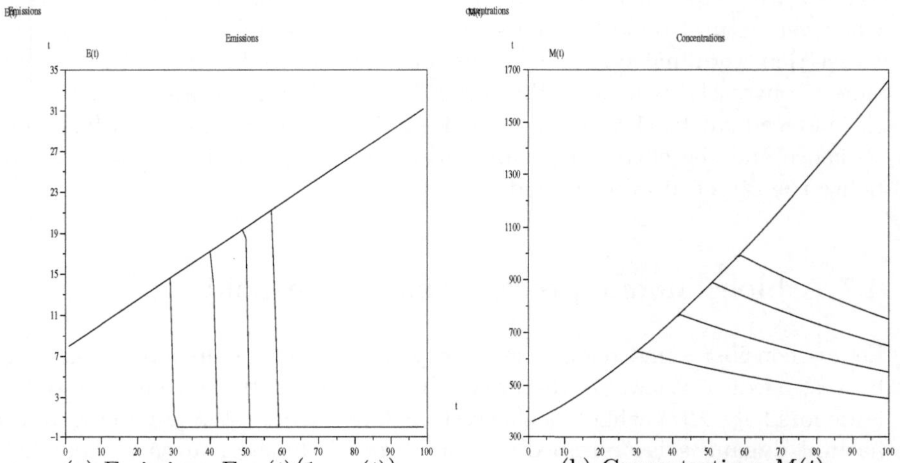


(a) Emissions $E_{\text{BAU}}(t)\big(1-a(t)\big)$ (b) Concentrations $M(t)$

**Fig. 4.3.** The increasing curves are the baseline $CO_2$ emissions $E_{\text{BAU}}(t) = \mathfrak{E}_{\text{BAU}}(Q(t))$ and concentrations $M_{\text{BAU}}(t)$ over the period $[2000, 2100]$. Viable emissions $E_{\text{BAU}}(t)\big(1-a(t)\big)$ and concentrations $M(t)$ are trajectories that coincide with the baseline until they decrease to reach the different concentration targets $M^\sharp = 450$, 550 and 650 ppm.

Let us now consider the problem introduced in Subsect. 4.2.2. Using viable dynamic programming, we compute explicitly the feasible solutions of the problem and, in particular, a viable concentration ceiling. We need to introduce the following maximal concentration values:

$$M^\sharp(t) := (M^\sharp - M_\infty)(1-\delta)^{t-T} + M_\infty \,. \tag{4.25}$$

These induced $M^\sharp(t)$ thresholds play the role of a tolerable ceiling.

**Result 4.11** *An effective (viable) policy exists if, and only if, the initial concentration $M(t_0)$ is smaller than $M^\sharp(t_0)$. In this case, the whole policy $a(t_0), a(t_0+1), \ldots, a(T-1)$ is effective if and only if associated concentrations $M(t)$ remain smaller than $M^\sharp(t)$. In other words, we have:*

$$\mathbb{V}\text{iab}(t) = ]-\infty, M^\sharp(t)] \times \mathbb{R}_+ \,.$$

*Viable feedback abatement rates belong to the interval*

$$A^{viab}(t, M, Q) = \left[\max\left\{0, \frac{(1-\delta)\big(M - M^\sharp(t)\big) + \mathfrak{E}_{\text{BAU}}(Q)}{\mathfrak{E}_{\text{BAU}}(Q)}\right\}, 1\right] \,.$$

The proof of this result is provided in Sect. A.2 in the Appendix.

Let us observe that we always have $M^\sharp(T) = M^\sharp$, which means that the terminal tolerable concentration is $M^\sharp$, as expected. As illustrated by Fig. 4.3, the expression (4.25) of the safety thresholds $M^\sharp(t)$ indicates that, whenever natural removal occurs ($\delta > 0$), the thresholds $M^\sharp(t)$ are strictly larger than terminal target $M^\sharp$: this allows for exceeding the target during time. Conversely, whenever the natural removal term disappears ($\delta = 0$), the induced safety thresholds coincide with final $M^\sharp$ along the whole time sequence and the effectiveness mitigation policies make it necessary to stay below the target at every period.

## 4.7 A bioeconomic precautionary threshold

Let us consider some regulating agency aiming at the sustainable use and harvesting of a renewable resource. The biomass of this resource at time $t$ is denoted by $B(t)$ while the harvesting level is $h(t)$. We assume that the natural dynamics is described by some growth function $g$ as in Sect. 2.2. Under exploitation, the following controlled dynamics is obtained

$$B(t+1) = g\big(B(t) - h(t)\big) \tag{4.26}$$

with the admissibility constraint

$$0 \leq h(t) \leq B(t) \; . \tag{4.27}$$

The policy goal is to guarantee at each time $t$ a minimal harvesting

$$h_{\mathrm{LIM}} \leq h(t) \; , \tag{4.28}$$

together with a non extinction level for the resource

$$B_{\mathrm{LIM}} \leq B(t) \; . \tag{4.29}$$

The combination of constraints (4.27) and (4.28) yields the state constraint

$$h_{\mathrm{LIM}} \leq B(t) \; . \tag{4.30}$$

For the sake of simplicity, we assume that $h_{\mathrm{LIM}} > B_{\mathrm{LIM}} > 0$, so that (4.29) is useless. We need to recall equilibrium concepts and notations already introduced in Chap. 3.

- We take the sustainable yield function $\sigma$ already defined by (3.6), giving

$$h = \sigma(B) \iff B = g(B-h) \quad \text{and} \quad 0 \leq h \leq B \; . \tag{4.31}$$

- The maximum sustainable biomass $B_{\mathrm{MSE}}$ and maximum sustainable yield $h_{\mathrm{MSE}}$ are defined by

$$h_{\mathrm{MSE}} = \sigma(B_{\mathrm{MSE}}) = \max_{B \geq 0} \sigma(B) \,. \tag{4.32}$$

It turns out that when $h_{\mathrm{LIM}} \leq h_{\mathrm{MSE}}$, the viability kernel $\mathbb{V}\mathrm{iab}$ depends on a floor level $B_{\mathrm{PA}}$ which is the solution of

$$B_{\mathrm{PA}} = \min_{B,\, h_{\mathrm{LIM}} = \sigma(B)} B \,. \tag{4.33}$$

**Result 4.12** *Assuming that $g$ is an increasing continuous function on $\mathbb{R}_+$, the viability kernel is given by*

$$\mathbb{V}\mathrm{iab} = \begin{cases} [B_{\mathrm{PA}}, +\infty[ & \textit{if} \quad h_{\mathrm{LIM}} \leq h_{\mathrm{MSE}} \\ \emptyset & \textit{otherwise.} \end{cases} \tag{4.34}$$

*For any stock $B \in \mathbb{V}\mathrm{iab}$, the viable catch feedbacks $\mathfrak{h}(B)$ lie within the set*

$$H^{viab}(B) = [h_{\mathrm{LIM}}, h_{\mathrm{PA}}(B)]$$

*where the ceiling catch $h_{\mathrm{PA}}(B)$ is*

$$h_{\mathrm{PA}}(B) = h_{\mathrm{LIM}} + B - B_{\mathrm{PA}} \,. \tag{4.35}$$

The complete proof of this result is provided in Sect. A.2 in the Appendix. The SCILAB code 8 makes it possible to examine and illustrate the consequences of this result with population dynamics governed by Beverton-Holt recruitment

$$g(B) = \frac{RB}{1 + bB} \,.$$

Let us notice that, in this case, the ceiling guaranteed catch given by (3.14) and (3.15) is

$$h_{\mathrm{MSE}} = \frac{(R - \sqrt{R})^2}{Rb} \,.$$

*The unsustainable case:* $\mathbb{V}\mathrm{iab} = \emptyset$.

In the following simulations, we choose a guaranteed harvesting $h_{\mathrm{LIM}}$ strictly larger than $h_{\mathrm{MSE}}$. Then, for several initial conditions $B(t_0)$, we compute different trajectories for the smallest admissible harvesting, namely $h(t) = h_{\mathrm{LIM}}$. As displayed by Fig. 4.4, it can be observed that the viability constraint (4.30) is violated in such a case. The situation is even more catastrophic with sequences of harvesting $h(t)$ larger than the guaranteed one $h_{\mathrm{LIM}}$. We illustrate this with the admissible feedbacks[7] $h(t, B) = \alpha(t) h_{\mathrm{LIM}} + (1 - \alpha(t))B$, where $(\alpha(t))_{t \in \mathbb{N}}$ is an i.i.d. sequence of uniform random variables in $[0, 1]$.

[7] One may also try $h(t, B) = B$ or $h(t, B) = \alpha h_{\mathrm{LIM}} + (1 - \alpha)B$ with $\alpha \in [0, 1]$.

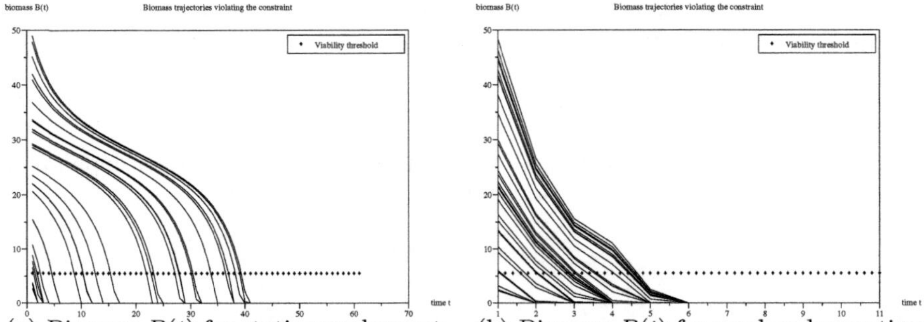


(a) Biomass $B(t)$ for stationary harvesting $h(t) = h_0$

(b) Biomass $B(t)$ for random harvesting $h(t)$

**Fig. 4.4.** Unsustainable case where $h_{\text{LIM}} > h_{\text{MSE}}$. Biomass trajectories $B(t)$ violating the viability constraint (horizontal line at $h_{\text{LIM}}$) in finite time whatever the initial condition $B_0$. Trajectories have been generated by SCILAB code 8.

*The sustainable case:* $\mathbb{V}\text{iab} \neq \emptyset$.

Now we choose a guaranteed harvesting $h_{\text{LIM}}$ strictly smaller than $h_{\text{MSE}}$. The value $B_{\text{PA}}$ is the smallest equilibrium solution of $\sigma(B) = h_{\text{LIM}}$, namely:

$$B = \frac{R(B - h_{\text{LIM}})}{1 + b(B - h_{\text{LIM}})} .$$

Hence, the value of $B_{\text{PA}}$ is

$$B_{\text{PA}} = \frac{(K + bh_{\text{LIM}})}{b} - \frac{\sqrt{\Delta}}{2b} ,$$

with $\Delta = (R - 1 + bh_{\text{LIM}})^2 - 4Rbh_{\text{LIM}}$ and $K = \frac{R-1}{b}$.

From (4.35), we know that any viable feedback $h(B)$ lies within the set $[h_{\text{LIM}}, h_{\text{PA}}(B)]$ where $h_{\text{PA}}(B) = h_{\text{LIM}} + B - B_{\text{PA}}$. In the simulations shown in Fig. 4.5, we compute different trajectories for a harvesting feedback of the form $h(t, B) = \alpha(t)h_{\text{LIM}} + (1 - \alpha(t))h_{\text{PA}}(B)$. The viability constraint (4.30) is now satisfied.

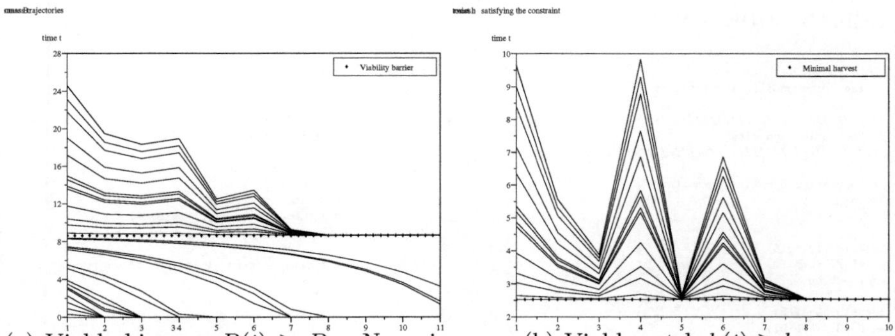


(a) Viable biomass $B(t) \geq B_{\text{PA}}$. Non viable biomass $B(t) < B_{\text{PA}}$

(b) Viable catch $h(t) \geq h_{\text{LIM}}$

**Fig. 4.5.** Sustainable case where $h_{\text{LIM}} \leq h_{\text{MSE}}$. Harvest trajectories satisfying the viability constraint (horizontal line at $h_{\text{LIM}}$). Biomass trajectories violating or satisfying the viability constraint, depending on whether the original biomass is lower or greater than the viability barrier (horizontal line at $B_{\text{PA}}$). Trajectories have been generated by Scilab code 8.

Scilab code 8.

```
//
// exec bioeconomic_viability.sce

/////////////////////////////////////////////
// Population dynamics
/////////////////////////////////////////////

// Population dynamics parameters
R=1.5;
b=0.01;

function [y]=g(B)
y=R*B./(1+b*B)
endfunction

function h=SY(B)
h= B-(B./(R-b*B))
endfunction

B_MSE=(R-sqrt(R))/b
h_MSE=SY(B_MSE)
K=(R-1)/b

xset("window",0); xbasc();
B=[0:1:K]; plot2d(B,SY(B))
xtitle('Sustainable yield curve associated to ...
   Beverton-Holt dynamics','biomass B')


// number of random initial conditions
nbsimul=30;


/////////////////////////////////////////////
// The unsustainable case
/////////////////////////////////////////////

h_min= h_MSE +0.5 // for instance

Horizon=60;
B=zeros(1,Horizon+1);


// // Stationary harvesting
xset("window",11); xbasc();

for i=1:nbsimul
B(1)=K*rand(1);
for t=1:Horizon
B(t+1)=max(0,g(B(t)-h_min));
end,
plot2d(1:(Horizon+1),B);
end,
plot2d(1:(Horizon+1),h_min * ones(1:(Horizon+1)),style=-4)
legends("Viability threshold",-4,'ur')

xtitle('Biomass trajectories violating the constraint',...
   'time t','biomass B(t)')


// // Random harvesting

function h=feedback(t,B,alpha)
h=max(h_min,alpha*h_min+(1-alpha)*B);
endfunction

Horizon=10;

alpha=rand(1,Horizon);
B=zeros(1,Horizon+1);

xset("window",12); xbasc();

for i=1:nbsimul
B(1)=K*rand(1);
for t=1:Horizon
h(t)=feedback(t,B(t),alpha(t));
B(t+1)=max(0,g(B(t)-h(t)));
end,
plot2d(1:(Horizon+1),B);
end,
plot2d(1:0.2:(Horizon+1),...
   h_min*ones(1:0.2:(Horizon+1)),style=-4)
legends("Viability threshold",-4,'ur')

xtitle('Biomass trajectories violating the constraint',...
   'time t','biomass B(t)')



/////////////////////////////////////////////
// The sustainable case
/////////////////////////////////////////////

h_min= h_MSE *0.5 // for instance

Horizon=10;

// numerical estimation of the viability barrier

function error=viabbarrier(B)
error=SY(B)-h_min
endfunction

B_V=fsolve(0,viabbarrier)



function h=h_max(B)
    h= B- ( B_V / (R - b*B_V) )
endfunction

function h=viab(t,B,alpha)
h=max(h_min,alpha*h_min+(1-alpha)*h_max(B))
endfunction

Horizon=10;

xset("window",21); xbasc();
xtitle("Biomass trajectories",'time t','biomass B');

xset("window",22);xbasc();
xtitle("Harvesting trajectories ...
   satisfying the constraint",'time t','harvest h');

alpha=rand(1,Horizon);
B=zeros(1,Horizon+1);
h=zeros(1,Horizon);

for i=1:nbsimul
B(1)=K*rand(1)/2;
for t=1:Horizon
h(t)=viab(t,B(t),alpha(t));
B(t+1)=max(0,g(B(t)-h(t)));
end,
xset("window",21);plot2d(1:(Horizon+1),B);
plot2d(1:0.2:(Horizon+1),...
   B_V*ones(1:0.2:(Horizon+1)),style=-4)
legends("Viability barrier",-4,'ur')
//
xset("window",22);plot2d(1:Horizon,h);
plot2d(1:0.2:Horizon,h_min*ones(1:0.2:Horizon),style=-4)
legends("Minimal harvest",-4,'ur')
end

//
```

## 4.8 The precautionary approach in fisheries management

Fisheries management agencies aim at driving resources on sustainable paths that should conciliate both ecological and economic goals. To manage fisheries and stocks, the ICES (*International Council for the Exploration of the Sea*) precautionary approach and advice rely on specific indicators and reference points such as spawning biomass or mean fishing mortality. In [9], the viable control approach is used to make the relationships between sustainability objectives and reference points for ICES advice on stock management explicit. It is shown how the age-structured dynamics can be taken into account for defining appropriate operational reference points, making it possible to reach the objective of keeping stock biomass above threshold $B_{\text{LIM}}$.

| Definition | Notation | *Anchovy* | *Hake* |
|---|---|---|---|
| Maximum age | $A$ | 3 | 8 |
| Mean weight at age (kg) | $(\upsilon_a)_a$ | $(16,\ 28,\ 36) \times 10^{-3}$ | (0.126, 0.2, 0.319, 0.583, 0.986, 1.366, 1.748, 2.42) |
| Maturity ogive | $(\gamma_a)_a$ | (1, 1, 1) | (0, 0, 0.23, 0.60, 0.90, 1, 1, 1) |
| Natural mortality | $M$ | 1.2 | 0.2 |
| Fishing mortality at age | $(F_a)_a$ | (0.4, 0.4, 0.4) | (0, 0, 0.1, 0.25, 0.22, 0.27, 0.42, 0.5, 0.5) |
| Presence of plus group | $\pi$ | 0 | 1 |
| fishing mortality precautionary RP | $F_{\text{PA}}$ | $1 - 1.2$ | 0.25 |
| *SSB* precautionary RP (tons) | $B_{\text{PA}}$ | 33 000 | 140 000 |
| fishing mortality limit RP | $F_{\text{LIM}}$ | / | 0.35 |
| *SSB* limit RP (tons) | $B_{\text{LIM}}$ | 21 000 | 100 000 |

**Table 4.1.** Parameter definitions and values for anchovy and hake. RP is for reference point.

### ICES indicators and reference points

Two indicators are used in the precautionary approach with associated limit reference points. Let us underline here that using limit reference points implies defining a boundary between unacceptable and admissible states, whereas it would be better to define desirable states using target reference points. The first indicator, denoted by *SSB* in (2.35), is the spawning stock biomass

$$SSB(N) = \sum_{a=1}^{A} \gamma_a \upsilon_a N_a \ ,$$

to which ICES associates the limit reference point $B_{\text{LIM}} > 0$. For management advice an additional precautionary reference point $B_{\text{PA}} > B_{\text{LIM}}$ is used, intended to incorporate uncertainty about the stock state.

The second indicator, denoted by $F$, is the mean fishing mortality over a pre-determined age range from $a_r$ to $A_r$, that is to say:

$$F(\lambda) := \frac{\lambda}{A_r - a_r + 1} \sum_{a=a_r}^{a=A_r} F_a \ . \tag{4.36}$$

The associated limit reference point is $F_{\text{LIM}}$ and a precautionary approach reference point $F_{\text{PA}} > 0$. Acceptable controls $\lambda$, according to this reference point, are those for which $F(\lambda) \leq F_{\text{LIM}}$, as higher fishing mortality rates might drive spawning stock biomass below its limit reference point.

### Acceptable configurations

To define sustainability, it can now be assumed that the decision maker can describe "acceptable configurations of the system," that is acceptable couples $(N, \lambda)$ of states and controls, which form a set $\mathbb{D} \subset \mathbb{X} \times \mathbb{U} = \mathbb{R}^A \times \mathbb{R}$, the acceptable set. In practice, the set $\mathbb{D}$ may capture ecological, economic and/or sociological requirements. Considering sustainable management within the precautionary approach, involving spawning stock biomass and fishing mortality indicators, we introduce the following ICES precautionary approach configuration set:

$$\mathbb{D}_{\text{LIM}} := \left\{(N, \lambda) \in \mathbb{R}_+^A \times \mathbb{R}_+ \mid \mathit{SSB}(N) \geq B_{\text{LIM}} \quad \text{and} \quad F(\lambda) \leq F_{\text{LIM}}\right\} . \quad (4.37)$$

### Interpreting the precautionary approach in light of viability

The precautionary approach can be sketched as follows: an estimate of the stock vector $N$ is made; the condition $\mathit{SSB}(N) \geq B_{\text{LIM}}$ is checked; if valid, the following usual advice (UA) is given:

$$\lambda_{\text{UA}}(N) = \max\{\lambda \in \mathbb{R}_+ \mid \mathit{SSB}(g(N, \lambda)) \geq B_{\text{LIM}} \quad \text{and} \quad F(\lambda) \leq F_{\text{LIM}}\} .$$

Here, the dynamics $g$ is the one introduced in (2.33), (2.34) and (2.36). However, the existence of such a fishing mortality multiplier for any stock vector $N$ such that $\mathit{SSB}(N) \geq B_{\text{LIM}}$ is tantamount to non-emptyness of a set of viable controls. This justifies the following definitions. Let us define the precautionary approach state set

$$\mathbb{V}_{\text{LIM}} := \left\{N \in \mathbb{R}_+^A \mid \mathit{SSB}(N) \geq B_{\text{LIM}}\right\} . \quad (4.38)$$

We shall say that the precautionary approach is sustainable if the precautionary approach state set $\mathbb{V}_{\text{LIM}}$ given by (4.38) is a viability domain for the acceptable set $\mathbb{D}_{\text{LIM}}$ under dynamics $g$.

Whenever the precautionary approach is sustainable, the set of viable controls is not empty. This observation implies the existence of a viable fishing mortality multiplier that allows the spawning stock biomass of the population to remain above $B_{\text{LIM}}$ along time. When $\mathbb{V}_{\text{LIM}}$ is not a viability domain of the acceptable set $\mathbb{D}_{\text{LIM}}$ under dynamics $g$, maintaining the spawning stock biomass above $B_{\text{LIM}}$ from year to year will not be sufficient to ensure the existence of controls which make it possible to remain indefinitely at this point. For example, in a stock with high abundance in the oldest age class and low abundances in the other age classes, spawning stock biomass would be above $B_{\text{LIM}}$ but would be at high risk of falling below $B_{\text{LIM}}$ the subsequent year, whatever the fishing mortality, if recruitment is low.

**Result 4.13** *If we suppose that the natural mortality is independent of age, that is* $M_a = M$*, and that the proportion* $\gamma_a$ *of mature individuals and the weight* $\upsilon_a$ *at age are increasing with age* $a$*, the precautionary approach is sustainable if, and only if, we obtain*

$$\inf_{B \in [B_{\text{LIM}}, +\infty[} \left[ \pi \mathrm{e}^{-M} B + \gamma_1 \upsilon_1 \varphi(B) \right] \geq B_{\text{LIM}} \, . \tag{4.39}$$

Notice that, when $\gamma_1 = 0$ (the recruits do not reproduce) condition (4.39) is never satisfied and the precautionary approach is not sustainable, whatever the value of $B_{\text{LIM}}$. That is, to keep spawning stock biomass above $B_{\text{LIM}}$ for an indefinite time, it is not enough to keep it there from year to year. Other conditions based upon more indicators have to be checked.

Condition (4.39) involves biological characteristics of the population and the stock recruitment relationship $\varphi$, as well as the threshold $B_{\text{LIM}}$. However, it is worth pointing out that condition (4.39) does not depend on the stock recruitment relationship $\varphi$ between 0 and $B_{\text{LIM}}$. It does not depend on $F_{\text{LIM}}$, either.

A constant recruitment is generally used for fishing advice, so the following simplified condition can be used. Assuming a constant recruitment $R$, the precautionary approach is sustainable if, and only if, we have $\pi \mathrm{e}^{-M} B_{\text{LIM}} + \gamma_1 \upsilon_1 R \geq B_{\text{LIM}}$, that is if, and only if,

$$R \geq \underline{R} \quad \text{where} \quad \underline{R} := \frac{1 - \pi \mathrm{e}^{-M}}{\gamma_1 \upsilon_1} B_{\text{LIM}} \, , \tag{4.40}$$

making thus of $\underline{R}$ a minimum recruitment required to preserve $B_{\text{LIM}}$.

The previous condition is easy to understand when there is no plus-group $\pi = 0$. Assuming a constant recruitment $R$ and no plus group, the precautionary approach is sustainable if, and only if,

$$\gamma_1 \upsilon_1 R \geq B_{\text{LIM}} \, . \tag{4.41}$$

Hence, in the worst case, where the whole population would spawn and die in a single time step, the resulting recruits would be able to restore the spawning biomass at the required level.

| stock recruitment relationship | | Condition | Parameter values | Threshold | Sustainable |
|---|---|---|---|---|---|
| Constant | (mean) | $R_{\text{MEAN}} \geq \underline{R}$ | $14\,016 \times 10^6$ | $1\,312 \times 10^6$ | yes |
| Constant | (geometric mean) | $R_{\text{GM}} \geq \underline{R}$ | $7\,109 \times 10^6$ | $1\,312 \times 10^6$ | yes |
| Constant | (2002) | $R_{2002} \geq \underline{R}$ | $3\,964 \times 10^6$ | $1\,312 \times 10^6$ | yes |
| Constant | (2004) | $R_{2004} \geq \underline{R}$ | $696 \times 10^6$ | $1\,312 \times 10^6$ | no |
| Linear | | $\gamma_1 \upsilon_1 r \geq 1$ | 0.84 | 1 | no |
| Ricker | | $\inf_{B \geq B_{\text{LIM}}} [\cdots] \geq B_{\text{LIM}}$ | 0 | 21 000 | no |

**Table 4.2.** Bay of Biscay anchovy: sustainability of advice based on the spawning stock biomass indicator for various stock recruitment relationships. The answer is given in the last column of the table.

**Case study: Northern hake**

For hake, the precautionary approach is never sustainable because $\gamma_1 = 0$.

**Case study: Bay of Biscay anchovy**

Because the first age class of Bay of Biscay anchovy accounts for *ca.* 80% of spawning stock biomass, the sustainability of the precautionary approach will depend on the relationship between the biomass reference point and the stock dynamics, mainly determined by the stock recruitment relationship as there is no plus-group. Assuming various stock recruitment relationships, and taking $\pi = 0$, since no plus group is present, it is determined whether the precautionary approach based on the current value of $B_{\mathrm{LIM}}$ is sustainable. The answer is given in the last column of Table 4.8. The second column contains an expression whose value is given in the third, and has to be compared, according to condition (4.39), to the threshold in the fourth.

## 4.9 Viable forestry management

Now we consider the problem introduced in Subsect. 4.2.4. For the sake of simplicity, we assume here that mortality and fertility parameters coincide in the sense that $m = \gamma$ in (4.9). In this case, we can verify that the total surface of the forest remains stationary:

$$\sum_{i=1}^{n} N_i(t) = S \; .$$

As described and proved in [20], the following felling threshold $h_{\mathrm{LIM}}$ is deciding to characterize the viability kernel:

$$h_{\mathrm{LIM}} := \begin{cases} S\dfrac{m(1-m)^{n-1}}{1-(1-m)^{n-1}} & \text{if} \quad m \neq 0 \; , \\ \dfrac{S}{n-1} & \text{if} \quad m = 0 \; . \end{cases} \tag{4.42}$$

The following result related to the vacuity of the viability kernel can be obtained.

**Result 4.14** *For any minimal harvesting level $h^\flat$ greater than $h_{\mathrm{LIM}}$, the viability kernel is empty:*

$$h^\flat > h_{\mathrm{LIM}} \Longrightarrow \mathbb{V}\mathrm{iab} = \emptyset \; .$$

Whenever it is not empty, the viability kernel $\mathbb{V}\mathrm{iab}$ is characterized as follows.

**Result 4.15** *Consider* $h^\flat \in [0, h_{\text{LIM}}]$. *Then the viability kernel is given by*

$$\mathbb{V}\text{iab} = \left\{ N \in \mathbb{R}_+^n \;\middle|\; \begin{array}{l} \sum_{l=1}^{n} N_l = S \\ \sum_{l=n-i}^{n} N_l \geq \frac{1-(1-m)^i}{m(1-m)^i} h^\flat \quad i = 1, \ldots, n-1 \end{array} \right\}.$$

The linear additional inequalities contained in the conditions advocate for the anticipation of potential crisis $N_n(t) < h^\flat$. Indeed, these additional inequalities impose other constraints on $N(t)$ to maintain sustainability in the long term horizon.

The viability kernel exhibits the states of the forest compatible with the constraints. The present step is to compute the sustainable management options (decisions or controls) associated with it. At any surface vector $N$ in viability kernel $\mathbb{V}\text{iab}$, we know that the viable regulation set is not empty.

**Result 4.16** *When* $h^\flat \in [0, h_{\text{LIM}}]$ *and for* $N \in \mathbb{V}\text{iab}$,

$$H^{viab}(N) = \left[ h^\flat \, , \, \min_{i=0,\ldots,n-2} \left\{ (1-m) \sum_{l=n-i-1}^{n} N_l - \frac{1-(1-m)^i}{m(1-m)^i} h^\flat \right\} \right] . \tag{4.43}$$

Two kinds of viable regulations have been used in the simulations hereafter.

- *Maximal viable harvesting.* The feedback $\mathfrak{h}_M$ consists in choosing at any $N \in \mathbb{V}\text{iab}$ the largest value $h$ allowed by the viability conditions:

$$\mathfrak{h}_M(N) = \max_{h \in H^{\text{viab}}(N)} h = \min_{i=0,\ldots,n-2} \left\{ (1-m) \sum_{l=n-i-1}^{n} N_l - \frac{1-(1-m)^i}{m(1-m)^i} h^\flat \right\}.$$

- *Inertial viable harvesting.* Given the resource state $N \in \mathbb{V}\text{iab}_{h^\flat}$ and a current felling level $h_c$, the inertial viable regulation $\mathfrak{h}_I$ consists in choosing a viable control $h$ minimizing decision's change $|h - h_c|$:

$$\mathfrak{h}_I(N) = \arg\min_{h \in H^{\text{viab}}(N)} |h - h_c| \, .$$

The interest of such a policy is to take into account rigidity of decisions or behaviors. Indeed, as long as the current decision is relevant, this policy is not modified.

Now the viability analysis is illustrated for five age classes $n = 5$, guaranteed catch $h^\flat = 4$, total surface $S = 20$ and the initial conditions $N(t_0) = (2, 6, 5, 3, 4)'$. Two cases for $m$ are distinguished.

i) $m = 7\%$. The largest sustainable value, computed from (4.42), is $h_{\text{LIM}} \simeq 4.2$. Maximal viable harvesting, provided by (4.43) is computed.

| $t$ | 0 | 1 | 2 | 3 | 4 | 5 | 6 | 7 | 8 | 9 | 10 | $\cdots$ | $+\infty$ |
|---|---|---|---|---|---|---|---|---|---|---|---|---|---|
| $N_5$ | 2.0 | 2.7 | 2.9 | 1.0 | 0.0 | 0.0 | 0.0 | 0.0 | 0.0 | 0.0 | 0.0 | $\rightarrow$ | 0.0 |
| $N_4$ | 6.0 | 4.6 | 2.6 | 3.2 | 5.0 | 4.3 | 4.3 | 4.3 | 4.8 | 4.4 | 4.4 | $\rightarrow$ | 4.5 |
| $N_3$ | 5.0 | 2.8 | 3.5 | 5.3 | 4.7 | 4.7 | 4.7 | 5.2 | 4.7 | 4.7 | 4.7 | $\rightarrow$ | 4.8 |
| $N_2$ | 3.0 | 3.7 | 5.7 | 5.0 | 5.0 | 5.0 | 5.5 | 5.0 | 5.0 | 5.0 | 5.5 | $\rightarrow$ | 5.2 |
| $N_1$ | 4.0 | 6.1 | 5.4 | 5.4 | 5.4 | 6.0 | 5.4 | 5.4 | 5.4 | 5.9 | 5.5 | $\rightarrow$ | 5.6 |
| $h_M$ | 4.7 | 4.0 | 4.0 | 4.0 | 4.6 | 4.0 | 4.0 | 4.0 | 4.5 | 4.0 | 4.0 | $\rightarrow$ | 4.2 |

We notice that viable trajectories converge asymptotically towards a stationary state.

ii) Again $m = 7\%$. This time, the inertial viable harvesting that fulfills (4.43) is applied.

| $t$ | 0 | 1 | 2 | 3 | 4 | 5 | 6 | 7 | 8 | 9 | 10 |
|---|---|---|---|---|---|---|---|---|---|---|---|
| $N_5$ | 2.0 | 3.0 | 2.9 | 1.0 | 0.0 | 0.41 | 0.56 | 0.56 | 0.56 | 0.56 | 0.56 |
| $N_4$ | 6.0 | 4.6 | 2.6 | 3.2 | 4.7 | 4.5 | 4.3 | 4.3 | 4.3 | 4.3 | 4.3 |
| $N_3$ | 5.0 | 2.8 | 3.5 | 5.1 | 4.8 | 4.7 | 4.7 | 4.7 | 4.7 | 4.7 | 4.7 |
| $N_2$ | 3.0 | 3.7 | 5.5 | 5.2 | 5.0 | 5.0 | 5.0 | 5.0 | 5.0 | 5.0 | 5.0 |
| $N_1$ | 4.0 | 6.0 | 5.6 | 5.4 | 5.4 | 5.4 | 5.4 | 5.4 | 5.4 | 5.4 | 5.4 |
| $h_I$ | 4.5 | 4.2 | 4.0 | 4.0 | 4.0 | 4.0 | 4.0 | 4.0 | 4.0 | 4.0 | 4.0 |

It has to be noticed that these viable trajectories become stationary in finite time.

iii) $m = 0$. The largest sustainable value is now $h_{\text{LIM}} = 5$. The maximal viable harvesting $h_M$ is chosen.

| $t$ | 0 | 1 | 2 | 3 | 4 | 5 | 6 | 7 | 8 | 9 | 10 |
|---|---|---|---|---|---|---|---|---|---|---|---|
| $N_5$ | 2.0 | 0.0 | 1.0 | 0.0 | 0.0 | 0.0 | 0.0 | 0.0 | 0.0 | 0.0 | 0.0 |
| $N_4$ | 6.0 | 5.0 | 3.0 | 4.0 | 8.0 | 4.0 | 4.0 | 4.0 | 8.0 | 4.0 | 4.0 |
| $N_3$ | 5.0 | 3.0 | 4.0 | 8.0 | 4.0 | 4.0 | 4.0 | 8.0 | 4.0 | 4.0 | 4.0 |
| $N_2$ | 3.0 | 4.0 | 8.0 | 4.0 | 4.0 | 4.0 | 8.0 | 4.0 | 4.0 | 4.0 | 8.0 |
| $N_1$ | 4.0 | 8.0 | 4.0 | 4.0 | 4.0 | 8.0 | 4.0 | 4.0 | 4.0 | 8.0 | 4.0 |
| $h_M$ | 8.0 | 4.0 | 4.0 | 4.0 | 8.0 | 4.0 | 4.0 | 4.0 | 8.0 | 4.0 | 4.0 |

It has to be stressed that these viable trajectories become $(n-1)$-cyclic in finite time.

## 4.10 Invariance or strong viability

It may happen that, for some initial states, the constraints do not really influence the evolution. This is the idea of invariance or strong viability.

### 4.10.1 Invariance kernel

The invariance kernel refers to the set of states from which any state trajectory generated by any *a priori* admissible decision sequence satisfies the constraints.

**Definition 4.17.** *The invariance kernel at time $s \in \{t_0, \ldots, T\}$ for the dynamics (4.1) and the constraints (4.2a)–(4.2b)-(4.2c) is the set denoted by $\mathbb{I}\mathrm{nv}(s)$ defined by:*

$$\mathbb{I}\mathrm{nv}(s) := \left\{ x \in \mathbb{X} \;\middle|\; \begin{array}{l} \textit{for every decision } u(\cdot) \\ \textit{and state } x(\cdot) \textit{ starting from } x \textit{ at time } s \\ \textit{satisfying for any time } t \in \{s, \ldots, T-1\} \\ \textit{dynamics (4.1) and constraint (4.2a)} \\ \textit{the constraints (4.2b) are satisfied} \\ \textit{for } t \in \{s, \ldots, T-1\} \\ \textit{and (4.2c) at final time } T. \end{array} \right\} . \tag{4.44}$$

Notice that

$$\mathbb{I}\mathrm{nv}(s) \subset \mathbb{A}(s) ,$$

because the constraint (4.2b) is satisfied at time $s$, and that the invariance kernel at horizon $T$ is the target set $\mathbb{A}(T)$:

$$\mathbb{I}\mathrm{nv}(T) = \mathbb{A}(T) .$$

### 4.10.2 Invariance and viability

Of course, the invariance requirement is more stringent than the viability one and the invariance kernel is contained in the viability kernel:

$$\mathbb{I}\mathrm{nv}(t) \subset \mathbb{V}\mathrm{iab}(t) .$$

Let us point out that the viability and invariance kernels coincide whenever the set of possible decisions or controls $\mathbb{B}(t, x)$ is reduced to a singleton and only one decision is available:

$$\mathbb{B}(t, x) = \{\overline{u}(t, x)\} \Longrightarrow \mathbb{I}\mathrm{nv}(t) = \mathbb{V}\mathrm{iab}(t) .$$

Moreover, once the "true constraints" $\mathbb{B}^{\mathrm{viab}}(t, x)$ in (4.14) and the viability kernels $\mathbb{V}\mathrm{iab}(t)$ associated to the dynamics (4.1) and the constraints (4.2a)–(4.2d) are known, the same dynamics can be considered but under these new constraints, namely:

$$\begin{cases} x(t_0) = x_0 \in \mathbb{V}\mathrm{iab}(t_0) , \\ x(t+1) = F\big(t, x(t), u(t)\big) , \; t = t_0, \ldots, T-1 , \\ u(t) \in \mathbb{B}^{\mathrm{viab}}\big(t, x(t)\big) , \quad t = t_0, \ldots, T-1 , \\ x(t) \in \mathbb{V}\mathrm{iab}(t) , \qquad t = t_0, \ldots, T . \end{cases} \tag{4.45}$$

It happens then that the domains $\mathbb{V}\mathrm{iab}(t)$ are the invariant kernels for this new problem related to the property of viable controls which ensure that the above problem is equivalent to:

$$\begin{cases} x(t_0) = x_0 \in \mathbb{V}\text{iab}(t_0) , \\ x(t+1) = F\big(t, x(t), u(t)\big) , t = t_0, \ldots, T-1 , \\ u(t) \in \mathbb{B}^{\text{viab}}\big(t, x(t)\big) , \quad t = t_0, \ldots, T-1 . \end{cases} \tag{4.46}$$

This situation may be useful in case of optimal intertemporal selection, as in Sect. 5. This case occurs in a cost-effectiveness perspective. In this case, *ex post* admissible decisions $\mathbb{B}^{\text{viab}}(t, x)$ are used instead of $\mathbb{B}(t, x)$ in the dynamic programming method.

### 4.10.3 Maximality and Bellman properties

We also recover Bellman properties and the dynamics programming principle for invariance. This principle holds true through both geometrical and functional formulations. We distinguish the infinite and finite horizons. The proofs of the following Propositions follow the ones given for viability in Sect. A.2 in the Appendix. A geometric characterization in the finite horizon case is given by the following Proposition.

**Proposition 4.18.** *The invariance kernel satisfies the backward induction, where $t$ runs from $T-1$ down to $t_0$:*

$$\begin{cases} \mathbb{I}\text{nv}(T) = \mathbb{A}(T) , \\ \mathbb{I}\text{nv}(t) = \{x \in \mathbb{A}(t) \mid \forall u \in \mathbb{B}(t, x) , \quad F(t, x, u) \in \mathbb{I}\text{nv}(t+1)\} . \end{cases} \tag{4.47}$$

Similarly, we obtain a functional dynamics programming characterization in the finite horizon case.

**Proposition 4.19.** *The extended function $V : (t, x) \mapsto \Psi_{\mathbb{I}\text{nv}(t)}(x)$ is the solution of the* Bellman equation, *or* dynamic programming equation, *where $t$ runs from $T-1$ down to $t_0$:*

$$\begin{cases} V(T, x) = \Psi_{\mathbb{A}(T)}(x) , \\ V(t, x) = \sup_{u \in \mathbb{B}(t,x)} \Big( \Psi_{\mathbb{A}(t)}(x) + V\big(t+1, F(t, x, u)\big) \Big) . \end{cases} \tag{4.48}$$

The autonomous case and infinite horizon framework provide the simplest characterization since time disappears from the statements. In this case, the invariance kernel does not depend on $t$ and we write it $\mathbb{I}\text{nv}$.

**Proposition 4.20.** *In the autonomous case (4.19) and with $T = +\infty$, the extended function $V : x \mapsto \Psi_{\mathbb{I}}\text{nv}(x)$ is the largest solution of the* Bellman equation, *or* dynamic programming equation,

$$\begin{cases} V(x) = \sup_{u \in \mathbb{B}(x)} V\big(F(x, u)\big) , \\ V(x) \geq \Psi_{\mathbb{A}}(x) . \end{cases} \tag{4.49}$$

A geometric characterization is given using invariant domains as follows.

**Definition 4.21.** *In the autonomous case (4.19), a subset* $\mathbb{V} \subset \mathbb{X}$ *is said to be an invariant (or strongly invariant) domain if*

$$\forall x \in \mathbb{V}\,, \quad \forall u \in \mathbb{B}(x)\,, \quad F(x,u) \in \mathbb{V}\,. \tag{4.50}$$

An equivalent functional characterization is $\Psi_{\mathbb{V}}(x) = \sup_{u \in \mathbb{B}(x)} \Psi_{\mathbb{V}}\big(F(x,u)\big)$.

**Proposition 4.22.** *In the autonomous case (4.19) and with* $T = +\infty$*, the invariance kernel* $\mathbb{I}\mathrm{nv}$ *of* $\mathbb{A}$ *is the largest invariant domain contained in* $\mathbb{A}$*.*

# References

[1] J.-P. Aubin. *Viability Theory.* Birkhäuser, Boston, 1991. 542 pp.

[2] C. Béné and L. Doyen. Storage and viability of a fishery with resource and market dephased seasonnalities. *Environmental Resource Economics*, 15:1–26, 2000.

[3] C. Béné, L. Doyen, and D. Gabay. A viability analysis for a bio-economic model. *Ecological Economics*, 36:385–396, 2001.

[4] R. P. Berrens. The safe minimum standard of conservation and endangered species: a review. *Environmental Conservation*, 28:104–116, 2001.

[5] T. Bruckner, G. Petschell-Held, F. L. Toth, H.-M. Fussel, C. Helm, M. Leimbach, and H. J. Schellnhuber. Climate change decision-support and the tolerable windows approach. In *Environmental Modeling and Assessment, Earth system analysis and management*, chapter 4, pages 217–234. H. J. Schellnhuber and F. L. Toth, 1999.

[6] F. H. Clarke, Y. S. Ledayev, R. J. Stern, and P. R. Wolenski. Qualitative properties of trajectories of control systems: a survey. *Journal of Dynamical Control Systems*, 1:1–48, 1995.

[7] P. Cury, C. Mullon, S. Garcia, and L. J. Shannon. Viability theory for an ecosystem approach to fisheries. *ICES Journal of Marine Science*, 62(3):577–584, 2005.

[8] M. De Lara, L. Doyen, T. Guilbaud, and M.-J. Rochet. Monotonicity properties for the viable control of discrete time systems. *Systems and Control Letters*, 56(4):296–302, 2006.

[9] M. De Lara, L. Doyen, T. Guilbaud, and M.-J. Rochet. Is a management framework based on spawning-stock biomass indicators sustainable? A viability approach. *ICES J. Mar. Sci.*, 64(4):761–767, 2007.

[10] L. Doyen and C. Béné. Sustainability of fisheries through marine reserves: a robust modeling analysis. *Journal of Environmental Management*, 69:1–13, 2003.

[11] L. Doyen, M. De Lara, J. Ferraris, and D. Pelletier. Sustainability of exploited marine ecosystems through protected areas: a viability model and

a coral reef case study. *Ecological Modelling*, 208(2-4):353–366, November 2007.

[12] L. Doyen, P. Dumas, and P. Ambrosi. Optimal timing of $CO_2$ mitigation policies for a cost-effectiveness model. *Mathematics and Computer Modeling*, in press.

[13] K. Eisenack, J. Sheffran, and J. Kropp. The viability analysis of management frameworks for fisheries. *Environmental Modeling and Assessment*, 11(1):69–79, February 2006.

[14] G. Heal. *Valuing the Future, Economic Theory and Sustainability.* Columbia University Press, New York, 1998.

[15] ICES. Report of the ices advisory committee on fishery management and advisory committee on ecosystems, 2004. Ices advice, 1, 2004. 1544 pp.

[16] IPCC. http://www.ipcc.ch/.

[17] V. Martinet and L. Doyen. Sustainable management of an exhaustible resource: a viable control approach. *Resource and Energy Economics*, 29(1):p.17–39, 2007.

[18] V. Martinet, L. Doyen, and O. Thébaud. Defining viable recovery paths toward sustainable fisheries. *Ecological Economics*, 64(2):411–422, 2007.

[19] W. F. Morris and D. F. Doak. *Quantitative Conservation Biology: Theory and Practice of Population Viability Analysis.* Sinauer Associates, 2003.

[20] A. Rapaport, J.-P. Terreaux, and L. Doyen. Sustainable management of renewable resource: a viability approach. *Mathematics and Computer Modeling*, 43(5-6):466–484, March 2006.

[21] H. J. Schellnhuber and V. Wenzel. *Earth System Analysis, Integrating Science for Sustainability.* Springer, 1988.

[22] M. Tichit, L. Doyen, J.Y. Lemel, and O. Renault. A co-viability model of grazing and bird community management in farmland. *Ecological Modelling*, 206(3-4):277–293, August 2007.

[23] M. Tichit, B. Hubert, L. Doyen, and D. Genin. A viability model to assess the sustainability of mixed herd under climatic uncertainty. *Animal Research*, 53(5):405–417, 2004.

[24] R. Vidal, S. Schaffert, J. Lygeros, and S. Sastry. Controlled invariance of discrete time systems. In *Hybrid Systems: Computation and Control*, volume 1790 of *Lecture Notes in Computer Science*, pages 437–450. Springer-Verlag, 2000.

# 5

# Optimal sequential decisions

Many decision problems deal with the intertemporal optimization of some criteria. Achieving a $CO_2$ concentration target in minimal time, maximizing the sum of discounted utility of consumption along time in an economic growth model, minimizing the cost of recovery for an endangered species, maximizing the discounted rent of a fishery are all examples of dynamic optimal control problems. Basic contributions for the optimal management or conservation of natural resources and bioeconomic modeling can be found in [3, 4, 5] together with [2] that contains optimality methods and models especially for the management of animal populations. Optimality approaches to address the sustainability issues and, especially, intergenerational equity and conservation issues, are exposed for instance in [8] in the continuous case. It includes in particular the maximin, Green Golden and Chichilnisky approaches.

An important debate for the optimal management of natural resources and the environment relates to the method of economic valuation of ecological or environmental outcomes. Such concerns have fundamental connections with "price versus quantity" issues pointed out by [14]. The usual economic approach of Cost-Benefit (CB) relies on the monetary assessment of all outcomes and costs involved in the management process. However some advantages or outputs are not easy to assess in monetary units. This is the case for non-use values or indicators of ecological services of biodiversity [13] such as aesthetic or existence values. Such is also the case for costs relying on damages of global changes [7] whose pricing turns out to be highly uncertain and controversial. Such difficulties advocate for the use of Cost-Effectiveness (CE) analysis which also measures costs but with outputs expressed in quantity terms. In mathematical terms, CE deals with optimal control under constraints. In the case of CB, the quantity constraints are directly taken into account in the criterion to optimize through adjoint variables including prices, shadow prices or existence values. Contingent valuation methods aim at pricing such non-use values.

For relevant mathematical references in the discrete time optimal control context, we refer to [1, 15]. In this textbook, we especially support the use of the dynamic programming method and Bellman equation since it turns out to be a relevant approach for many optimal control problems including additive and maximin criterion as well as viability and invariance problems. Moreover, it is shown that the other well known approach termed *Pontryagin maximum principle* to cope with optimal control can be derived from the Bellman equation in a variational setting under an appropriate hypothesis. Furthermore, the dynamic programming can be expanded to the uncertain case as will be exposed in the chapters to come. By *dynamic programming*, the dynamic optimality decision problem is solved sequentially: one starts at the final time horizon and then applies some backward induction mechanism at each time step. We detail this process and ideas in what follows.

The chapter is organized as follows. Section 5.1 provides the general problem formulation for several intertemporal optimality criteria. The dynamic programming method for the additive criterion case is exposed in Sect. 5.2. Then various examples illustrate the concepts and results for natural resource management. The so-called "maximum principle" is presented in Sect. 5.8, with an application to the Hotelling rule in Sect. 5.9. Section 5.11 focuses on the final payoff problem related to the Green Golden issues. Section 5.14 presents the maximin case, with an illustration of the management of an exhaustible resource in Sect. 5.15.

## 5.1 Problem formulation

Let us briefly recall the mathematical material already introduced in Sect. 2.9 for optimal and feasible sequential decisions in discrete time.

### 5.1.1 Dynamics and constraints

We consider again the following nonlinear dynamical system, as in Sect. 2.9,

$$x(t+1) = F\big(t, x(t), u(t)\big) \; , \quad t = t_0, \ldots, T-1 \; , \quad \text{with} \quad x(t_0) = x_0 \; , \qquad (5.1)$$

where $x(t) \in \mathbb{X} = \mathbb{R}^n$ is the state, $u(t) \in \mathbb{U} = \mathbb{R}^p$ is the control or decision, $T$ corresponds to the time horizon which may be finite or infinite, while $x_0$ and $t_0$ stand for the initial state and time conditions.

In addition, some admissibility or effectiveness constraints are to be satisfied as underlined in Chap. 4 focusing on invariance and viability. These constraints include state and control constraints respected at any time

$$\begin{cases} x(t) \in \mathbb{A}(t) \subset \mathbb{X} \; , & t = t_0, \ldots, T \; , \\ u(t) \in \mathbb{B}\big(t, x(t)\big) \subset \mathbb{U} \; , & t = t_0, \ldots, T-1 \; . \end{cases} \qquad (5.2)$$

### 5.1.2 The criterion and the evaluation of the decision

Now, one may aim at ranking the different feasible decision paths $u(\cdot)$ according to some specific indicator of performance or *criterion* $\pi$, representing the total gain, payoff, utility, cost or cost-benefit over $T+1$ stages. Of particular interest is the selection of a policy optimizing (minimizing or maximizing) this performance or criterion $\pi$. *We shall consider maximization problems where the criterion is a payoff.*

A *criterion* $\pi$ is a function

$$\pi : \mathbb{X}^{T+1-t_0} \times \mathbb{U}^{T-t_0} \to \mathbb{R} \tag{5.3}$$

which assigns a real number to a state and control trajectory. Different criteria have been detailed in Sect. 2.9.4 and we shall here concentrate on two main cases.

- **General additive criterion.** The *additive criterion* with final payoff in the finite horizon is

$$\pi\big(x(\cdot), u(\cdot)\big) = \sum_{t=t_0}^{T-1} L\big(t, x(t), u(t)\big) + M\big(T, x(T)\big) \, . \tag{5.4}$$

  Function $L$ is referred to as the system's *instantaneous performance* (or profit, benefit, payoff, etc.) while $M$ is the *final performance.* Such payoff includes usual discounted cases as well as Green Golden Rule or Chichilnisky performances. In our discrete time framework when the horizon $T$ is finite, we shall label *Green Golden* a criterion of the form

$$\pi\big(x(\cdot), u(\cdot)\big) = M\big(T, x(T)\big) \, . \tag{5.5}$$

- **The Maximin.** The *Rawlsian* or *maximin* form in the finite horizon is

$$\pi\big(x(\cdot), u(\cdot)\big) = \min_{t=t_0,\ldots,T-1} L\big(t, x(t), u(t)\big) \, , \tag{5.6}$$

  and, with a final payoff, the expression is somewhat heavier to write:

$$\pi\big(x(\cdot), u(\cdot)\big) = \min\left( \min_{t=t_0,\ldots,T-1} L\big(t, x(t), u(t)\big), M\big(T, x(T)\big) \right) . \tag{5.7}$$

### 5.1.3 The general problem of optimal control under constraints

The constraint (5.2) specified beforehand combined with the dynamic (5.1) settle the set of all possible and feasible state and decision trajectories. Such a feasibility set, denoted by $\mathcal{T}^{ad}(t_0, x_0) \subset \mathbb{X}^{T+1-t_0} \times \mathbb{U}^{T-t_0}$, has already been introduced in Subsect. 2.9.5 and studied in Chap. 4:

$$\mathcal{T}^{ad}(t_0,x_0) = \left\{ (x(\cdot),u(\cdot)) \;\middle|\; \begin{array}{ll} x(t_0) = x_0 \,, & \\ x(t+1) = F\big(t,x(t),u(t)\big) \,, & t = t_0,\ldots,T-1 \\ u(t) \in \mathbb{B}\big(t,x(t)\big) \,, & t = t_0,\ldots,T-1 \\ x(t) \in \mathbb{A}(t) \,, & t = t_0,\ldots,T \end{array} \right\} . \tag{5.8}$$

This is the set of admissible trajectories which visit $x_0$ at time $t_0$ while respecting both the constraints and the dynamics after time $t_0$.

When the assessment $\pi$ measures payoff (or profit, benefit, utility, etc.), the problem reads[1]

$$\pi^\star(t_0,x_0) := \sup_{\big(x(\cdot),u(\cdot)\big)\in\mathcal{T}^{ad}(t_0,x_0)} \pi\big(x(\cdot),u(\cdot)\big) \,. \tag{5.9}$$

Abusively, we shall in practice abbreviate (5.9) in

$$\pi^\star(t_0,x_0) := \sup_{u(\cdot)} \pi\big(x(\cdot),u(\cdot)\big) \,. \tag{5.10}$$

Whenever feasibility is impossible, *i.e.* $\mathcal{T}^{ad}(t_0,x_0) = \emptyset$, by convention we set the optimal criterion to minus infinity: $\pi^\star(t_0,x_0) = -\infty$.

**Definition 5.1.** *The optimal value $\pi^\star(t_0,x_0)$ in (5.9) is called the* optimal performance. *Any path $\big(x^\star(\cdot),u^\star(\cdot)\big) \in \mathcal{T}^{ad}(t_0,x_0)$ such that*

$$\pi^\star(t_0,x_0) = \max_{\big(x(\cdot),u(\cdot)\big)\in\mathcal{T}^{ad}(t_0,x_0)} \pi\big(x(\cdot),u(\cdot)\big) = \pi\big(x^\star(\cdot),u^\star(\cdot)\big) \tag{5.11}$$

*is a feasible* optimal trajectory *or an* optimal path.

In fact, the feasible set $\mathcal{T}^{ad}(t_0,x_0)$ is related to the viability kernel $\mathbb{V}\mathrm{iab}(t_0)$ in (4.10) and to viable decisions examined in Chap. 4 (see Definition 4.1 and (4.14)). As pointed out in that chapter (Proposition 4.4), the "true" state constraints are captured by the viability kernels and $\mathcal{T}^{ad}(t_0,x_0) \subset \mathbb{X}^{T+1-t_0} \times \mathbb{U}^{T-t_0}$ is given equivalently by

$$\begin{aligned}
(x(\cdot),u(\cdot)) \in \mathcal{T}^{ad}(t_0,x_0) \iff & \left\{ \begin{array}{ll} x(t_0) = x_0 & \\ x(t+1) = F\big(t,x(t),u(t)\big), & t = t_0,\ldots,T-1 \\ u(t) \in \mathbb{B}^{\mathrm{viab}}(t,x(t)), & t = t_0,\ldots,T-1 \\ x(t) \in \mathbb{V}\mathrm{iab}(t), & t = t_0,\ldots,T \,, \end{array} \right. \\
\iff & \left\{ \begin{array}{ll} x(t_0) = x_0 \in \mathbb{V}\mathrm{iab}(t_0) & \\ x(t+1) = F\big(t,x(t),u(t)\big), & t = t_0,\ldots,T-1 \\ u(t) \in \mathbb{B}^{\mathrm{viab}}(t,x(t)), & t = t_0,\ldots,T-1 \,. \end{array} \right.
\end{aligned}$$

[1] We shall only consider maximization problems. To cope with the minimization problem $\inf_{\big(x(\cdot),u(\cdot)\big)\in\mathcal{T}^{ad}(t_0,x_0)} \pi\big(x(\cdot),u(\cdot)\big)$, one should simply change the sign of $\pi$ since $\inf_{z\in Z} f(z) = -\sup_{z\in Z}\big(-f(z)\big)$.

Let us stress the fact that, whenever the viability kernel $\mathbb{V}\mathrm{iab}(t_0)$ is empty, it is worthwhile to put a hold on the computation of any optimal solution as no feasible solution exists:

$$\mathbb{V}\mathrm{iab}(t_0) = \emptyset \iff \mathcal{T}^{ad}(t_0, x_0) = \emptyset\,, \quad \forall x_0 \in \mathbb{X}\,.$$

Let us also point out that an interesting case corresponds to an invariant state constraint set as explained in detail in Sect. 4.10. An invariant state constraint set means that, for any time $t$, the invariance kernel $\mathbb{I}\mathrm{nv}(t)$ of Definition 4.17 coincides with the initial state constraint set $\mathbb{A}(t_0)$, thus revealing that every *a priori* admissible state $x(\cdot)$ and all decision $u(\cdot)$ trajectories will satisfy the constraints along time. In other words, the constraints do not really reduce the choice of the decisions and only the initial state constraint $\mathbb{A}(t_0)$ prevails. Such a convenient configuration occurs in the special case without state constraint, namely when $\mathbb{A}(t) = \mathbb{X}$.

### 5.1.4 Cost-Benefit versus Cost-Effectiveness

Two specific classes of problems are worth distinguishing: Cost-Benefit versus Cost-Effectiveness. Such concerns have connections with "price versus quantity" debates. The usual economic approach of Cost-Benefit (CB) relies on the monetary assessment of all outcomes and costs involved in the management process. However some advantages or outputs are not easy to assess in monetary units. This is the case for non-use values of ecological services of biodiversity such as aesthetic or existence values. Such is also the case for costs relying on damages of global changes whose pricing turns out to be highly unknown. Such difficulties justify the use of Cost-Effectiveness (CE) analysis which also measures costs but with outputs expressed in quantity terms. In mathematical terms, CE deals with optimal control under constraints. In the case of CB, the quantity constraints are directly taken into account in the criterion to optimize through adjoint variables including prices, shadow prices or existence values.

*Cost-Benefit*

In this case, no state constraints bind the optimal solution and, generally, the objective is the intertemporal discounted profit, namely the difference between benefits $B$ and costs $C$ in monetary units

$$\pi^\star(t_0, x_0) = \sup_{\big(x(\cdot),u(\cdot)\big) \in \mathcal{T}^{ad}(t_0,x_0)} \sum_{t=t_0}^{T-1} \rho^t \Big( B\big(x(t), u(t)\big) - C\big(x(t), u(t)\big) \Big)\,, \tag{5.12}$$

where $\rho \in [0, 1]$ is a discount factor.

*Cost-Effectiveness*

In this case, there is generally an additional state-control constraint, in vectorial form $Y\big(x(t), u(t)\big) \geq 0$, capturing the effectiveness of the decisions, and replacing, in a sense, the monetary evaluation of benefits. For this reason the objective is now intertemporal discounted costs minimization:

$$\pi^\star(t_0, x_0) = \inf_{\begin{cases} \big(x(\cdot), u(\cdot)\big) \in \mathcal{T}^{ad}(t_0, x_0) \\ Y\big(x(t), u(t)\big) \geq 0 \end{cases}} \sum_{t=t_0}^{T-1} \rho^t C\big(x(t), u(t)\big) . \quad (5.13)$$

## 5.2 Dynamic programming for the additive payoff case

The additive payoff case plays a prominent role in optimal control for the separable form which allows for the classical Bellman equation and for its economic interpretation as an intertemporal sum of payoffs.

### 5.2.1 The optimization problem

In this section, we focus on the maximization problem for additive and separable forms in the finite horizon. A general form is considered which combines an instantaneous payoff $L$ together with a final requirement $M$:

$$\pi^\star(t_0, x_0) = \sup_{\big(x(\cdot), u(\cdot)\big) \in \mathcal{T}^{ad}(t_0, x_0)} \sum_{t=t_0}^{T-1} L\big(t, x(t), u(t)\big) + M\big(T, x(T)\big) . \quad (5.14)$$

We shall display that the following *value function* is the solution of a backward induction equation by observing that we can split the maximization operation into two parts for the following reasons: the criterion $\pi$ is additive[2], the dynamic is a first order difference equation and, finally, constraints at time $t$ depend only on time $t$ and state $x(t)$.

### 5.2.2 Additive value function

The additive value function $V$ at time $t$ and for state $x$ represents the optimal value of the criterion over $T - t$ periods, given that the state of the system $x(t)$ at time $t$ is $x$.

---

[2] This is a traditional assumption. However, it is not essential for dynamic programming to apply it in the deterministic case, as pointed out in [15]. See the proofs in Sect. A.3.

**Definition 5.2.** *For the maximization problem (5.14) with dynamic (5.1) and under constraints (5.2), we define a function* $V(t,x)$ *of time* $t$ *and state* $x$*, named* additive value function *or* Bellman function*as follows: for* $t = t_0, \ldots, T-1$*, for* $x \in \mathbb{V}\mathrm{iab}(t)$

$$V(t,x) := \sup_{\big(x(\cdot),u(\cdot)\big)\in\mathcal{T}^{ad}(t,x)} \sum_{s=t}^{T-1} L\big(s,x(s),u(s)\big) + M\big(T,x(T)\big) , \qquad (5.15)$$

*with the convention that* $V(t,x) := -\infty$ *for* $x \notin \mathbb{V}\mathrm{iab}(t)$.

Therefore, given the initial state $x_0$, the optimal sequential decision problem (5.1)-(5.2)-(5.14) refers to the particular value $V(t_0,x_0)$ of Bellman function:

$$V(t_0,x_0) = \pi^\star(t_0,x_0) .$$

### 5.2.3 Dynamic programming equation

We shall now see the principle of dynamic programming which consists in replacing the optimization problem (5.14) over the trajectories space $\mathbb{X}^{T+1-t_0} \times \mathbb{U}^{T-t_0}$ by a sequence of $T+1-t_0$ interconnected optimization problems (5.17) over the decision space $\mathbb{U}$.

The common proof of the three following Propositions 5.3, 5.4 and 5.5 can be found in Sect. A.3 in the Appendix.

#### Case without state constraint

We first distinguish the case without state constraints, which is simpler to formulate.

**Proposition 5.3.** *In the case without state constraint where* $\mathbb{A}(t) = \mathbb{X}$ *in (5.2), the value function in (5.15) is the solution of the following backward* dynamic programming equation *(or* Bellman equation*), where* $t$ *runs from* $T-1$ *down to* $t_0$*:*

$$\begin{cases} V(T,x) = M(T,x) , \\ V(t,x) = \sup_{u\in\mathbb{B}(t,x)} \Big( L(t,x,u) + V\big(t+1, F(t,x,u)\big)\Big) . \end{cases} \qquad (5.16)$$

Thus, the value function is given by a *backward* induction, starting from the final term $M$ in (5.14) and proceeding with a sequence of maximizations over $u \in \mathbb{U}$, for all $(t,x)$. This approach generally renders dynamic programming numerically untractable for high dimensions of the state (*curse of dimensionality*).

**Case with state constraints**

Now, the case taking into account state constraints is presented. Of course, it is more complicated than the previous one since viability and optimality conditions are combined.

**Proposition 5.4.** *The value function defined in (5.15) is the solution of the following backward* dynamic programming equation *(or* Bellman equation*), where $t$ runs from $T-1$ down to $t_0$,*

$$\begin{cases} V(T,x) = M(T,x) \;, & \forall x \in \mathbb{V}\mathrm{iab}(T) \;, \\ V(t,x) = \sup\limits_{u \in \mathbb{B}^{viab}(t,x)} \Big( L(t,x,u) + V\big(t+1, F(t,x,u)\big) \Big) \;, & \forall x \in \mathbb{V}\mathrm{iab}(t) \;, \end{cases} \tag{5.17}$$

*where $\mathbb{V}\mathrm{iab}(t)$ is given by the backward induction (4.11)*

$$\begin{cases} \mathbb{V}\mathrm{iab}(T) = \mathbb{A}(T) \;, \\ \mathbb{V}\mathrm{iab}(t) = \{x \in \mathbb{A}(t) \mid \exists u \in \mathbb{B}(t,x) \;, \quad F(t,x,u) \in \mathbb{V}\mathrm{iab}(t+1)\} \;, \end{cases} \tag{5.18}$$

*and where the supremum in (5.17) is over viable controls in $\mathbb{B}^{viab}(t,x)$ given by (4.14), namely*

$$\mathbb{B}^{viab}(t,x) = \{u \in \mathbb{B}(t,x) \mid F(t,x,u) \in \mathbb{V}\mathrm{iab}(t+1)\} \;. \tag{5.19}$$

### 5.2.4 Optimal feedback

As we have seen, the backward equation of dynamic programming (5.17) enables us to compute the value function $V(t,x)$ and, thus, the optimal payoff $\pi^\star$. In fact, a stronger result is obtained since Bellman induction optimization reveals relevant feedback controls. Indeed, assuming the additional hypothesis that the supremum is achieved in (5.17) for at least one decision, if we denote by $\mathfrak{u}^\star(t,x)$ a value[3] $u \in \mathbb{B}(t,x)$ which achieves the maximum in (5.17), then $\mathfrak{u}^\star(t,x)$ defines an optimal feedback for the optimal control problem in the following sense.

The following Proposition 5.5 claims that the dynamic programming equation exhibits optimal viable feedbacks.

**Proposition 5.5.** *For any time $t$ and state $x \in \mathbb{V}\mathrm{iab}(t)$, assume the existence of the following feedback decision*

---

[3] See the footnote 13 in Sect. 2.10.

$$\mathrm{u}^\star(t,x) \in \mathop{\arg\max}_{u \in \mathbb{B}^{viab}(t,x)} \Big( L(t,x,u) + V\big(t+1, F(t,x,u)\big) \Big) . \quad (5.20)$$

*Then* u *is an* optimal viable feedback *for the maximization problem (5.14) under constraints (5.8) in the sense that, when* $x_0 \in \mathbb{V}\mathrm{iab}(t_0)$*, the trajectory* $\big(x^\star(\cdot), u^\star(\cdot)\big)$ *generated by*

$$x^\star(t_0) = x_0 \, , \quad x^\star(t+1) = F(t, x^\star(t), u^\star(t)) \, , \quad u^\star(t) = \mathrm{u}^\star(t, x(t)) \, , \quad (5.21)$$

*for* $t = t_0, \ldots, T-1$*, belongs to* $\mathcal{T}^{ad}(t_0, x_0)$ *given by (5.8) and is an optimal feasible trajectory, that is,*

$$\max_{\big(x(\cdot),u(\cdot)\big) \in \mathcal{T}^{ad}(t_0,x_0)} \pi\big(x(\cdot), u(\cdot)\big) = \pi\big(x^\star(\cdot), u^\star(\cdot)\big) . \quad (5.22)$$

Notice that the feedback found is appropriate for *any* initial time $t_0$ and initial state $x_0 \in \mathbb{V}\mathrm{iab}(t_0)$.

## 5.3 Intergenerational equity for a renewable resource

Let us return to the renewable resource management problem presented in Sect. 2.2

$$\sup_{h(\cdot)} \Big( \sum_{t=t_0}^{T-1} \rho^t L\big(h(t)\big) + \rho^T L\big(B(T)\big) \Big) ,$$

under the dynamic and constraints

$$B(t+1) = R\big(B(t) - h(t)\big) \, , \quad 0 \le h(t) \le B(t) \, .$$

For the sake of simplicity, we consider the particular case of $T = 2$ periods and we suppose as well that the discount factor $\rho$ equals the growth rate of the resource

$$\rho = \frac{1}{R} \quad \text{with} \quad R > 1 \, . \quad (5.23)$$

The utility function $L$ is supposed sufficiently smooth (twice continuously differentiable, for instance), and strictly concave, that is to say $L'' < 0$. The Bellman equation (5.16) implies

$$\begin{cases} V(2,B) = \rho^2 L(B) \\ V(1,B) = \displaystyle\sup_{0 \le h \le B} \{\rho L(h) + V(2, RB - h))\} \\ \qquad\quad = \displaystyle\sup_{0 \le h \le B} \{\rho L(h) + \rho^2 L(R(B-h))\} \\ \qquad\quad = \displaystyle\sup_{0 \le h \le B} v(h) \end{cases}$$

where $\upsilon(h) = \rho L(h) + \rho^2 L\big(R(B-h)\big)$. Assuming that the last supremum is achieved for an interior solution $h^\star \in ]0, B[$, it must necessarily satisfy the so-called *first order optimality condition*

$$\frac{d\upsilon}{dh}(h^\star) = 0 \,.$$

We then have $L'(h^\star) = \rho R L'\big(R(B-h^\star)\big) = L'\big(R(B-h^\star)\big)$, by (5.23). Thus, as $L'$ is a strictly decreasing function, hence injective, one has $h^\star = R(B-h^\star)$ so that the optimal feedback is linear in $B$

$$\mathfrak{h}^\star(1, B) = \frac{R}{1+R} B$$

and we check that $0 < h^\star = \frac{R}{1+R} B < B$. The value function is of the same form as $L$:

$$V(1, B) = \rho L(\frac{R}{1+R} B) + \rho^2 L(\frac{R}{1+R} B) = \rho(1+\rho) L(\frac{R}{1+R} B) \,.$$

Likewise, we check that

$$\begin{aligned} V(0, B) &= \sup_{0 \le h \le B} \big\{ L(h) + V\big(1, R(B-h)\big) \big\} \\ &= \sup_{0 \le h \le B} \Big\{ L(h) + \rho(1+\rho) L(\frac{R}{1+R} R(B-h)) \Big\} \,. \end{aligned}$$

The first order optimality condition now implies

$$L'(h^\star) = L'(\frac{R}{1+R} R(B-h^\star)) \,.$$

Thus, $h^\star = \dfrac{R}{1+R} R(B-h^\star)$, so that the optimal feedback is still linear

$$\mathfrak{h}^\star(0, B) = \frac{R^2}{1+R+R^2} B$$

and the value function is of the same form as $L$. Going forward with $h^\star(0) = \mathfrak{h}^\star(0, B_0)$ and $B^\star(1) = R(B_0 - h^\star(0))$, we easily find that the optimal catches $h^\star(0) = \mathfrak{h}^\star(0, B_0)$ and $h^\star(1) = \mathfrak{h}^\star(1, B^\star(1))$ are stationary:

$$h^\star(0) = h^\star(1) = \frac{R^2}{1+R+R^2} B_0 \,.$$

In this sense, the optimal harvests promote intergenerational equity as may be seen in Fig. 5.3. This is closely related to the assumption $\rho R = 1$ as shown by the following example of an exhaustible resource management.

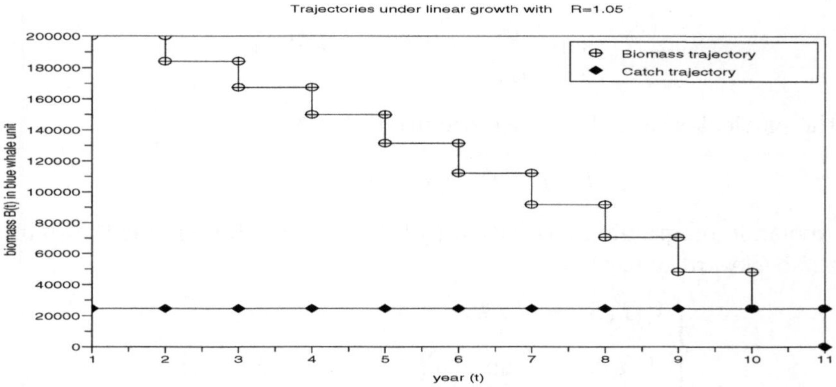


**Fig. 5.1.** Biomass $B(t)$ and "equity" catch $h(t)$ trajectories for a linear dynamic model of whale growth with yearly growth rate $R = 1.05$. Time is measured in years while population and catches is in blue whale unit. Trajectories are computed with the SCILAB code 9.

SCILAB CODE 9.

```
//
// exec intergen_equity.sce

R_whale=1.05 ;
// per capita productivity
R=R_whale ;
K_whale = 400000;
// carrying capacity  (BWH, blue whale unit)
K=K_whale ;

// LINEAR DYNAMICS
function [y]=linear(B), y=R*B , endfunction;

Horizon= 10;
years=1:Horizon;
yearss=1:(Horizon+1);

Binit= K/2 ;
// initial condition
trajectory_whale=zeros(yearss) ;
// vector will contain the trajectory B(1),...,B(Horizon+1)
catch_whale=zeros(years) ;
// vector will contain the catches h(1),...,h(Horizon)

trajectory_whale(1)=Binit ;
// initialization of vector B(1),...,B(Horizon+1)
for t=years
  catch_whale(t)=R^{Horizon-t+1}/ ...
  sum(R.^{Horizon-years(t:$)+1})*...
  trajectory_whale(t);
  trajectory_whale(t+1)=...
  linear(trajectory_whale(t)-catch_whale(t));
end

// Graphic display

xset("window",20+1); xbasc(20+1);
plot2d2(yearss,[trajectory_whale ; [catch_whale 0]]' );
xtitle('Trajectories under linear growth with...
   R='+string(R_whale),'year (t)',...
   'biomass B(t) in blue whale unit')
plot2d2(yearss,[trajectory_whale ; [catch_whale 0]]',...
style=-[3,4]);
legends(['Biomass trajectory';'Catch trajectory'],-[3,4],'ur')

//
```

## 5.4 Optimal depletion of an exhaustible resource

Consider the model presented in Sect. 2.1

$$S(t+1) = S(t) - h(t) , \quad 0 \leq h(t) \leq S(t)$$

for the optimal management of an exhaustible resource

$$\sup_{h(\cdot)} \Big( \sum_{t=t_0}^{T-1} \rho^t h(t)^\eta + \rho^T S(T)^\eta \Big)$$

in the particular case of an isoelastic utility function

$$L(h) = h^\eta \quad \text{with} \quad 0 < \eta < 1 \, .$$

By dynamic programming equation (5.16), the value function is the solution of the backward induction:

$$\begin{cases} V(T,S) = \rho^T S^\eta \, , \\ V(t,S) = \sup_{0 \le h \le S} \big\{ \rho^t h^\eta + V(t+1, S-h) \big\} \, . \end{cases}$$

Notice the “existence” or “inheritance” term $M(T,S(T)) = \rho^T S(T)^\eta$ enhancing the resource.

**Result 5.6** *Using dynamic programming, one can prove by induction that*

$$\begin{cases} V(t,S) = \rho^t b(t)^{\eta-1} S^\eta \, , \\ h^\star(t,S) = b(t) S \, , \end{cases}$$

*where*

$$b(t) = \frac{a-1}{a-a^{t-T}} \quad \textit{with} \quad a := \rho^{\frac{1}{\eta-1}} \, .$$

Thus, by stock dynamic $S^\star(s+1) = S^\star(s) - h^\star(s, S^\star(s))$ as in (5.21), we get the optimal paths in open-loop terms from initial time $t_0 = 0$ as follows:

$$\begin{cases} S^\star(t) = \frac{a^{T+1-t}-1}{a^{T+1}-1} S_0 \, , \\ h^\star(t) = \frac{a-1}{a-a^{t-T}} \frac{a^{T+1-t}-1}{a^{T+1}-1} S_0 \, . \end{cases}$$

Let us also deduce that the profile of optimal consumptions basically depends on the discount factor $\rho$ through $a = \rho^{\frac{1}{\eta-1}}$ since

$$h^\star(t+1) = a^{-1} h^\star(t) \, .$$

In particular, the optimal extractions strictly decrease with time for usual values of the discount factor strictly smaller than 1. This fact illustrates the preference for the present of such a discounted framework. How does this alter sustainability of both the exploitation and the resource? At a first glance, some sustainability remains since the optimal final stock is strictly positive, $S^\star(T) > 0$, contributing to the final inheritance or existence value of the resource $\rho^T L(S^\star(T)) = \rho^T (S^\star(T))^\eta > 0$. However, let us stress the fact that, for an infinite horizon $T = +\infty$, the final optimal stock $S^\star(T)$ and consumption $h^\star(T)$ vanish for usual values of the discount factor $\rho \in [0,1[$, thus pointing out how sustainability is threatened in such an optimality context.

**Result 5.7** *Assume that the discount factor $\rho$ is strictly smaller than 1 ($0 < \rho < 1$). Then sustainability disappears in the sense*

$$\begin{cases} h^\star(t+1) < h^\star(t) , \\ \lim_{T \to +\infty} S^\star(T) = 0 , \\ \lim_{T \to +\infty} h^\star(T) = 0 . \end{cases}$$

Furthermore, the optimal control problem

$$\sup_{h(\cdot)} \sum_{t=t_0}^{T-1} \rho^t h(t)^\eta$$

under the same dynamics and constraints is the same as the previous one, except for the absence of an inheritance value of the stock. A similar computation shows that $V(T-1, S) = \rho^{T-1} S^\eta$ and that all results above may be used with $T$ replaced by $T+1$. We find that $S^\star(T) = 0$: without "inheritance" value of the stock, it is optimal to totally deplete the resource at the final period. Hence, not surprisingly, sustainability of the resource is altered more in such a case.

## 5.5 Over-exploitation, extinction and inequity

Here we follow the material introduced in Sect. 2.2. Let us consider the exploitation of a renewable resource whose growth is linear $g(B) = RB$. It is assumed that the utility function is also linear $L(h) = h$, which means that price is normalized at $p = 1$ while harvesting costs are not taken into account, that is to say $C(h, B) = 0$. Hence, the optimality problem over $T$ periods corresponds to

$$\sup_{h(t_0), h(t_0+1), \ldots, h(T-1)} \sum_{t=t_0}^{T-1} \rho^t h(t) ,$$

under the dynamics and constraints

$$B(t+1) = R(B(t) - h(t)) , \quad B(t_0) = B_0 \quad \text{and} \quad 0 \leq h(t) \leq B(t) .$$

By dynamic programming equation (5.16), it can be proved that the optimal catches $h^\star(t)$ are computed for any initial state $B_0$ as follows.

**Result 5.8** *Optimal catches depend on $\rho R$ as follows.*

1. *If $\rho R > 1$, then $0 = h^\star(t_0) = h^\star(t_0+1) = \cdots = h^\star(T-2)$ and $h^\star(T-1) = B_0 R^{T-1-t_0}$.*
2. *If $\rho R < 1$, then $B_0 = h^\star(t_0)$ and $0 = h^\star(t_0+1) = \cdots = h^\star(T-1)$.*

3. *If $\rho R = 1$, many solutions exist. One optimal solution is given by* $h^\star(t_0) = h^\star(t_0+1) = \cdots = h^\star(T-1) = B_0/(T-t_0)$.

The sustainability issues captured by such an assertion are the following and depend critically upon the product

$$\underbrace{\rho}_{\text{economic discount factor}} \times \underbrace{R}_{\text{biological growth factor}} \tag{5.24}$$

which mixes economic and biological characteristics of the problem.

- Intergenerational equity is a central problem: whenever $\rho R \neq 1$, no guaranteed harvesting occurs. Either the resource is completely depleted at the first period and thus catches vanish along the remaining time (case when $\rho R < 1$), or catches are maintained at zero until the last period when the whole biomass is harvested ($\rho R > 1$).
- It can be optimal to destroy the resource. Especially if biomass growth $R$ is weak, it is efficient to destroy the whole biomass at the initial date. An extinction problem occurs as pointed out by [3].
- The discount factor $\rho$ plays a basic role in favoring the future or the present or providing equity through $\rho R = 1$.

An illustrative example is provided by blue whales whose growth rate is estimated to be lower than 5% a year [10]. Such a low rate might explain the overexploitation of whales during the twentieth century that led regulating agencies to a moratorium in the sixtees [3].

Another example with the utility function $L(h) = 1 - e^{-h}$ is examined in Scilab code 10. The solutions are displayed in Fig. 5.2. Although the optimal paths are smoother, the intergenerational equity problems again occur in a similar qualitative way depending on the critical value $\rho R$. Preference for the future or present basically depends on $\rho R$.

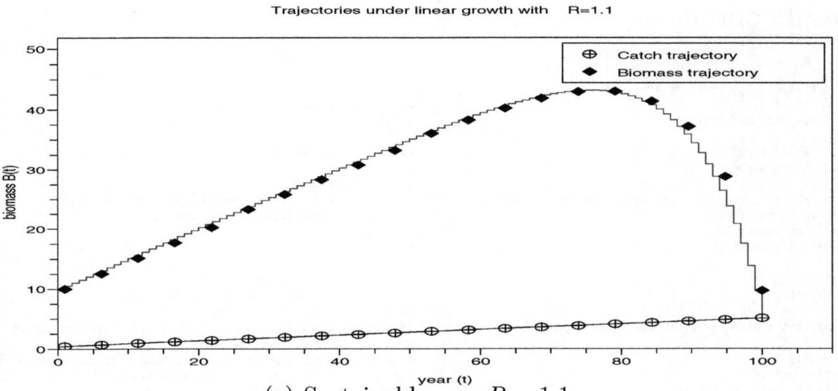


(a) Sustainable case $R = 1.1$

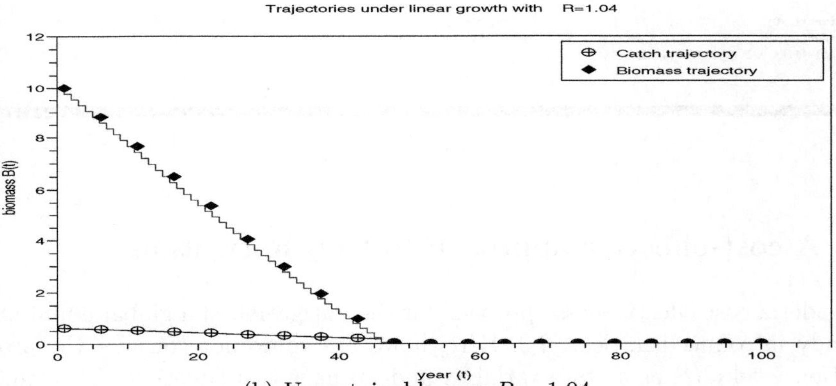


(b) Unsustainable case $R = 1.04$

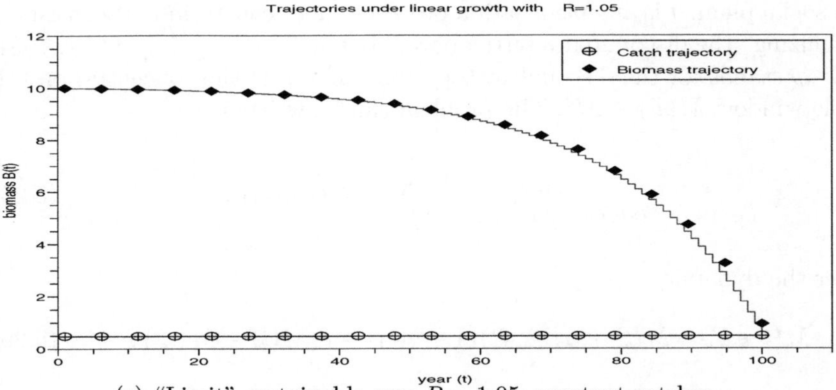


(c) "Limit" sustainable case $R = 1.05$: constant catches

**Fig. 5.2.** Optimal biomass $B^{\star}(t)$ (in ♦) and optimal catches $h^{\star}(t)$ (in $\oplus$) for different productivity $R$ of the resource $B(t)$ with $T = 100$, initial biomass $B_0 = 10$ and discount factor $\rho = 1/1.05$ for the utility function $L(h) = 1 - e^{-h}$. Figures are generated by SCILAB code 10.

SCILAB CODE 10.

```
//
// exec opti_certain_resource.sce

Horizon=100; // time Horizon

r0=0.05;rho=1/(1+r0);
// discount rate r0=5%

R=1/rho; // limit between sustainable and unsustainable
R=1.04; // unsustainable
R=1.1; // sustainable
growth=[1.04 1/rho 1.1];

for i=1:3
  R=growth(i);

// Construction of b and f through dynamic programming
b=[];
b(Horizon+1)=1;
for t=Horizon:-1:1
    b(t)=R*b(t+1)/(1+ R*b(t+1));
    end;

f=[];
f(Horizon+1)=0;
for t=Horizon:-1:1
    f(t)=(f(t+1)-log(rho*R))/(1+R*b(t+1));
    end;

// Optimal catches and stocks
opt=[];
hopt=[];
P0=10; // initial biomass
Popt(1)=P0;
for t=1:Horizon
     hopt(t)=min(Popt(t),max(0,b(t)*Popt(t)+f(t)));
    Popt(t+1)=max(0,R*(Popt(t)-hopt(t)));
    end

// Graphics
xset("window",10+i);xbasc();
plot2d2(0:(Horizon)',[[hopt; Popt($)] Popt],...
rect=[0,0,Horizon*1.1,max(Popt)*1.2]);
// drawing diamonds, crosses, etc. to identify the curves
abcisse=linspace(1,Horizon,20);
plot2d(abcisse,[ hopt(abcisse) Popt(abcisse) ],style=-[3,4]);
legends(['Catch trajectory';'Biomass trajectory'],-[3,4],'ur')
xtitle('Trajectories under linear growth with...
   R='+string(R),'year (t)','biomass B(t)')

end

//
```

## 5.6 A cost-effective approach to $CO_2$ mitigation

Consider a cost effectiveness approach for the mitigation of a global pollutant already introduced in Sect. 2.3. Here, however, we do not consider the production level $Q(t)$ as a state variable, and we have cost function $C(t,a)$ and baseline emissions $E_{\text{BAU}}(t)$ directly dependent upon time $t$. The problem faced by a social planner is an optimization problem under constraints. It consists in minimizing[4] the discounted intertemporal abatement cost $\sum_{t=t_0}^{T-1} \rho^t C\big(t,a(t)\big)$, where $\rho$ stands for the discount factor, while achieving the concentration tolerable window $M(T) \leq M^\sharp$. The problem can be written

$$\inf_{a(t_0),a(t_0+1),..,a(T-1)} \sum_{t=t_0}^{T-1} \rho^t C\big(t,a(t)\big) , \tag{5.25}$$

under the dynamic

$$M(t+1) = M(t) + \alpha E_{\text{BAU}}(t)\big(1-a(t)\big) - \delta(M(t) - M_{-\infty}) , \tag{5.26}$$

and constraint

$$M(T) \leq M^\sharp . \tag{5.27}$$

[4] This differs from the general utility maximization approach followed thus far in the book.

We denote by $a^\star(t_0), a^\star(t_0+1), .., a^\star(T-1)$ an effective optimal solution of this problem whenever it exists. We assume that BAU emissions increase with time, namely, when $E_{\mathrm{BAU}}$ is regular enough

$$\frac{dE_{\mathrm{BAU}}(t)}{dt} > 0 . \tag{5.28}$$

For the sake of simplicity, we assume that the abatement costs $C(t,a)$ are linear with respect to abatement rate $a$ in the sense that

$$C(t,a) = c(t)a . \tag{5.29}$$

We do not specify the marginal cost function, allowing again for non linear processes. We just assume that the abatement cost $C(t,a)$ increases with $a$ which implies

$$\frac{\partial C(t,a)}{\partial a} = c(t) > 0 . \tag{5.30}$$

Furthermore, following for instance [7], we assume that growth lowers marginal abatement costs. This means that the availability and costs of technologies for fuel switching improve with growth. Thus if the marginal abatement cost $c(\cdot)$ is regular enough, it decreases with time in the sense

$$\frac{\partial^2 C(t,a)}{\partial t \partial a} = \frac{dc(t)}{dt} < 0 . \tag{5.31}$$

As a result, the costs of reducing a ton of carbon decline.

Using backward dynamic programming equation (5.16), the optimal and feasible solutions of the cost-effectiveness problem can be computed explicitly as in [6]. However, let us mention that the proofs are not obvious and require the use of a generalized gradient. Indeed, the value function and feedback controls display some non smooth shapes because of kink solutions and active constraints. Numerical solutions are displayed in Fig. 5.3 and can also be obtained using any scientific software with optimizing routines.

As explained in detail in Sect. 4.6 of Chap. 4, a viability result brings the following maximal concentration values to light:

$$M^\sharp(t) := (M^\sharp - M_\infty)(1-\delta)^{t-T} + M_\infty . \tag{5.32}$$

It turns out that an optimal cost-effective policy exists if and only if the initial concentration $M_0$ is smaller than $M^\sharp(t_0)$. In this case, the whole abatement rates sequence $a(t_0)$, $a(t_0+1)$, ..., $a(T-1)$ is effective if and only if associated concentrations $M(t)$ remain lower than $M^\sharp(t)$.

Now, under the previous existence and effectiveness assumption, we obtain the optimal policy in terms of feedback depending on the current state $M$ of the system. A proof based on Bellman dynamic programming under constraint can be found in [6].

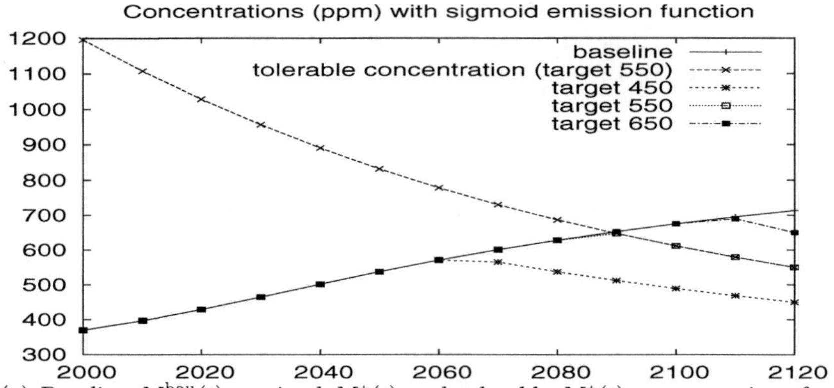


(a) Baseline $M^{\text{bau}}(t)$, optimal $M^{\star}(t)$ and tolerable $M^{\star}(t)$ concentrations for different target $M^{\sharp}$ of 450, 550 and 650 ppm.

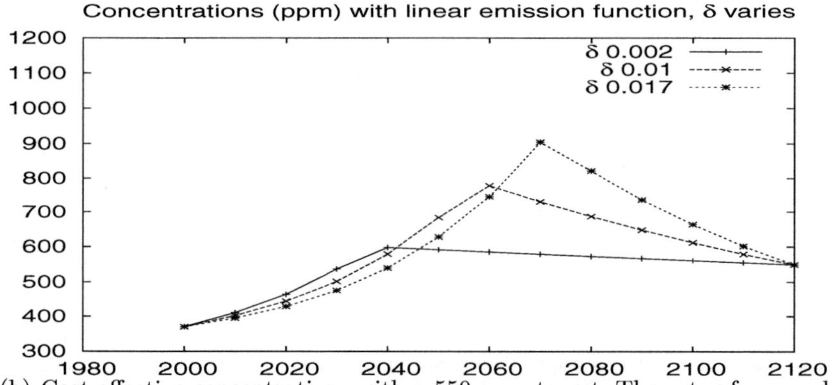


(b) Cost-effective concentrations with a 550 ppm target. The rate of removal $\delta$ varies.

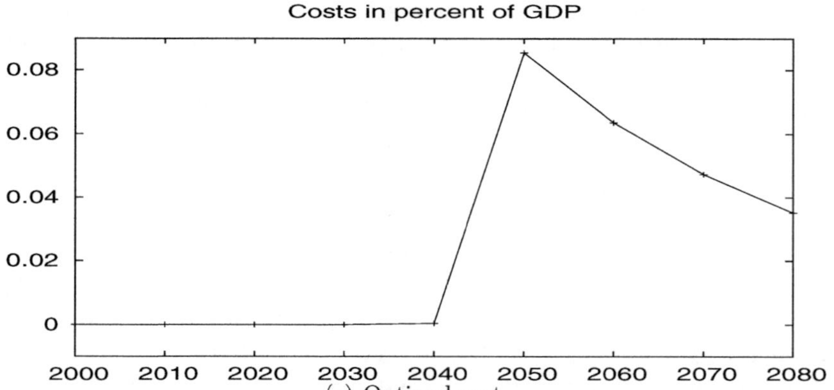


(c) Optimal cost

**Fig. 5.3.** Cost-effective concentration trajectories $M^{\star}(t)$ over the time window [2000, 2120]. The BAU emission function is sigmoid and the optimal concentrations are plotted for different concentration targets. The baseline concentrations $M^{\text{bau}}(t)$, and the tolerable concentrations $M^{\star}(t)$ for a 550 ppm target are also shown.

**Result 5.9** *Consider a tolerable initial situation $M(t_0) \leq M^\sharp(t_0)$. If assumptions for emissions and cost functions (5.28) and (5.31) hold true, then the optimal effective mitigation policy is defined by the feedback abatement*

$$\mathfrak{a}^\star(t, M) = \max\left(0, \frac{(1-\delta)(M - M^\sharp(t)) + E_{\text{BAU}}(t)}{E_{\text{BAU}}(t)}\right).$$

Let us point out that the abatement $\mathfrak{a}^\star(t, M)$ reduces to zero when condition $(1-\delta)\big(M - M^\sharp(t)\big) + E_{\text{BAU}}(t)$ is negative which corresponds to the case where the violation of the tolerable threshold $M^\sharp(t)$ is not at risk even with BAU emissions. We also emphasize that the case of total abatement where $\mathfrak{a}^\star(t, M) = 1$ occurs when the current concentration $M$ coincides with maximal tolerable concentration $M^\sharp(t)$.

Using the optimal feedback abatement above, we obtain the following monotonicity result which favors a non precautionary policy in the sense that the reduction of emissions is more intensive at the end of period than at the beginning.

**Result 5.10** *Consider a tolerable situation $M(t_0) \leq M^\sharp(t_0)$. If assumptions for emissions and cost functions (5.28) and (5.31) hold true, then the optimal abatement rates sequence $a^\star(t) = \mathfrak{a}^\star(t, M^\star(t))$ is increasing with time in the sense that*

$$a^\star(t_0) \leq a^\star(t_0 + 1) \leq \cdots \leq a^\star(T-1).$$

At this stage, let us point out that the previous qualitative results depend on neither the discount factor $\rho \leq 1$ nor the specific form of the emission and marginal abatement cost functions. This fact emphasizes the generality of the assertions. In other words, only a change in the described behavior of the emission function or the use of a non linear cost function could justify another abatement decision profile on the grounds of this simple optimality model.

## 5.7 Discount factor and extraction path of an open pit mine

Consider an open pit mine supposed to be made of blocks of ore with different values [11]. Each block is a two-dimensional rectangle identified by its horizontal position $i \in \{1, \ldots, N\}$ and by its vertical position $j \in \{1, \ldots, H\}$ (see Fig. 5.4). In the sequel, it will also be convenient to see the mine as a collection of columns indexed by $i \in \{1, \ldots, N\}$, each column containing $H$ blocks.

We assume that blocks are extracted sequentially under the following hypothesis:

- it takes one time unit to extract one block;

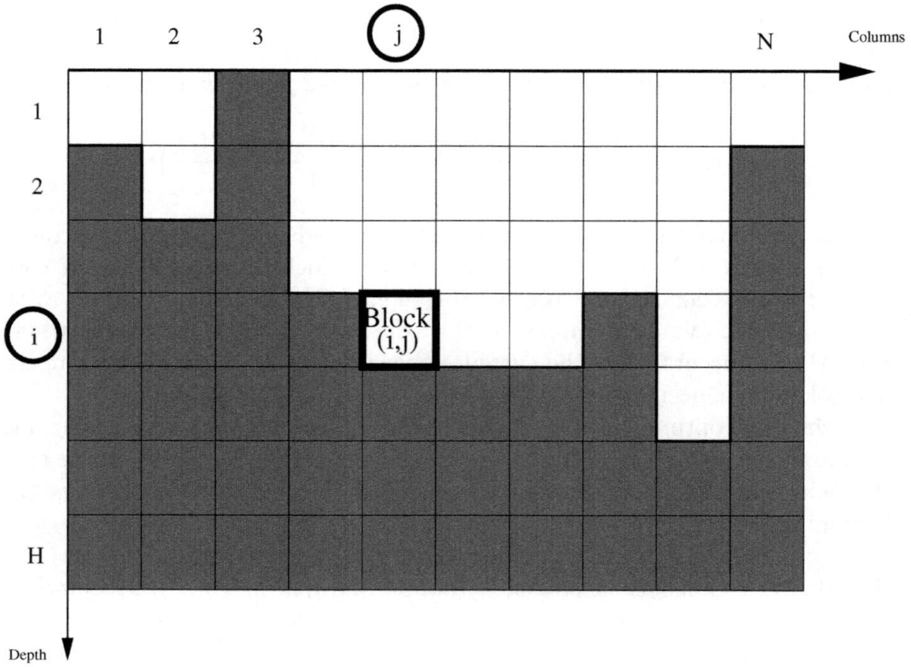


**Fig. 5.4.** An extraction profile in an open pit mine

- only blocks at the surface may be extracted;
- a block cannot be extracted if the slope made with one of its two neighbors is too high, due to physical requirements;
- a retirement option is available where no block is extracted.

## States and admissible states

Denote discrete time by $t = 0, 1, \ldots, T$, where an upper bound for the number of extraction steps is obviously the number $N \times H$ of blocks (it is in fact strictly lower due to slope constraints). At time $t$, the *state* of the mine is a *profile* $x(t) = \big(x_1(t), \ldots, x_N(t)\big) \in \mathbb{X} = \mathbb{R}^N$ where $x_i(t) \in \{1, \ldots, H+1\}$ is the vertical position of the top block with horizontal position $i \in \{1, \ldots, N\}$ (see Fig. 5.4). The initial profile is $x(0) = (1, 1, \ldots, 1)$ while the mine is totally exhausted in state $x = (H+1, H+1, \ldots, H+1)$.

An admissible profile is one that respects local angular constraints at each point, due to physical requirements. A state $x = (x_1, \ldots, x_N)$ is said to be *admissible* if the slope constraints are respected in the sense that

$$\begin{cases} x_1 = 1 \text{ or } 2 \text{ (border slope)} \\ |x_{i+1} - x_i| \leq 1\,, \quad \text{for}\, i = 1, \ldots, N-1 \\ x_N = 1 \text{ or } 2 \text{ (border slope).} \end{cases} \tag{5.33}$$

Denote by $\mathbb{A} \subset \{1, \ldots, H+1\}^N$ the set of admissible states satisfying the above slope constraints (5.33).

## Controlled dynamics

A decision is the selection of a column in $\{1, \ldots, N\}$ whose top block will be extracted, or the retirement option that we shall identify with a column $N+1$. Thus a decision $u$ is an element of the set $\mathbb{B} = \{1, \ldots, N, N+1\} \subset \mathbb{U} = \mathbb{R}$. At time $t$, if a column $u(t) \in \{1, \ldots, N\}$ is chosen at the surface of the open pit mine, the corresponding block is extracted and the profile $x(t) = \big(x_1(t), \ldots, x_N(t)\big()$ becomes

$$x_j(t+1) = \begin{cases} x_j(t) - 1 & \text{if} \quad j = u(t) \\ x_j(t) & \text{else.} \end{cases}$$

In case of retirement option $u(t) = N+1$, then $x(t+1) = x(t)$ and the profile does not change. In other words, the dynamics is given by $x(t+1) = F(x, u)$ where

$$F_j(x, u) = \begin{cases} x_j - 1 & \text{if } j = u \in \{1, \ldots, N\} \\ x_j & \text{if } j \neq u \text{ or } j = N+1\,. \end{cases} \tag{5.34}$$

Indeed, the top block of column $j$ is no longer at altitude $x_j(t)$ but at $x_j(t)-1$, while all other top blocks remain. Of course, not all decisions $u(t) = j$ are possible either because there are no blocks left in column $j$ ($x_j = H+1$) or because of slope constraints.

## Intertemporal profit maximization

The optimal mining problem consists in finding a sequence of admissible blocks which maximizes an intertemporal discounted extraction profit. It is assumed that the value of blocks differs in altitude and column because richness of the mine is not uniform among the zones as well as costs of extraction. For instance Figs. 5.5 (a) and (b) display high value levels in darker (black and red) blocks. The net value a each block is denoted by $\mathcal{R}(i, j)$. By convention $\mathcal{R}(i, N+1) = 0$ when the retirement option is selected. Selecting a square $u(t) \in \mathbb{B}$ at the surface of the open pit mine, and extracting the corresponding block[5] at depth $x_{u(t)}(t)$ yields the value $\mathcal{R}\big(x_{u(t)}(t), u(t)\big)$. With discounting $0 < \rho < 1$, the

[5] When $u(t) = N+1$, there is no corresponding block and the following notation $x_{u(t)}(t) = x_{N+1}(t)$ is meaningless, but this is without incidence since the value $\mathcal{R}\big(x_{N+1}(t), N+1\big) = 0$.

optimization problem is $\sup_{u(\cdot)} \sum_{t=0}^{+\infty} \rho^t \mathcal{R}\big(x_{u(t)}(t), u(t)\big)$. Notice that the sum is in fact finite, bounded above by the number of blocks. Thus, we shall rather consider[6]

$$\sup_{u(0),\dots,u(T-1)} \sum_{t=0}^{T-1} \rho^t \mathcal{R}\big(x_{u(t)}(t), u(t)\big) . \tag{5.35}$$

## Dynamic programming equation

The value function $V(t, x)$ solves $V(T, x) = 0$ and[7]

$$V(t, x) = \max_{u \in \mathbb{B}} \Big( \rho^t \mathcal{R}(x_u, u) - \Psi_{\mathbb{A}}\big(F(x, u)\big) + V\big(t, F(x, u)\big) \Big) . \tag{5.36}$$

The maximization problem (5.35) is solved by a numerical dynamic programming corresponding[8] to SCILAB codes 11-12. Figs. 5.5 exhibit optimal extraction mine profiles for $t = 1$, $t = 18$ and final time for two discount values. The left column is for discount factor $\rho = 0.95$, while the right one is for $\rho = 0.99$. It is shown how a large discount factor $\rho = 0.99$ leads to a larger exploitation of the mine than a lower $\rho = 0.95$. The high preference for present neglects potential use of the mine at stronger depths while a complete admissible exploitation occurs with a more important account of future incomes through a larger discount factor.

---

[6] If we account for transportation costs, we may subtract to $\mathcal{R}$ a term proportional to $\delta(u(t), u(t-1))$, measuring the distance between two subsequent extraction columns.

[7] We have $-\Psi_{\mathbb{A}}(F(x, u)) = 0$ if $F(x, u) \in \mathbb{A}$, while $-\Psi_{\mathbb{A}}(F(x, u)) = -\infty$ if $F(x, u) \notin \mathbb{A}$. This is how we capture the fact that a decision $u$ is admissible.

[8] However, due to numerical considerations and curse of dimensionality, a more parsimonious state is introduced in the SCILAB code 11 before applying the dynamic programming algorithm. Indeed, to give a flavor of the complexity of the problem, notice that 4 columns ($N = 4$) each consisting of nine blocks ($H = 9$) give 10 000 states ($(H + 1)^N = 10^4$), while this raises to $10^{10\ 000}$ if we assume that the surface consists of $100 \times 100$ columns ($N = 10\ 000$). However, the set $\mathbb{A}$ of acceptable states has a cardinal $\text{card}(\mathbb{A})$ which is generally much smaller than $(H + 1)^N$. To see this, let us introduce the mapping $x = (x_1, \dots, x_N) \mapsto \varphi(x) = (x_1, x_2 - x_1, \dots, x_N - x_{N-1})$. Let $x \in \mathbb{A}$ and $y = \varphi(x)$. Since $x$ satisfies the admissibility condition (5.33), $y$ necessarily satisfies $y_1 \in \{1, 2\}$ and $\sup_{i=2,\dots,N} |y_i| \leq 1$. Hence, $\text{card}(\mathbb{A}) \leq 2 \times 3^{N-1}$ and the dynamic programming algorithm will be released with the new state $y = (y_1, \dots, y_N) \in \{1, 2\} \times \{-1, 0, 1\}^{N-1}$ corresponding to the increments of the state $x$.

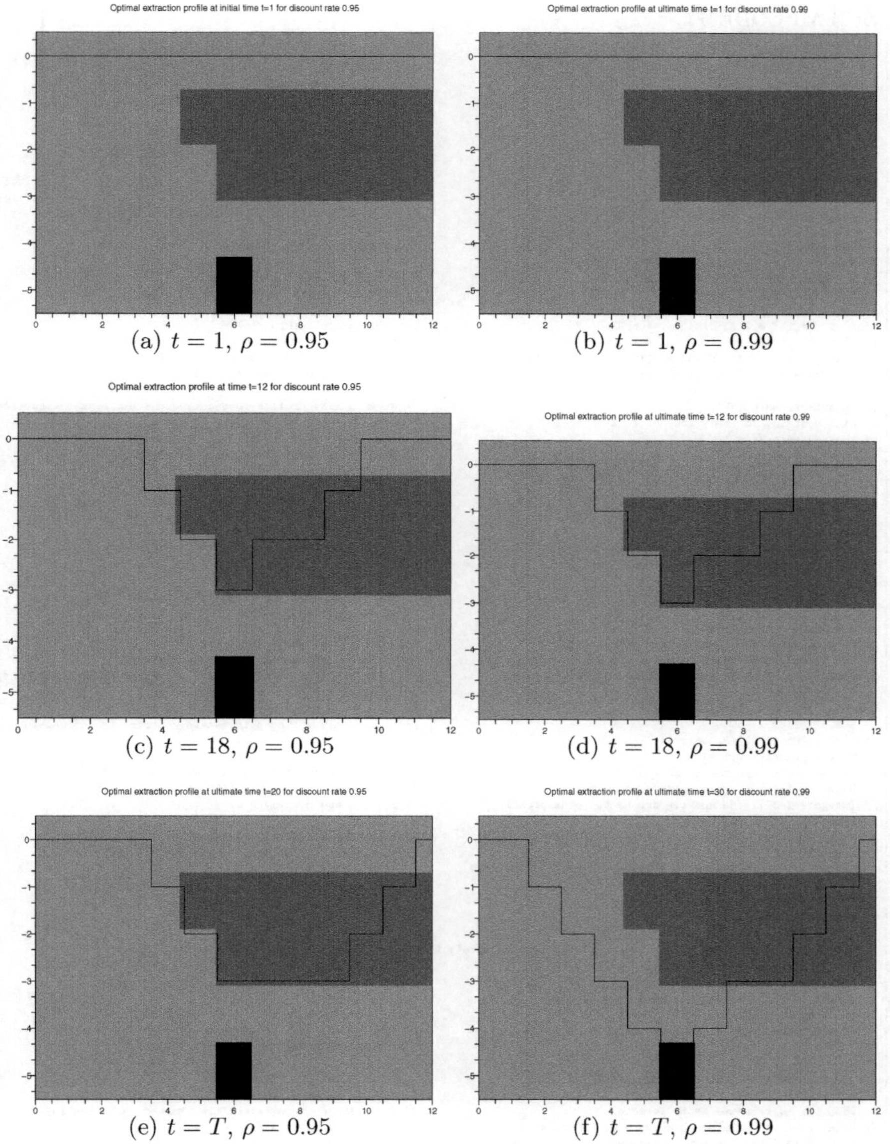


(a) $t = 1$, $\rho = 0.95$ (b) $t = 1$, $\rho = 0.99$

(c) $t = 18$, $\rho = 0.95$ (d) $t = 18$, $\rho = 0.99$

(e) $t = T$, $\rho = 0.95$ (f) $t = T$, $\rho = 0.99$

**Fig. 5.5.** Optimal extraction mine profiles for $t = 1$, $t = 18$ and final time. The left column is for discount factor $\rho = 0.95$, while the right one is for $\rho = 0.99$. The darker the zone, the more valuable. Trajectories are computed with the SCILAB codes 11-12.

## Scilab code 11.

```
//
stacksize(2*10^8);

// ----------------------------------------
// LABELLING STATES AND DYNAMICS
// ----------------------------------------

MM=3^{NN} ; // number of states

Integers=(1:MM)' ;  // labelled states

State=zeros(MM,NN);
// Will contain, for each integer z in Integers,
// a sequence s=(s_1,...,s_NN)  with
// s_1 \in \{1,2\}
// and s_k \in \{0,1,2\} for k \geq 2, such that
// 1) z = s_1 + s_2*3^{1} + ... + s_NN*3^{NN-1}
// 2) a mine profile p_1,...,p_NN is given by
// p_k = y_1 + ... + y_k where y_1= s_1-1 \in \{0,1\}
// and y_j = s_j - 1 \in \{-1,0,1\} for j>1.

Increments=zeros(MM,NN);
// Will contain, for each integer z in Integers,
// the sequence y=(y_1,...,y_NN).

// The initial profile is supposed to be p(0)=(0,0,...,0)
// to which corresponds y(0)=(0,0,...,0) and
// s(0)=(1,1,...,1) and z(0)=1+ 3^{1} + ... + 3^{NN-1}.

Partial_Integer=zeros(MM,NN);
// Will contain, for each integer z in Integers,
// a lower aproximation of z in the basis
// 1, 3^{1},..., 3^{NN-1}
// Partial_Integer(z,k)=s_1 + s_2*3^{1} +...+ s_k*3^{k-1}.

Future_Integer=zeros(MM,NN);
// Will contain, for each integer z in Integers,
// the image by the dynamics under the control consisting in
// extracting block in the corresponding column.

State(:,1)=pmodulo(Integers, 3^{1}) ;
// State(z,1)=s_1
Partial_Integer(:,1)=State(:,1)  ;
// Partial_Integer(z,1)=s_1
Future_Integer(:,1)= maxi(1,Integers+1-3^{1}) ;
// Dynamics (with a "maxi" because some integers in
// Integers+1-3^{1} do not correspond to "mine profiles").
Increments(:,1)=State(:,1)-1;

for k=2:NN
  remainder= ( Integers-Partial_Integer(:,k-1) ) / 3^{k-1} ;
  // s_{k} + s_{k+1}*3 + ...
  State(:,k)=pmodulo(remainder, 3) ;
  // State(:,k)=s_{k}
  Increments(:,k)=State(:,k)-1;
  Partial_Integer(:,k)=Partial_Integer(:,k-1)+3^{k-1}*State(:,k);
// Partial_Integer(z,k)=s_1+s_2*3^{1}+...+s_k*3^{k-1}
Future_Integer(:,k)= maxi(1,Integers+3^{k-1}-3^{k}) ;
// Dynamics (with a "maxi" because some integers
// in Integers+3^{k-1}-3^{k}
// do not correspond to "mine profiles")
end

Future_Integer(:,NN)= mini(MM,Integers+3^{NN-1}) ;
// Correction for the dynamics
// when the last column NN is selected.
// Dynamics (with a "mini" because some integers in
// Integers+3^{NN-1} do not correspond to "mine profiles").



// ----------------------------------------
// FROM PROFILES TO INTEGERS
// ----------------------------------------

function z=profile2integer(p)
  // p : profile as a row vector
yy= p-[ 0 p(1:($-1)) ] ;
ss=yy+1 ;
z=sum( ss .* [1 3.^{[1:(NN-1)]}] ) ;
endfunction
```

```
// ----------------------------------------
// ADMISSIBLE INTEGERS
// ----------------------------------------

// Mine profiles are those for which
// HH \geq p_1 \geq 0,..., HH \geq p_NN \geq 0
// that is, HH \geq y_1 + ... + y_k \geq 0 for all k
// Since, starting from the profile p(0)=(0,0,...,0)),  the
// following extraction rule will always give "mine profiles",
// we shall not exclude other unrealistic profiles.

Admissible=zeros(MM,NN);

Profiles= cumsum (Increments , "c" ) ;
// Each line contains a profile, realistic or not.

adm_bottom=bool2s(Profiles<HH) ;
// A block at the bottom cannot be extracted:
// an element in adm_bottom is one if and only if
// the top block of the column is not at the bottom.

// Given a mine profile, extracting one block at the surface
// is admissible if the slope is not too high.
//
// Extracting block in column 1 is admissible
// if and only if p_1=0.
//
// Extracting block in column j 1<j<NN) is not admissible
// if and only if y_j=1 or y_{j+1}=-1 that is,
// (s_j-1)=1 or (s_{j+1}-1)=-1.
// Extracting block in column j is admissible
// if and only if s_j < 2 and s_{j+1} > 0.
//
// Extracting block in column NN is admissible
// if and only if p_{NN}=0.

Admissible(:,1)=bool2s(Profiles(:,1)==0);
Admissible(:,NN)=bool2s(Profiles(:,NN)==0);
// Corresponds to side columns 1 and NN, for which only the
// original top block may be extracted:
// an element in columns 1 and NN of Admissible is one
// if and only if the pair (state,control) is admissible.

Admissible(:,2:($-1))=...
bool2s( State(:,2:($-1))<2 & State(:,3:$)>0 ) ;
// An element in column 1<j<NN of AA is one if and only
// s_j < 2 and s_{j+1} > 0.

Admissible=Admissible .* adm_bottom ;
// An element in column j of admissible is zero
// if and only if
// extracting block in column j is not admissible,
// else it is one.

Stop_Integers=Integers( prod(1-Admissible,"c") ==1 ) ;
// Labels of states for which no decision is admissible,
// hence the decision is the retirement option


// ----------------------------------------
// INSTANTANEOUS GAIN
// ----------------------------------------

Forced_Profiles= mini(HH,  maxi(1, Profiles) ) ;
// Each line contains a profile, forced to be realistic.
// This trick is harmless and useful
// to fill in the instantaneous gain matrix.

instant_gain=zeros(MM,NN);

for uu=1:NN
  instant_gain(:,uu)= Admissible(:,uu) .* ...
  richness(Forced_Profiles(Future_Integer(:,uu),uu) , uu )...
  + (1-Admissible(:,uu)) *bottom;
end
// When the control uu is admissible,
// instant_gain is the richness of the top block of column uu.
// When the control uu is not admissible, instant_gain
// has value "bottom", approximation of -infinity.

//
```

SCILAB CODE 12.

```
//

// ----------------------------------------
// DYNAMIC PROGRAMMING ALGORITHM
// ----------------------------------------

VV=zeros(MM,T); // value functions in a matrix
// The final value function is zero.
UUopt=(NN+1)*ones(MM,T); // optimal controls in a matrix

for t=(T-1):(-1):1 // backward induction
  loc=[];
  // will contain the vector to be maximized
  loc_bottom=mini(VV(:,t+1));
  // The value attributed to the value function VV(:,t)
  // when a control is not admissible.
  //
  for uu=1:NN // columns 1 to NN selected
loc=[loc, Admissible(:,uu) .* ( ...
   discount^t * instant_gain(:,uu) + ...
  VV( Future_Integer(:,uu) , t+1 ) ) +...
+ (1-Admissible(:,uu)) .* (discount^t * bottom + loc_bottom) ] ;
  end
// When the control uu is admissible,
// loc is the usual DP expression.
// When the control uu is not admissible,
// loc is the  DP expression
// with both terms at the lowest values.
 //
  loc=[loc, VV(:,t+1) + discount^t *0] ;
// Adding an extra control/column which provides zero
// instantaneous gain and does not modify the state:
// retire option.
//
  [lhs,rhs]= maxi(loc,"c") ; // DP equation
  VV(:,t)=lhs;
  UUopt(:,t)=rhs;
//
UUopt(Stop_Integers,t)=(NN+1)*ones(Stop_Integers) ;
// retire option
end


// ----------------------------------------
// OPTIMAL TRAJECTORIES
// ----------------------------------------

xx=zeros(T,NN);
zz=zeros(T,1);
uu=(NN+1)*ones(T,1);
vv=0;

xx(1,:)=zeros(1,NN);
// initial profile

xset("window",1) ;
xset("colormap",hotcolormap(64));
// table of colors
xbasc();plot2d2(-0.5+0:(NN+2),-[0 xx(1,:) 0 0],...
rect=[0,-HH-0.5,NN+1,0.5])
Matplot1(rich_color,[0,-HH-0.5,NN+1,0.5]);

t=1 ; last_control=0 ;
//
while last_control<NN+1 do
  zz(t)=profile2integer(xx(t,:)) ;
  uu(t)=UUopt(zz(t),t) ;
  xx(t+1,:)=xx(t,:);
  if uu(t)<NN+1 then
  xx(t+1,uu(t))=xx(t,uu(t))+1;
  vv=vv+discount^t*richness(xx(t+1,uu(t)),uu(t)) ;
  end
//  halt()
  xbasc();plot2d2(-0.5+0:(NN+2),-[0 xx(t+1,:) 0 0],...
  rect=[0,-HH-0.5,NN+1,0.5]); xpause(400000);

  last_control=uu(t);
t=t+1 ;
end
xtitle("Optimal extraction profile for discount rate "...
   +string(discount))


function []=display_profile(time)
  // Displays graphics of mine profiles
xset("window",time) ;
xset("colormap",hotcolormap(64));
xbasc();
plot2d2(-0.5+0:(NN+2),-[0 xx(time,:) 0 0],...
rect=[0,-HH-0.5,NN+1,0.5]);
// Just to to set the frame
Matplot1(rich_color,[0,-HH-0.5,NN+1,0.5]);
// The value of the mine blocks in color.
// Will be in background.
plot2d2(-0.5+0:(NN+2),-[0 xx(time,:) 0 0],...
rect=[0,-HH-0.5,NN+1,0.5]);
xtitle("Optimal extraction profile at ultimate time t="...
   +string(time) +" for discount rate "+string(discount))
endfunction

time=t;
display_profile(time)

time=1;
display_profile(time)

time=ceil(T/3);
display_profile(time)

//
```

## 5.8 Pontryaguin's maximum principle for the additive case

A dual approach to solve intertemporal optimization problems emphasizes the role played by the adjoint state, Lagrangian or Kuhn and Tucker multipliers in optimal control problems. We present this so-called *Pontryaguin's maxi-*

*mum principle*[9] using a Hamiltonian formulation. The maximum principle is a *necessary* condition for optimality. Here, we basically deduce the maximum principle from the Bellman backward induction equation. We could as well deduce the principle from the existence of Lagrange multipliers in optimization problems with constraints on $\mathbb{R}^N$, as is often introduced in the literature.

### 5.8.1 Hamiltonian formulation without control and state constraints

We assume here the absence of control and state constraints:

$$\mathbb{B}(t,x) = \mathbb{U} = \mathbb{R}^p \quad \text{and} \quad \mathbb{A}(t) = \mathbb{X} = \mathbb{R}^n \; . \tag{5.37}$$

The maximum principle may be expressed in a compact manner introducing the so-called Hamiltonian of the problem.

**Definition 5.11.** *The* Hamiltonian *associated to the maximization problem (5.14) is the following function*[10]

$$\begin{aligned} \mathcal{H}(t,x,q,u) &:= \sum_{i=1}^{n} q_i F_i(t,x,u) + L(t,x,u) \\ &= F(t,x,u)'q + L(t,x,u) \; . \end{aligned} \tag{5.38}$$

*The new variable $q$ is usually called* adjoint state, adjoint variable, *or* multiplier.

Thus, to form the Hamiltonian, simply multiply the dynamics by the adjoint variable (scalar product in dimension more than one), and subtract the instantaneous payoff.

The Hamiltonian conditions involving the so-called optimal adjoint state $q^\star(\cdot)$ together with the optimal state $x^\star(\cdot)$ and decision $u^\star(\cdot)$ are described in the following Proposition, called the *maximum principle.* It basically stems from first order optimality conditions under constraints involving first order derivatives. One proof can be derived from a Lagrangian formulation. The other way that we have chosen is to apply the Bellman principle in a marginal version as exposed in Proposition 5.13. The proof can be found in Sect. A.3 in the Appendix.

**Proposition 5.12.** *Consider the maximization problem (5.14) under constraints (5.8) without state constraints as in (5.37). Assume that instantaneous utility $L$, final utility $M$ and dynamic $F$ are continuously differentiable*

[9] This is an abuse of language. Indeed, in the discrete time case, the optimality condition is not necessarily a *maximum* but may be a *minimum* or neither of both, which is not the case in continuous time.

[10] Recall the notations for transpose vectors: if $p$ and $q$ are two column vectors of $\mathbb{R}^n$, the scalar product of $p$ and $q$ is denoted indifferently $\langle p,q\rangle = p'q = q'p = \langle q,p\rangle$, where $'$ denotes the *transpose* operator.

*in the state variable $x$. Let the trajectory $\big(x^\star(\cdot), u^\star(\cdot)\big) \in \mathbb{X}^{T+1-t_0} \times \mathbb{U}^{T-t_0}$ be a solution of the maximization problem (5.14). Then, there exists a sequence $q^\star(\cdot) = (q^\star(t_0), \dots, q^\star(T-1)) \in \mathbb{X}^{T-t_0}$, called* adjoint state, *such that, for any $i = 1, \dots, n$ and $j = 1, \dots, p$, we have*

$$\begin{cases} x_i^\star(t+1) = \dfrac{\partial \mathcal{H}}{\partial q_i}\big(t, x^\star(t), q^\star(t), u^\star(t)\big) \ , \ t = t_0, \dots, T-1 \ , \\ q_i^\star(t-1) = \dfrac{\partial \mathcal{H}}{\partial x_i}\big(t, x^\star(t), q^\star(t), u^\star(t)\big) \ , \ t = t_0+1, \dots, T-1 \ , \\ 0 = \dfrac{\partial \mathcal{H}}{\partial u_j}\big(t, x^\star(t), q^\star(t), u^\star(t)\big) \ , \ t = t_0, \dots, T-1 \ , \end{cases} \tag{5.39}$$

*together with boundary conditions*

$$\begin{cases} x_i^\star(t_0) = x_{i0} \ , \\ q_i^\star(T-1) = \dfrac{\partial M}{\partial x_i}\big(t, x^\star(t)\big) \ . \end{cases} \tag{5.40}$$

Writing the previous Hamiltonian conditions, we may hope to simultaneously reveal the optimal state, decisions and adjoints. For this reason the principle proves useful and is applied for the analysis of many optimal control problems. Conditions (5.39) can be equivalently depicted in a more compact form using vectors and transposed as follows

$$\begin{cases} x^\star(t+1) = (\dfrac{\partial \mathcal{H}}{\partial q})'\big(t, x^\star(t), q^\star(t), u^\star(t)\big) \ , \\ q^\star(t-1) = (\dfrac{\partial \mathcal{H}}{\partial x})'\big(t, x^\star(t), q^\star(t), u^\star(t)\big) \ , \\ 0 = (\dfrac{\partial \mathcal{H}}{\partial u})'\big(t, x^\star(t), q^\star(t), u^\star(t)\big) \ . \end{cases} \tag{5.41}$$

### 5.8.2 The adjoint state as a marginal value

The adjoint state can be interpreted in terms of prices and marginal value: it appears indeed as the derivative of the value function with respect to the state variable along an optimal trajectory. Similarly, the adjoint state is also related to the Lagrangian multipliers associated with the dynamics seen here as an equality constraint between state trajectory, control trajectory and time.

**Proposition 5.13.** *Suppose that the value function $V(t,x)$ associated to the maximization problem (5.14) is smooth with respect to $x$. Assume that there exists an optimal trajectory $x^\star(\cdot)$ such that $\mathfrak{u}^\star\big(t, x^\star(t)\big)$ in (5.20) is unique for all $t = t_0, \dots, T-1$. Then, the sequence $q^\star(\cdot)$ defined by*

$$q_i^\star(t) := \frac{\partial V}{\partial x_i}\big(t+1, x^\star(t+1)\big) \,, \quad t = t_0, \dots, T-1 \,, \tag{5.42}$$

*is a solution of (5.39) and (5.40).*

In a more vectorial form, we write (5.42) equivalently as

$$q^\star(t) := \Big(\frac{\partial V}{\partial x}\Big)'\big(t+1, x^\star(t+1)\big) \,. \tag{5.43}$$

## 5.9 Hotelling rule

Let us now apply the maximum principle to a simple example. Let us handle the "cake-eating" model coping with an exhaustible resource presented at Sect. 2.1, where we ignore the constraints, giving

$$\sup_{h(t_0),\dots,h(T-1)} \sum_{t=t_0}^{T-1} \rho^t L\big(h(t)\big) + \rho^T L\big(S(T)\big) \tag{5.44}$$

with dynamics $S(t+1) = S(t) - h(t)$ and $S(t_0)$ given. We define the Hamiltonian as in (5.38)

$$\mathcal{H}(t, S, h, q) := \rho^t L(h) + q(S - h) \,.$$

From Proposition 5.12, a trajectory $\big(S^\star(\cdot), h^\star(\cdot)\big)$ is the solution of the optimization problem (5.44) if there exists an adjoint state trajectory $q^\star(\cdot)$ such that the following conditions are satisfied

$$\begin{cases} \dfrac{\partial \mathcal{H}}{\partial h}\big(t, S^\star(t), h^\star(t), q^\star(t)\big) = 0 \\ \dfrac{\partial \mathcal{H}}{\partial q}\big(t, S^\star(t), h^\star(t), q^\star(t)\big) = S^\star(t+1) \\ \dfrac{\partial \mathcal{H}}{\partial S}\big(t, S^\star(t), h^\star(t), q^\star(t)\big) = q^\star(t-1) \\ \qquad\qquad \dfrac{\partial M}{\partial S}\big(T, S^\star(T)\big) = q^\star(T-1) \end{cases}$$

giving

$$\begin{cases} \rho^t L'\big(h^\star(t)\big) - q^\star(t) = 0 \\ \qquad S^\star(t) - h^\star(t) = S^\star(t+1) \\ \qquad\qquad q^\star(t) = q^\star(t-1) \\ \rho^T L'\big(S^\star(T)\big) = q^\star(T-1) \,. \end{cases} \tag{5.45}$$

Thus, the multiplier $q^\star(t)$ is stationary and we obtain the following interesting result that

$$L'\big(h^\star(t)\big) = q^\star(t_0)\rho^{t_0-t} \,. \tag{5.46}$$

We may write the above relation as

$$\frac{L'\big(h^\star(t+1)\big)}{L'\big(h^\star(t)\big)} = \frac{1}{\rho} \,. \tag{5.47}$$

The *marginal utility* $L'\big(h^\star(t)\big)$ is interpreted as the price of the resource at time $t$, that we denote by

$$p(t) := L'\big(h^\star(t)\big) \,.$$

From (5.47), we derive that the price of the resource grows as

$$p(t+1) = (1+r_f)p(t) \quad \text{with} \quad r_f := \frac{1}{\rho} - 1 \,,$$

where the discount factor $\rho$ is usually related to the *interest rate* $r_f$ through the relation $\rho(1+r_f) = 1$. Thus the price growth rate coincides with the interest rate: this is the so-called *Hotelling rule* [9].

Moreover, writing

$$\log\left(\frac{L'\big(h^\star(t+1)\big)}{L'\big(h^\star(t)\big)}\right) = \int_{h^\star(t)}^{h^\star(t+1)} \frac{L''}{L'}(h)dh$$

and introducing the *elasticity of the marginal utility of consumption*

$$\eta(h) := -\frac{hL''(h)}{L'(h)} \,,$$

equation (5.47) becomes[11]

$$\int_{h^\star(t)}^{h^\star(t+1)} \frac{\eta(h)}{h} dh = \log\rho \,.$$

Thus, the greater the rate of discount $\rho$, the greater the rate of extraction, and a decrease in the elasticity of the marginal utility of consumption increases the rate of extraction. From (5.46), we obtain that

$$h^\star(t) = (L')^{-1}(q^\star(t_0)\rho^{t_0-t}) \,,$$

and the optimal extraction $h^\star(t)$ is stricly decreasing with time $t$ if marginal utility $L'$ is taken to be stricly decreasing ($L$ stricly concave) and since $\rho \leq 1$.

When $T \to +\infty$ and the marginal utility at zero is infinite $L'(0) = +\infty$, we observe that the extraction $h^\star(t)$ goes toward zero. Since the stock $S^\star(t)$

[11] In continuous time, the Hotelling rule is stated as $\frac{\dot{h}}{h} = -\frac{r_f}{\eta}$, where $r_f = -\log\rho$ is the *discount rate*.

decreases and is nonnegative, it also goes down to a limit. By the equality $\rho^T L'(S^\star(T)) = q^\star(t_0)$, we see that $S^\star(T) \searrow (L')^{-1}(+\infty) = 0$. Hence, both the extraction $h^\star(t)$ and the stock $S^\star(t)$ go toward zero under reasonable assumptions.

Let us summarize these results as follows.

**Result 5.14** *Assume that the concave utility $L$ satisfies $L'' < 0$ and the interest rate $r_f$ is strictly positive. Then*

- *the rate of return (financial) of the resource is the interest rate;*
- *optimal consumption decreases along time $h^\star(t+1) < h^\star(t)$;*
- *optimal stock and extraction are exhausted in the long term:*

$$\lim_{T\to+\infty} S^\star(T) = \lim_{T\to+\infty} h^\star(T) = 0\ .$$

### Optimal depletion without existence value

We have ignored the constraints $0 \leq h(t) \leq S(t)$ above. However, one should be cautious in doing so. Consider indeed the same problem but without "inheritance" value:

$$\sup_{h(t_0),\dots,h(T-1)} \sum_{t=t_0}^{T-1} \rho^t L\big(h(t)\big)\ .$$

The optimality equations are the same as in (5.45) except for $q^\star(T-1) = L'\big(S^\star(T)\big)\rho^T$ which now would become $q^\star(T-1) = 0$, giving $q^\star(t) = 0$ and a contradiction with $L'\big(h^\star(t)\big)\rho^t = q^\star(t) = 0$ and $L' > 0$ in general.

## 5.10 Optimal management of a renewable resource

We first present a sustainable exploitation when $\rho R = 1$, then expose the so called fundamental equation of renewable resource.

### 5.10.1 Sustainable exploitation

Let us return to the management problem presented at Sect. 2.2

$$\sup_{h(t_0),\dots,h(T-1)} \Big( \sum_{t=t_0}^{T-1} \rho^t L\big(h(t)\big) + \rho^T L\big(B(T)\big) \Big)$$

with dynamic $B(t+1) = R\big(B(t) - h(t)\big)$. We illustrate the maximum approach, temporarily ignoring the constraints $0 \leq h(t) \leq B(t)$; we shall verify afterwards that the optimal solution is an interior solution in the sense that it satisfies $0 < h^\star(t) < B(t)$. For the sake of simplicity, we consider the particular case of $T = 2$ periods and we suppose that the discount factor satisfies relation (5.23) namely, $\rho R = 1$.

The utility function $L$ is taken to be sufficiently regular (twice continuously differentiable, for instance), and strictly concave ($L'' < 0$).

Using the general notations of Sect. 5.1, we have $M(2, B) = \rho^2 L(B)$. The Hamiltonian as in (5.38) is given by

$$\mathcal{H}(t, B, q, h) = qR(B - h) + \rho^t L(h) .$$

The maximum principle (5.39) implies $q^\star(1) = \rho^2 L'\big(B^\star(2)\big)$ and

$$q^\star(t-1) = q^\star(t)R,\ t = 1, 2 \quad \text{and} \quad 0 = q^\star(t)R + \rho^t L'\big(h^\star(t)\big),\ t = 0, 1, 2 .$$

This yields

$$q^\star(1) = \rho^2 L'(B^\star(2)) \ , q^\star(0) = \rho L'(B^\star(2)) \ , L(h^\star(0)) = L'(h^\star(1)) = L'(B^\star(2)) .$$

Since $L$ is a strictly concave function, its derivative is strictly decreasing and one can deduce that we are dealing with stationary optimal consumptions $h^\star(0)$ and $h^\star(1)$, both equal to $B^\star(2)$. According to the dynamic equation, we deduce that

$$B^\star(2) = R(B^\star(1) - h^\star(1)) = R\Big(R(B_0 - h^\star(0)) - h^\star(1)\Big) = R\Big(R(B_0 - B^\star(2)) - B^\star(2)\Big) ,$$

implying that optimal catches are

$$h^\star(0) = h^\star(1) = \frac{R^2}{1 + R + R^2} B_0 .$$

We check that $0 < h^\star(t) < B^\star(t)$.

### 5.10.2 A new bioeconomic equilibrium

Following [4], we may derive from the maximum principle at equilibrium a well known relationship displaying a new bioeconomic equilibrium based on a long term steady state. Such equilibrium accounts for the interest rate $r_f$ or risk free asset, the growth of the resource $g(B)$, the sustainable yield $\sigma(B)$ defined in (3.6) of Chap. 3 and the rent $\mathcal{R}$ in marginal terms, namely

$$g_B\big(B - \sigma(B)\big) = (1 + r_f) \frac{\mathcal{R}_h(\sigma(B), B) + \mathcal{R}_B(\sigma(B), B)}{\mathcal{R}_h(\sigma(B), B)} . \tag{5.48}$$

Such equilibrium has to be compared with characterizations of MSE or PPE equilibria in Subsect. 3.3.1 of Chap. 3.

To achieve assertion (5.48), we consider the dynamic optimization problem

$$\max_{h(t_0),\dots,h(T-1)} \Big( \sum_{t=t_0}^{T-1} \rho^t \mathcal{R}\big(h(t), B(t)\big) + \rho^T M\big(B(T)\big) \Big) ,$$

where the discount $\rho$ refers to the interest rate through $\rho(1 + r_f) = 1$, under the dynamic

$$B(t+1) = g\big(B(t) - h(t)\big) .$$

The Hamiltonian is

$$\mathcal{H}(t, B, q, h) := qg(B - h) + \rho^t \mathcal{R}(h, B) .$$

Under regularity assumptions on the functions $\mathcal{R}$ and $g$, the maximum principle reads

$$\begin{cases} 0 = -q(t) g_B\big(B(t) - h(t)\big) + \rho^t \mathcal{R}_h\big(h(t), B(t)\big) , \\ B(t+1) = g\big(B(t) - h(t)\big) , \\ q(t-1) = q(t) g_B\big(B(t) - h(t)\big) + \rho^t \mathcal{R}_B\big(h(t), B(t)\big) . \end{cases}$$

Setting $l(t) := q(t)\rho^{-t}$, the conditions become

$$\begin{cases} 0 = \mathcal{R}_h\big(h(t), B(t)\big) - l(t) g_B\big(B(t) - h(t)\big) , \\ B(t+1) = g\big(B(t) - h(t)\big) , \\ l(t-1) = \rho\Big(\mathcal{R}_B\big(h(t), B(t)\big) + l(t) g_B\big(B(t) - h(t)\big)\Big) . \end{cases}$$

A steady state $(B^\star, h^\star, l^\star)$ associated with these two dynamics satisfies

$$\begin{cases} 0 = \mathcal{R}_h(h^\star, B^\star) - l^\star g_B(B^\star - h^\star) , \\ h^\star = \sigma(B^\star) , \\ l^\star = \rho\big(\mathcal{R}_B(h^\star, B^\star) + l^\star g_B(B^\star - h^\star)\big) . \end{cases}$$

Combining the equations yields the desired result (5.48).

Despite the numerous assumptions underlying such reasoning, this equation has been called the *fundamental equation of renewable resource*. It has an interesting economic interpretation. On the left hand side, the first term $g_B\big(B^\star - \sigma(B^\star)\big)$ is the marginal productivity of the resource (at equilibrium) while the second term involving the marginal stock effect measures the marginal value of the stock relative to the marginal value of harvest.

Such equilibrium questions the long term sustainability of an optimal management. In particular, we may wonder whether this equilibrium can take negative values. Let us consider the illustrative case where

- the resource dynamic is of the Beverton-Holt type $g(B) = \frac{RB}{1+bB}$;
- rent is of the form $\mathcal{R}(h, B) = ph - c$.

It has been already shown that

$$g_B(B) = \frac{R}{(1+bB)^2} \quad \text{and} \quad \sigma(B) = B\left(1 - \frac{1}{R-bB}\right).$$

Moreover we obtain

$$\mathcal{R}_B(h, B) = 0 \quad \text{and} \quad \mathcal{R}_h(h, B) = p.$$

Thus, the long term equilibrium satisfies

$$1 + r_f = \frac{R}{\left(1 + b\left(\frac{B^\star}{R - bB^\star}\right)\right)^2} = \frac{(R - bB^\star)^2}{R}.$$

Thus, whenever $R\rho \leq 1$, the unique solution $B^\star$ is a negative biomass.

**Result 5.15** *Consider the Beverton-Holt dynamics and fixed harvesting costs. If $R\rho \leq 1$ then extinction is optimal in the sense that the long term equilibrium is negative $B^\star \leq 0$.*

More general conditions for "optimal extinction" are exposed in [3]. Again, such assertions stress the fact that major economic rationales exist for non conservation and non sustainability of renewable resources.

## 5.11 The Green Golden rule approach

As already been pointed out in the introduction, the *Green Golden* criterion sheds interesting light on the sustainability issue from the point of view of the future preferences. Indeed, by focusing on the decisions that favor the final payoff, this approach promotes the distant future. For the management of natural resources, it turns out that such an approach may reduce consumption to zero along time and promote the stock resource. In this sense, it constitutes a dictatorship of the future. From the methodological viewpoint, this framework turns out to be a particular instance of the additive case of Sect. 5.2 without instantaneous payoff. Thus, both the dynamic programming and maximum principle apply with major simplifications.

Again, the *Green Golden rule value function* or *Bellman function* $V(t, x)$ at time $t$ and for state $x$ represents the optimal value of the criterion over $T - t$ remaining periods, given that the state of the system at time $t$ is $x$, namely

$$V(t, x) := \sup_{\big(x(\cdot), u(\cdot)\big) \in \mathcal{T}^{ad}(t,x)} M\big(T, x(T)\big), \tag{5.49}$$

where $\mathcal{T}^{ad}(t, x)$ is given by (5.8).

An induction equation appears for the Green Golden rule value function $V(t,x)$. The proof of the following assertion is a particular instance of the additive case of Sect. 5.2. We describe it in the case with the state constraint involving viability conditions.

The value function defined by (5.49) is the solution of the following *dynamic programming equation* (or *Bellman equation*), where $t$ runs from $T-1$ down to $t_0$,

$$\begin{cases} V(T,x) = M(T,x) , & \forall x \in \mathbb{V}\mathrm{iab}(T) , \\ V(t,x) = \sup_{u \in \mathbb{B}^{\mathrm{viab}}(t,x)} V\big(t+1, F(t,x,u)\big) , & \forall x \in \mathbb{V}\mathrm{iab}(t) , \end{cases} \tag{5.50}$$

where $\mathbb{V}\mathrm{iab}(t)$ is again given by the backward induction (5.18) and where the supremum in (5.50) is over viable controls in $\mathbb{B}^{\mathrm{viab}}(t,x)$ as in (5.19).

As in the additive case, Bellman induction optimization in the Green Golden rule case also reveals relevant feedback controls. Indeed, assuming the additional hypothesis that the supremum is achieved in (5.50) for at least one decision, if we denote by $\mathfrak{u}^\star(t,x)$ a value $u \in \mathbb{B}(t,x)$ which achieves the maximum in equation (5.50), then $\mathfrak{u}^\star(t,x)$ is an optimal feedback for the optimal control problem.

## 5.12 Where conservation is optimal

Now, let us examine an application of the Green Golden rule for the following renewable resource management problem

$$B(t+1) = g\big(B(t) - h(t)\big) , \quad 0 \le h(t) \le B(t) ,$$

with criterion

$$\sup_{h(t_0),\dots,h(T-1)} M\big(T, B(T)\big) ,$$

where $M$ is some payoff function with respect to resource state $B$ and $g$ represents the resource dynamic. The case where $g(B) = B$ stands for the exhaustible issue. It can be proved that, under general and simple assumptions, the optimal catches are reduced to zero.

**Result 5.16** *Assume that the final payoff $M$ and resource productivity $g$ are increasing functions. Then, the value function and optimal feedbacks are given by*

$$V(t,B) = M\big(T, g^{(T-t)}(B)\big) , \quad \mathfrak{u}^\star(t,B) = 0 , \tag{5.51}$$

*where* $g^{(i)}(B) = \overbrace{g(g(\cdots g(B)\cdots))}^{i \text{ times}}$ *denotes the $i$-th composition of function $g$ ($g^{(1)} = g$, $g^{(2)} = g \circ g$, etc.).*

Notice that such an optimal viable rule provides zero extraction or catches along time and thus favors the resource which takes its whole value from the final payoff. In this sense, such an approach promotes ecological and environmental dimensions. Although such a Green Golden rule solution displays intergenerational equity since depletion is stationary along time and favors the resource, it is not a satisfying solution for sustainability as soon as consumption is reduced to zero along the generations.

Why is it so mathematically? From (5.50), the inductive relation (5.51) holds true at final time $T$. Assume now that the relation (5.51) is satisfied at time $t+1$. From dynamic programming equation (5.50), we infer that

$$\begin{aligned} V(t,B) &= \sup_{0 \le h \le B} V\big(t+1, g(B-h)\big) \\ &= \sup_{0 \le h \le B} \rho^T M\Big(T, g^{T-t-1}\big(g(B-h)\big)\Big) \,. \end{aligned}$$

Since payoff $M$ and dynamic $g$ increase with resource $B$, we deduce that the optimal feedback control is given by its lower admissibility boundary $\mathfrak{u}^\star(t,B) = 0$ and consequently

$$V(t,B) = M\Big(T, g^{T-t-1}\big(g(B)\big)\Big) = M\big(T, g^{T-t}(B)\big) \,,$$

which is the desired statement.

## 5.13 Chichilnisky approach for exhaustible resources

The optimality problem proposed by Chichilnisky to achieve sustainability relies on a mix of the Green Golden rule and discounted approaches. The basic idea is to exhibit a trade-off between preferences for the future and the resource underlying the Green Golden approach and preferences for the present and consumption generated by the discounted utility criterion. The general problem of Chichilnisky performance can be formulated as a specific case of additive performance with a scrap payoff where the discount is time dependent. We focus here on the case of exhaustible resource $S(t+1) = S(t) - h(t)$ as in Sect. 2.1 with criterion

$$\max_{h(\cdot)} \Big( \theta \sum_{t=t_0}^{T-1} \rho^t L\big(h(t)\big) + (1-\theta) L\big(S(T)\big) \Big) \,. \tag{5.52}$$

Here, $\theta \in [0,1]$ stands for the coefficient of present preference. Note that the extreme case where $\theta = 1$ corresponds to the usual sum of discounted utility of consumptions, while $\theta = 0$ is the Green Golden rule payoff. Hence comparison of the three contexts can be achieved only by changing $\theta$, the coefficient of present preference. Applying dynamic programming, the following assertions can be proved.

**Result 5.17** *Using dynamic programming, one can prove by induction that*

$$\begin{cases} V(t,S) = \rho(t)b(t)^{\eta-1}S^{\eta}\,, \\ \mathfrak{h}^{\star}(t,S) = b(t)S\,, \end{cases} \tag{5.53}$$

*where $b(\cdot)$ is the solution of backward induction*

$$b(t) = \frac{\rho(t+1)^{\mu}b(t+1)}{\rho(t)^{\mu}+\rho(t+1)^{\mu}b(t+1)}\,, \quad b(T)=1\,,$$

*with $\mu = \frac{1}{\eta-1}$ and* $\rho(t) = \begin{cases} \theta\rho^{t} & \textit{if} \quad t = t_0,\dots,T-1\,, \\ (1-\theta) & \textit{if} \quad t = T\,. \end{cases}$

At final time $t=T$, the relation holds true as

$$V(T,S) = (1-\theta)S^{\eta} = \rho(T)b(T)^{\eta-1}S^{\eta}\,.$$

Now, assume that (5.53) is satisfied at time $t+1$. By dynamic programming equation (5.16), the value function is the solution of the backward induction

$$\begin{aligned} V(t,S) &= \sup_{0\le h\le S}\left\{\theta\rho^{t}h^{\eta} + V(t+1,S-h)\right\}\,, \\ &= \sup_{0\le h\le S}\left\{\theta\rho^{t}h^{\eta} + \rho(t+1)b(t+1)^{\eta-1}(S-h)^{\eta}\right\}. \end{aligned}$$

Applying first order optimality conditions, we deduce that

$$\rho^{t}\theta(h^{\star})^{\eta-1} = \rho(t+1)b(t+1)^{\eta-1}(S-h^{\star})^{\eta-1}\,.$$

Consequently, the optimal feedback is linear in $S$,

$$\mathfrak{h}^{\star}(t,S) = \frac{\rho(t+1)^{\mu}b(t+1)}{\rho(t)^{\mu}+\rho(t+1)^{\mu}b(t+1)}S\,,$$

which is similar to the desired result. Inserting the optimal control $\mathfrak{h}^{\star}(t,S)$ into the Bellman relation, it can be claimed that $V(t,S) = \rho(t)b(t)^{\eta-1}S^{\eta}$. Indeed, we derive that for any time $t = t_0,\dots,T$

$$\begin{aligned} V(t,S) &= \theta\rho^{t}h^{\star}(t,S)^{\eta} + V(t+1,S-h^{\star}(t,S)) \\ &= \theta\rho^{t}(b(t)S)^{\eta} + \rho(t+1)b(t+1)^{\eta-1}(S-b(t)S)^{\eta} \\ &= S^{\eta}\left(\rho(t)b(t)^{\eta} + \rho(t+1)b(t+1)^{\eta-1}(1-b(t))^{\eta}\right) \\ &= S^{\eta}\frac{\rho(t)\rho(t+1)^{\mu\eta}b(t+1)^{\eta} + \rho(t+1)b(t+1)^{\eta-1}\rho(t)^{\mu\eta}}{\left(\rho(t)^{\mu}+\rho(t+1)^{\mu}b(t+1)\right)^{\eta}} \\ &= S^{\eta}\frac{\rho(t)\rho(t+1)^{1+\mu}b(t+1)^{\eta} + \rho(t+1)b(t+1)^{\eta-1}\rho(t)^{1+\mu}}{\left(\rho(t)^{\mu}+\rho(t+1)^{\mu}b(t+1)\right)^{\eta}} \\ &= S^{\eta}\rho(t)\rho(t+1)b(t+1)^{\eta-1}\frac{\rho(t+1)^{\mu}b(t+1)+\rho(t)^{\mu}}{\left(\rho(t)^{\mu}+\rho(t+1)^{\mu}b(t+1)\right)^{\eta}} \\ &= S^{\eta}\rho(t)\frac{\left(\rho(t+1)^{\mu}b(t+1)\right)^{\eta-1}}{\left(\rho(t)^{\mu}+\rho(t+1)^{\mu}b(t+1)\right)^{\eta-1}} \\ &= S^{\eta}\rho(t)b(t)^{\eta-1}\,. \end{aligned}$$

It is worth pointing out that the feedback control $\mathfrak{h}(t,S)$ is smaller than the stock $S$ since coefficient $b(t)$ remains smaller than 1.

By stock dynamics $S^\star(s+1) = S^\star(s) - \mathfrak{h}^\star(s, S^\star(s))$ as in (5.21), we get the optimal paths in open-loop terms from initial time $t_0$. The behavior of optimal consumption $h^\star(t) = \mathfrak{h}^\star(t, S^\star(t))$ can be described as follows:

$$\begin{aligned}\frac{h^\star(t+1)}{h^\star(t)} &= \frac{b(t+1)S^\star(t+1)}{b(t)S^\star(t)} \\ &= \frac{b(t+1)S^\star(t)(1-b(t))}{b(t)S^\star(t)} \\ &= b(t+1)\frac{\rho(t)^\mu}{\rho(t+1)^\mu b(t+1)} \\ &= \left(\frac{\rho(t)}{\rho(t+1)}\right)^\mu .\end{aligned}$$

Hence, we recover the discounted case for periods before final time since optimal consumptions for $t = t_0, \dots, T-1$ change at a constant rate

$$h^\star(t+1) = \tau h^\star(t) \quad \text{with} \quad \tau = \rho^{-\mu} \quad \text{for} \quad t = t_0, \dots, T-1 .$$

Now, let us compute the final resource level $S^\star(T)$ generated by optimal extractions $h^\star(t)$. From Result 5.17, it can be claimed by induction that

$$S^\star(T) = S_0 \prod_{t=t_0}^{T-1} (1 - b(t)) .$$

Since $1 - b(t) = b(t)\left(\frac{1}{b(t)} - 1\right)$ and $\frac{1}{b(t)} = \left(\frac{\rho(t)}{\rho(t+1)}\right)^\mu \frac{1}{b(t+1)} + 1$, we deduce that

$$1 - b(t) = \left(\frac{\rho(t)}{\rho(t+1)}\right)^\mu \frac{b(t)}{b(t+1)} .$$

Consequently,

$$\prod_{t=t_0}^{T-1} (1 - b(t)) = \left(\frac{\rho(t_0)}{\rho(T)}\right)^\mu \frac{b(t_0)}{b(T)} .$$

Moreover, by virtue of the relation $\frac{1}{b(t)} = \left(\frac{\rho(t)}{\rho(t+1)}\right)^\mu \frac{1}{b(t+1)} + 1$ and by $b(T) = 1$, we also obtain that

$$\frac{1}{b(t_0)} = \sum_{t=t_0}^{T} \left(\frac{\rho(t_0)}{\rho(t)}\right)^\mu .$$

Combining these results, we write

$$\prod_{t=t_0}^{T-1} (1 - b(t) = \left(\frac{\rho(t_0)}{\rho(T)}\right)^\mu \left(\sum_{t=t_0}^{T} \left(\frac{\rho(t_0)}{\rho(t)}\right)^\mu\right)^{-1} = \left(\sum_{t=t_0}^{T} \left(\frac{\rho(T)}{\rho(t)}\right)^\mu\right)^{-1} .$$

Therefore, the optimal final resource state is characterized by

$$S^\star(T) = S_0 \left( \sum_{t=t_0}^{T} \left( \frac{\rho(T)}{\rho(t)} \right)^{\mu} \right)^{-1} .$$

Replacing the discount factors $\rho(t)$ by their specific value for the Chichilnisky performance gives

$$S^\star(T) = S_0 \left( 1 + \left( \frac{1-\theta}{\theta} \right)^{\mu} \sum_{t=t_0}^{T-1} \rho^{-\mu t} \right)^{-1} = S_0 \left( 1 + \left( \frac{1-\theta}{\theta} \right)^{\mu} \frac{\rho^{-\mu(T-t_0)} - 1}{\rho^{-\mu} - 1} \right)^{-1} .$$

Therefore, whenever the discount factor is strictly smaller than one $\rho < 1$, the Chichilnisky criterion exhibits a guaranteed resource $S^\star(+\infty) > 0$ together with a decreasing consumption that vanishes with the time horizon.

**Result 5.18** *Assume the utility function is isoelastic in the sense that* $L(h) = h^\eta$ *with* $0 < \eta < 1$*. If the discount factor is strictly smaller than unity (*$\rho < 1$*) and the present preference coefficient is strictly positive,* $\theta > 0$*, then the Chichilnisky criterion yields*

- *guaranteed stock:* $\lim_{T \to +\infty} S^\star(T) = S_0 \left( 1 + \left( \frac{1-\theta}{\theta} \right)^{\mu} \frac{1}{1-\rho^{-\mu}} \right)^{-1} > 0$*;*
- *consumption decreasing toward zero:* $h^\star(t+1) < h^\star(t)$ *and* $\lim_{T \to +\infty} h^\star(T) = 0$.

At this stage, a comparison can be achieved between the DU (discounted), GGR (Green Golden rule) and CHI (Chichilnisky) approaches as shown by Fig. 5.6. Each criterion can be characterized through the parameter $\theta$. It is worth noting that for DU ($\theta = 1$) and CHI frameworks, consumptions $h^\star(t)$ are decreasing toward 0 while it remains zero along time for GGR ($\theta = 0$). This observation emphasizes the unsustainable feature of such optimal solutions. However, as far as resource $S(t)$ is concerned, it turns out that both CHI and GGR provide conservation of the resource. In this sense, the CHI framework represents an interesting trade-off for sustainability as it allows both for consumption and conservation.

## 5.14 The "maximin" approach

As already pointed out in Subsect. 5.1.2, the maximin or Rawls criterion sheds an interesting light on the sustainability issue. Indeed, by focusing on the worst output of decisions along generations and time, this approach promotes more intergenerational equity than the additive criterion which often yields strong preferences for the present. From the methodological point of view, this framework turns out to be more complicated to handle because a "max"

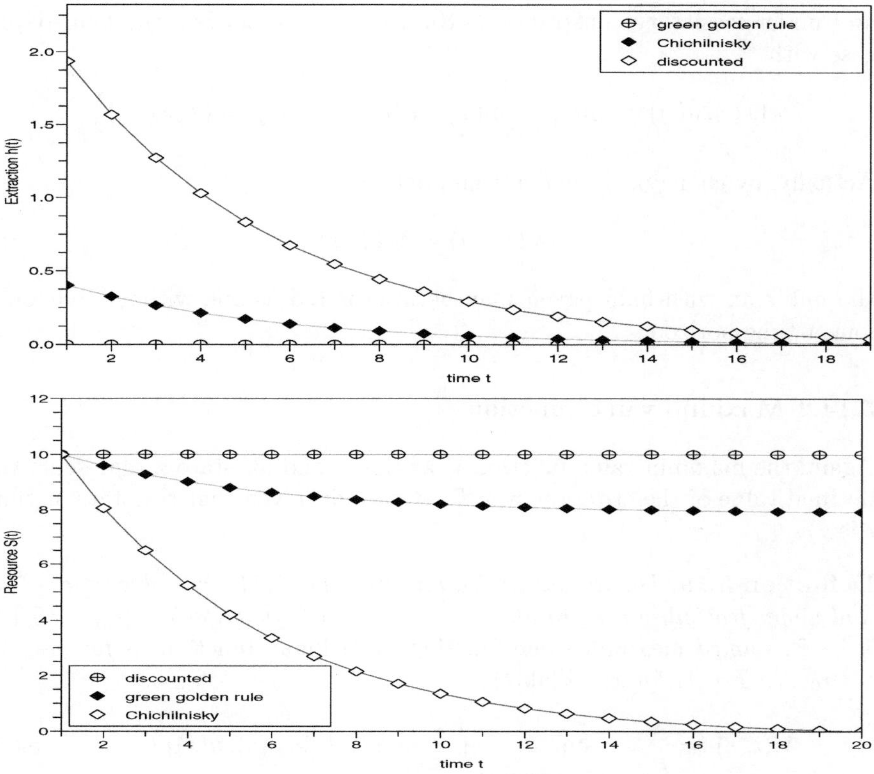


**Fig. 5.6.** A comparison between discounted ($\diamond$, $\theta = 1$), Green Golden rule ($\oplus$, $\theta = 0$) and Chichilnisky ($\blacklozenge$, $\theta = 0.2$) approaches for optimal stock $S(t)$ and consumption $h(t)$. Here the utility parameter is $\eta = 0.5$, the discount factor is $\rho = 0.9$ and the final horizon is $T = 20$.

operator is less regular than an addition (in an algebraic sense that we do not treat here). For instance, the maximum principle no longer holds for the maximin framework, at least in simple formulations. However, the dynamic programming principle and Bellman equation still hold true, although some adaptations are needed. Strong links between maximin and viability or weak invariance approaches are worth being outlined.

### 5.14.1 The optimization problem

In this section, we focus on the maximization of the worst result along time of an instantaneous payoff in the finite horizon ($T < +\infty$)

$$\pi^\star(t_0, x_0) = \sup_{\big(x(\cdot),u(\cdot)\big)\in\mathcal{T}^{ad}(t_0,x_0)} \min_{t=t_0,\dots,T-1} L\big(t, x(t), u(t)\big) \tag{5.54}$$

and under feasibility constraints (5.8). We can also address the final payoff case with

$$\pi\big(x(\cdot),u(\cdot)\big) = \min\Big(\min_{t=t_0,\dots,T-1} L\big(t,x(t),u(t)\big), M\big(T,x(T)\big)\Big) .$$

Actually, by changing $T$ in $T+1$ and defining

$$L(T,x,u) = M\big(T,x\big) ,$$

the minimax with final payoff may be interpreted as one without, but on a longer time horizon.

### 5.14.2 Maximin value function

Again, the maximin value function $V$ at time $t$ and for state $x$ represents the optimal value of the criterion over $T-t$ periods, given that the state at time $t$ is $x$.

**Definition 5.19.** *For the maximization problem (5.54) with dynamics (5.1) and under feasibility constraints (5.2), we define a function* $V : \{t_0,\dots,T\} \times \mathbb{X} \to \mathbb{R}$, *named* maximin value function *or* Bellman function *as follows: for* $t = t_0,\dots,T-1$, *for* $x \in \mathbb{V}\mathrm{iab}(t)$

$$V(t,x) := \sup_{\big(x(\cdot),u(\cdot)\big)\in\mathcal{T}^{ad}(t,x)} \Big(\min_{s=t,\dots,T-1} L\big(s,x(s),u(s)\big)\Big) , \tag{5.55}$$

*with the convention that, for* $x \in \mathbb{V}\mathrm{iab}(T) = \mathbb{A}(T)$,

$$V(T,x) := +\infty . \tag{5.56}$$

Therefore, given the initial state $x_0$, the optimal sequential decision problem (5.1)-(5.2)-(5.54) refers to the particular value $V(t_0,x_0)$ of the Bellman function:

$$V(t_0,x_0) = \pi^\star(t_0,x_0) .$$

### 5.14.3 Maximin dynamic programming equation

As in the additive case of Sect. 5.2, a backward induction equation appears for the maximin value function $V(t,x)$ when we note that the maximization operation can again be split up into two parts. However, this separation has to be adapted to the maximin case. The proof of the following Proposition 5.20 can be found in Sect. A.3 in the Appendix.

**Proposition 5.20.** *The value function defined by (5.55)–(5.56) is the solution of the following backward* dynamic programming equation *(or* Bellman equation*), where* $t$ *runs from* $T-1$ *down to* $t_0$,

$$\begin{cases} V(T,x) = +\infty \,, & \forall x \in \mathbb{V}\text{iab}(T) \,, \\ V(t,x) = \sup\limits_{u \in \mathbb{B}^{viab}(t,x)} \min\Big(L(t,x,u), V\big(t+1, F(t,x,u)\big)\Big) \,, & \forall x \in \mathbb{V}\text{iab}(t) \,, \end{cases} \tag{5.57}$$

*where* $\mathbb{V}\text{iab}(t)$ *is again given by the backward induction (5.18) and where the supremum in (5.57) is over viable controls in* $\mathbb{B}^{viab}(t,x)$ *as in (5.19).*

Thus, as in the additive case of Sect. 5.2, to compute the maximin value $V(t,x)$, we begin with final payoff $V(T,x) = +\infty$ and then apply a backward induction.

### 5.14.4 Optimal feedback

As in the additive case, Bellman induction optimization (5.57) in the maximin case also reveals relevant feedback controls. Indeed, assuming the additional hypothesis that the supremum is achieved in (5.57) for at least one decision, if we denote by $\mathfrak{u}^\star(t,x)$ a value $u \in \mathbb{B}(t,x)$ which achieves the maximum in equation (5.57), then $\mathfrak{u}^\star(t,x)$ is an optimal feedback for the optimal control problem in the following sense. The proof of the following Proposition 5.21 follows from the proof of Proposition 5.20 in Sect. A.3 in the Appendix.

**Proposition 5.21.** *For any time* $t$ *and state* $x$*, assume the existence of the following maximin feedback decision*

$$\mathfrak{u}^\star(t,x) \in \mathop{\arg\max}_{u \in \mathbb{B}^{viab}(t,x)} \min\Big(L(t,x,u), V\big(t+1, F(t,x,u)\big)\Big) \,, \tag{5.58}$$

*where the maximin value function* $V$ *is given in (5.55). Then an optimal trajectory* $\big(x^\star(\cdot), u^\star(\cdot)\big)$ *for the maximization problem (5.54) is given for any* $t = t_0, \ldots, T-1$ *by*

$$x^\star(t+1) = F\big(t, x^\star(t), u^\star(t)\big) \,, \quad u^\star(t) = \mathfrak{u}^\star\big(t, x^\star(t)\big) \,. \tag{5.59}$$

### 5.14.5 Are maximin and viability approaches equivalent?

Here we examine the links between the maximin framework and viability approaches. Such a connection has already been emphasized in [12] in the context of exhaustible resource management. In particular, it turns out that the maximin value function corresponds to a static optimality problem involving the viability kernel. We need to consider a specific viability kernel associated with the additional constraint requiring a guaranteed payoff $L$, namely:

$$L\big(t, x(t), u(t)\big) \geq L^\flat \ .$$

Hence, we introduce the viability kernel

$$\mathbb{V}\mathrm{iab}(t, L^\flat) := \left\{ x \in \mathbb{X} \;\middle|\; \begin{array}{l} \exists \big(x(\cdot), u(\cdot)\big) \text{ such that } \forall s = t, \ldots, T \\ x(s+1) = F\big(s, x(s), u(s)\big) \\ x(t) = x \\ x(s) \in \mathbb{A}(s),\ u(s) \in \mathbb{B}(s, x(s)) \\ L(s, x(s), u(s)) \geq L^\flat \end{array} \right\}$$

which depends on the guaranteed payoff $L^\flat$. The main result is a characterization of the maximin value function $V(t,x)$ through a static optimization problem involving the viability kernel.

The proof of the following Proposition 5.22 can be found in in Sect. A.3 in the Appendix.

**Proposition 5.22.** *The maximin value function $V$ in (5.55) satisfies:*

$$V(t,x) = \sup\{L^\flat \in \mathbb{R} \mid x \in \mathbb{V}\mathrm{iab}(t, L^\flat)\} \ .$$

Such a result points out, on the one hand, the interest of the viable control framework to deal with equity issues and, on the other hand, the viability property related to the maximin approach.

## 5.15 Maximin for an exhaustible resource

Consider the exhaustible resource management

$$S(t+1) = S(t) - h(t) \ , \quad 0 \leq h(t) \leq S(t) \ , \quad S(t_0) = S_0 \ ,$$

in the maximin perspective

$$\sup_{h(t_0),\ldots,h(T-1)} \ \min_{t=t_0,..,T-1} L\big(h(t)\big) \ ,$$

where $L$ is an increasing utility function. We shall show that the value function and optimal feedbacks are given by

$$V(t,S) = L\big(\frac{S}{T-t}\big) \ , \quad \mathfrak{h}^\star(t,S) = \frac{S}{T-t} \ , \quad t = t_0, \ldots, T-1 \ .$$

From (5.56), we have $V(T,S) = +\infty$ since there is no $M\big(T, x(T)\big)$ final term and, thus, the above formula holds true at final time $T$. Assume that it is satisfied at time $t+1$. From dynamic programming equation (5.57), we infer that

$$
\begin{aligned}
V(t,S) &= \sup_{0\leq h\leq S} \min\big(L(h), L(\frac{S-h}{T-t-1})\big) \\
&= \max\left\{ \sup_{\frac{S}{T-t}\leq h\leq S} \min\big(L(h), L(\frac{S-h}{T-t-1})\big), \right. \\
&\qquad \left. \sup\nolimits_{0\leq h\leq \frac{S}{T-t}} \min\big(L(h), L(\tfrac{S-h}{T-t-1})\big) \right\} \\
&= \max\left\{ \sup_{\frac{S}{T-t}\leq h\leq S} L(\frac{S-h}{T-t-1}), \sup_{0\leq h\leq \frac{S}{T-t}} L(h) \right\} \\
&= \max\left\{ L(\tfrac{S-\frac{S}{T-t}}{T-t-1}), L(\tfrac{S}{T-t}) \right\} \\
&= L(\tfrac{S}{T-t}) \, ,
\end{aligned}
$$

and that the optimal feedback control is unique and given by $\mathfrak{h}^\star(t,S) = \frac{S}{T-t}$. Therefore the solution at time $t_0$ is given by

$$
V(t_0, S_0) = L(\frac{S_0}{T-t_0})
$$

and we deduce the optimal path as follows.

**Result 5.23** *Assume that the instantaneous payoff $L$ is increasing. Then the maximin optimal extraction path $h^\star(\cdot)$ is stationary and*

$$
S^\star(t) = \frac{S_0(T-t+t_0)}{T} \, , \quad h^\star(t) = \mathfrak{h}^\star\big(t, S^\star(t)\big) = \frac{S_0}{T-t_0} \, .
$$

Such a maximin solution promotes intergenerational equity in the sense that depletion $h^\star(t)$ is stationary. The “cake” $S_0$ is shared in equal parts eaten along time.

# References


[1] D. P. Bertsekas. *Dynamic Programming and Optimal Control.* Athena Scientific, Belmont, Massachusets, second edition, 2000. Volumes 1 and 2.

[2] K. W. Byron, J. D. Nichols, and M. J. Conroy. *Analysis and Management of Animal Populations.* Academic Press, 2002.

[3] C. W. Clark. *Mathematical Bioeconomics.* Wiley, New York, second edition, 1990.

[4] J. M. Conrad. *Resource Economics.* Cambridge University Press, 1999.

[5] J. M. Conrad and C. Clark. *Natural Resource Economics.* Cambridge University Press, 1987.

[6] L. Doyen, P. Dumas, and P. Ambrosi. Optimal timing of $CO_2$ mitigation policies for a cost-effectiveness model. *Mathematics and Computer Modeling*, in press.

[7] L. H. Goulder and K. Mathai. Optimal $CO_2$ abatement in the presence of induced technical change. *Journal of Environmental Economics and Management*, 39:1–38, 2000.

[8] G. Heal. *Valuing the Future, Economic Theory and Sustainability.* Columbia University Press, New York, 1998.

[9] H. Hotelling. The economics of exhaustible resources. *Journal of Political Economy*, 39:137–175, april 1931.

[10] M. Kot. *Elements of Mathematical Ecology.* Cambridge University Press, Cambridge, 2001.

[11] H. Lerchs and I. F. Grossman. Optimum design of open-pit mines. *Transactions, CIM*, 68:17–24, 1965.

[12] V. Martinet and L. Doyen. Sustainable management of an exhaustible resource: a viable control approach. *Resource and Energy Economics*, 29(1):p.17–39, 2007.

[13] MEA. *Millennium Ecosystem Assessment.* 2005. Available on http://www.maweb.org/en/index.aspx.

[14] M. L. Weitzman. Prices vs. quantities. *Review of Economic Studies*, 41(4):477–491, October 1974.

[15] P. Whittle. *Optimization over Time: Dynamic Programming and Stochastic Control*, volume 1. John Wiley & Sons, New York, 1982.

# 6

# Sequential decisions under uncertainty

The issue of uncertainty is fundamental in environmental management problems. Decision makers have to take sequential decisions without precise knowledge about phenomena that can affect the system state and evolution [3, 4]. Such uncertainties are involved, for instance, in natural mechanisms including demographic or environmental stochasticity for population dynamics [10, 12, 14], or the natural removal rate in the carbon cycle. Uncertainties also occur at more anthropic stages, as for example with scenarios such as the global economic growth rate or demographic evolution along with the uncontrollability of catches for harvesting of renewable resources, to name a few. Errors in measurement and data also constitute sources of uncertainty.

By fundamental, we mean that the issue of uncertainty is inherent to environmental problems in the sense that it cannot be easily removed by scientific investigation, although it can be reduced by it. For instance, the scientific resolution of uncertainty for global warming will, if it is ever achieved, occur much later than the date at which decisions must be taken. For fish harvesting, stock evaluations will remain poor, although recognizing this problem in no way negates the necessity of taking decisions. We simply underline here that such irreducible uncertainties open the way for controversies.

In this context, the methods of decision-making under uncertainty are useful [11]. We can distinguish several kinds of uncertainty. First, there is *risk*, which covers events with *known* probabilities. To deal with risk, policy makers may refer to *risk assessment*, which can be useful when the probability of an outcome is known from experience and statistics. In the framework of dynamic decision-making under uncertainty, the usual approach applied is the expected or mean value of utility or cost-benefits. The general method is *stochastic control* [1, 2, 15].

On the other hand, there are cases presenting *ambiguity* or "Knightian" uncertainty [9] with unknown probability or with no probability at all. Most precaution and environmental problems involve ambiguity in the sense of controversies, beliefs, irreducible scientific uncertainties. In this sense, by dealing with ambiguity, multiprior models [7] may appear relevant alternatives for the precaution issue. Similarly, pessimistic, worst case, total risk-averse or guaranteed and robust control frameworks may also shed interesting light. As an initial step in these directions, the present textbook proposes to introduce ambiguity through the use of "total" uncertainty and *robust control* [6].

This chapter is devoted to the extension of the concepts and results of deterministic control to the uncertain framework. Dynamics, constraints and optimality have to be carefully expanded. The principal difficulties stem from the fact that we are still manipulating states and controls, but they are now dependent on external variables (disturbance, noise, etc.). As a first consequence, the open loop approach – consisting in designing control laws dependent only upon time – is no longer relevant: now, the decision at each point in time must rely at least on the uncertain available information (current state for instance) to display required adaptive properties. Assuming perfect information, in the sense that the state is observed by the decision maker, we shall focus on state feedback policies. Another difficulty lies in optimal criterion. Since states and controls are now uncertain, the criterion also becomes uncertain. This fact opens the door to different options: taking the mathematical expectation leads to *stochastic control*, while minimax operations lead to *robust control*. A similar type of difficulty arises for constraints in *stochastic or robust viability* that will be handled in Chap. 7.

For the sake of simplicity, we consider control dynamical systems with perturbations. This is a natural extension of deterministic control systems which cover a large class of situations. This context makes it possible to treat robust and stochastic approaches simultaneously. In both cases, the interest of dynamic programming is emphasized.

The chapter is organized as follows. The notion of dynamics introduced in Chap. 2 is extended to the uncertain case in Sect. 6.1. Then, due to uncertainties, trajectories are no longer unique in contrast to the deterministic case, and we are led to define solution maps and feedback strategies in Sect. 6.2. Specific treatment is devoted to the probabilistic or stochastic case in Sect. 6.3. Different options for the criteria are presented in Sect. 6.4. The remaining Sections are devoted to examples.

## 6.1 Uncertain dynamic control system

Now, the dynamic control system which has been the basic model in the previous chapters is no longer deterministic. For instance, some parameters are not under direct control in the dynamics and may vary along time. Perturbations disturb the system and yield uncertainties on the state paths whatever the decision applied. In this context, decisions and controls are now to be selected to display reasonable performance for the system despite these uncertainties. The assessments rely on mean performance in the stochastic case while worst case criteria are adapted to the robust perspective.

## Uncertain dynamics

Extending the state equation (2.48), the uncertain dynamic model is described in discrete time by the state equation:

$$x(t+1) = F\big(t, x(t), u(t), w(t)\big) \;, \quad t = t_0, \ldots, T-1 \quad \text{with} \quad x(t_0) = x_0 \quad (6.1)$$

where again $F$ is the so-called *dynamics* function representing the system's evolution, the *horizon* $T \in \mathbb{N}^*$ or $T = +\infty$ stands for the term, $x(t) \in \mathbb{X} = \mathbb{R}^n$ represents the system's *state* vector at time $t$, $x_0 \in \mathbb{X}$ is the *initial condition* at *initial time* $t_0 \in \{0, \ldots, T-1\}$, $u(t) \in \mathbb{U} = \mathbb{R}^p$ represents the *decision* or *control* vector. The new element $w(t)$ in the dynamic stands for the *uncertain variable*[1], or *disturbance*, *noise*, taking its values from a given set $\mathbb{W} = \mathbb{R}^q$.

## Scenarios

We assume that

$$w(t) \in \mathbb{S}(t) \subset \mathbb{W} \,, \quad (6.2)$$

so that the sequences

$$w(\cdot) := \big(w(t_0), w(t_0+1), \ldots, w(T-1), w(T)\big) \quad (6.3)$$

belonging to

$$\Omega := \mathbb{S}(t_0) \times \cdots \times \mathbb{S}(T) \subset \mathbb{W}^{T+1-t_0} \quad (6.4)$$

capture the idea of possible *scenarios* for the problem. A scenario is an *uncertainty trajectory.* The deterministic or certain case is recovered as soon as the alternatives are reduced to one choice, namely when $\Omega$ is a singleton $\Omega = \{\overline{w}(\cdot)\}$.

Probabilistic assumptions on the uncertainty $w(\cdot)$ may also be added as we shall see in Sect. 6.3.

## Constraints and viability

As in the certain case, we may require state and decision constraints to be satisfied. However, since state trajectories are no longer unique, the following requirements depend upon the scenarios $w(\cdot) \in \Omega$ in a way that we shall specify later, in Chap. 7. In particular, it is worth distinguishing robust and stochastic approaches for handling such invariance or viability issues. *The assertions below are thus to be taken in a loose sense at this stage.*

- The *control constraints* are respected at any time $t$:

$$u(t) \in \mathbb{B}\big(t, x(t)\big) \subset \mathbb{U} \,. \quad (6.5a)$$

[1] Notice that we do not include here the initial condition in the uncertain variables. This issue is addressed in Chap. 9.

- The *state constraints* are respected at any time $t$:

$$x(t) \in \mathbb{A}(t) \subset \mathbb{X} . \tag{6.5b}$$

- The final state achieves a fixed *target* $\mathbb{A}(T) \subset \mathbb{X}$:

$$x(T) \in \mathbb{A}(T) . \tag{6.5c}$$

### Criteria to optimize

The *criterion* $\pi$ of Subsect. 5.1.2 now depends upon the scenarios $w(\cdot)$: this point raises issues as to how to turn this family of values (one per scenario) into a single one to be optimized. A *criterion* $\pi$ is a function $\pi : \mathbb{X}^{T+1-t_0} \times \mathbb{U}^{T-t_0} \times \mathbb{W}^{T+1-t_0} \to \mathbb{R}$ which assigns a real number to a state, control and uncertainty trajectory. Extending the different forms exposed for the deterministic case in Sect. 2.9.4, we distinguish the following cases.

- The *additive* and *separable* form in finite horizon is

$$\pi\big(x(\cdot), u(\cdot), w(\cdot)\big) = \sum_{t=t_0}^{T-1} L\big(t, x(t), u(t), w(t)\big) + M\big(T, x(T), w(T)\big) \tag{6.6}$$

  in which the function $L$ again specifies the *instantaneous payoff* (or gain, profit, benefit, utility, etc.) when the criterion $\pi$ is maximized. The final performance is measured through function $M$ which can depend on time and terminal uncertainty $w(T)$. Such a general additive framework encompasses more specific performance as the Green Golden or the Chichilnisky form.
  The so-called *Green Golden* form focuses on the final performance:

$$\pi\big(x(\cdot), u(\cdot), w(\cdot)\big) = M\big(T, x(T), w(T)\big) . \tag{6.7}$$

  An intermediary form (Chichilnisky type) may be obtained by adding instantaneous and scrap performances (with $0 \leq \theta \leq 1$):

$$\pi\big(x(\cdot), u(\cdot), w(\cdot)\big) = \theta \sum_{t=t_0}^{T-1} L\big(t, x(t), u(t), w(t)\big) + (1-\theta) M\big(T, x(T), w(T)\big) . \tag{6.8}$$

- The *Rawls* or *maximin* form in the finite horizon is (without final term)

$$\pi\big(x(\cdot), u(\cdot), w(\cdot)\big) = \min_{t=t_0,\ldots,T-1} L\big(t, x(t), u(t), w(t)\big) , \tag{6.9}$$

  and, with a final payoff, the expression is somewhat heavier to write:

$$\pi\big(x(\cdot), u(\cdot), w(\cdot)\big) = \min\Big(\min_{t=t_0,\ldots,T-1} L\big(t, x(t), u(t), w(t)\big), M\big(T, x(T), w(T)\big)\Big) .$$

These distinct ways[2] of taking time into account for assessing the intertemporal performance have major impacts on sustainability (equity, conservation...) as previously discussed in the deterministic case.

## 6.2 Decisions, solution map and feedback strategies

Decision issues are much more complicated than in the certain case. In the uncertain context, we must drop the idea that the knowledge of decisions $u(\cdot)$ induces one single path of sequential states $x(\cdot)$. Open loop controls $u(t)$ depending only upon time $t$ are no longer relevant, in contrast to closed loop or feedback controls $\mathfrak{u}\big(t, x(t)\big)$ which display more adaptive properties by taking the uncertain state evolution $x(t)$ into account.

We shall assume in the present chapter that, at time $t$, the whole state $x(t)$ is observed and available for control design. This *perfect information* case corresponds to *state feedback* controls $\mathfrak{u} : \mathbb{N} \times \mathbb{X} \to \mathbb{U}$ which may be any mapping apart from general measurability assumptions in the stochastic case. Extensions to the context of imperfect information are partly handled in Chap. 9.

We define a *feedback* as an element of the set of all functions from the couples time-state towards the controls:

$$\mathfrak{U} := \{\mathfrak{u} : (t, x) \in \mathbb{N} \times \mathbb{X} \mapsto \mathfrak{u}(t, x) \in \mathbb{U}\} . \tag{6.11}$$

At this level of generality, no measurability assumptions are made. However, in the probabilistic setting, $\sigma$-algebras will be introduced with respect to which feedbacks will be supposed measurable.

Let us mention that, in the stochastic context, a feedback decision is also termed a *pure Markovian strategy*[3]. Markovian means that the current state contains all the sufficient information of past system evolution to determine the statistical distribution of future states. Thus, only current state $x(t)$ is needed in the feedback loop among the whole sequence of past states $x(t_0)$,..., $x(t)$.

Hereafter, for the sake of clarity, we restrict the notation $u(t)$ for a control *variable* belonging to $\mathbb{U}$, $u(t) \in \mathbb{U}$, while we denote by $\mathfrak{u} \in \mathfrak{U}$ a feedback

[2] For the mathematical proofs, the *multiplicative form* is also used:

$$\pi\big(x(\cdot), u(\cdot), w(\cdot)\big) = \prod_{t=t_0}^{T-1} L\big(t, x(t), u(t), w(t)\big) \times M\big(T, x(T), w(T)\big) . \tag{6.10}$$

[3] A pure strategy (or policy) at time $t$ is a rule which assigns to $(x(t_0),\dots, x(t), u(t_0),\dots, u(t-1))$ a control $u(t) \in \mathbb{U}$. Such a policy is said to be *pure* by opposition to a *mixed strategy* that assigns to $(x(t_0),\dots, x(t), u(t_0),\dots, u(t-1))$ a probability law on $\mathbb{U}$: in a mixed strategy, the decision maker draws randomly a decision $u(t)$ on $\mathbb{U}$ according to this latter law.

*mapping*, with $\mathfrak{u}(t,x) \in \mathbb{U}$ (see the footnote 13 in Sect. 2.10). The terminology *unconstrained case* covers the situation where all feedbacks in $\mathfrak{U}$ are allowed. The *control constraints case* restricts feedbacks to *admissible feedbacks* as follows

$$\mathfrak{U}^{ad} = \{\mathfrak{u} \in \mathfrak{U} \mid \mathfrak{u}(t,x) \in \mathbb{B}(t,x) \ , \quad \forall (t,x)\} \ , \tag{6.12}$$

corresponding to control constraints (6.5a).

The *viability case* covers control and state constraints as in (6.5a)-(6.5b)-(6.5c). However, its definition depends upon the context, either robust or stochastic. Specific definitions are given hereafter accordingly.

At this stage, we need to introduce some notations which will appear quite useful in the sequel: the *state map* $x_F[t_0, x_0, \mathfrak{u}, w(\cdot)]$ and the *control map* $u_F[t_0, x_0, \mathfrak{u}, w(\cdot)]$.

Given a feedback $\mathfrak{u} \in \mathfrak{U}$, a scenario $w(\cdot) \in \Omega$ and an initial state $x_0$ at time $t_0 \in \{0, \ldots, T-1\}$, the solution state $x_F[t_0, x_0, \mathfrak{u}, w(\cdot)]$ is the state path $x(\cdot) = (x(t_0), x(t_0+1), \ldots, x(T))$ solution of dynamic

$$x(t+1) = F\big(t, x(t), \mathfrak{u}(t, x(t)), w(t)\big) \ , \quad t = t_0, \ldots, T-1 \quad \text{with} \quad x(t_0) = x_0$$

starting from the initial condition $x(t_0) = x_0$ at time $t_0$ and associated with feedback control $\mathfrak{u}$ and scenario $w(\cdot)$. The solution control $u_F[t_0, x_0, \mathfrak{u}, w(\cdot)]$ is the associated decision path $u(\cdot) = (u(t_0), u(t_0+1), \ldots, u(T-1))$ where $u(t) = \mathfrak{u}\big(t, x(t)\big)$.

It should be noticed that, with straightforward notations,

$$\begin{cases} x_F[t_0, x_0, \mathfrak{u}, w(\cdot)](t_0) = x_0 \ , \\ x_F[t_0, x_0, \mathfrak{u}, w(\cdot)](t) \ = x_F[t_0, x_0, \mathfrak{u}, (w(t_0), \ldots, w(t-1))](t) \\ \qquad\qquad\qquad\qquad\qquad \text{for} \quad t \geq t_0 + 1 \ , \end{cases} \tag{6.13}$$

thus expressing a *causality* property: the future state $x_F[t_0, x_0, \mathfrak{u}, w(\cdot)](t)$ depends upon the disturbances $(w(t_0), \ldots, w(t-1))$ and not upon all $w(\cdot) = (w(t_0), \ldots, w(T-1), w(T))$. This property will be used in Sect. A.4 in the Appendix.

Once a state feedback law is selected, state and control variables are functions of the disturbances $w(\cdot)$ and in this sense they become *uncertain variables*[4]. Thus, *what characterizes uncertain systems is the non uniqueness of trajectories.*

## 6.3 Probabilistic assumptions and expected value

Probabilistic assumptions on the uncertainty $w(\cdot) \in \Omega$ *may* be added, providing a stochastic nature to the problem. Mathematically speaking, we equip

[4] We reserve the term *random variable* to the case where the set $\mathbb{W}^{T+1-t_0}$ is equipped with a $\sigma$-field and a probability measure, and $w(\cdot)$ is identified with identity mapping.

the domain of scenarios $\Omega \subset \mathbb{W}^{T+1-t_0} = \mathbb{R}^q \times \cdots \times \mathbb{R}^q$ with a $\sigma$-field[5] $\mathcal{F}$ and a *probability* $\mathbb{P}$: thus, $(\Omega, \mathcal{F}, \mathbb{P})$ constitutes a *probability space*. The sequences

$$w(\cdot) = (w(t_0), w(t_0+1), \ldots, w(T-1), w(T)) \in \Omega$$

now become the *primitive random variables.*[6]

The notation $\mathbb{E}$ refers to the *mathematical expectation* over $\Omega$ under probability $\mathbb{P}$. To be able to perform mathematical expectations, we are led to consider measurability assumptions. In the stochastic setting, all the objects considered will be implicitly equipped with appropriate measurability properties[7]. Once a feedback $\mathfrak{u}$ is picked up in $\mathfrak{U}^{ad}$ defined in (6.11), the state and control *variables* $x$ and $u$ become *random variables*[8] defined over $(\Omega, \mathcal{F}, \mathbb{P})$ by means of the relations

$$x(t) = x_F[t_0, x_0, \mathfrak{u}, w(\cdot)](t) \quad \text{and} \quad u(t) = \mathfrak{u}\big(t, x(t)\big) ,$$

which refer to state and control solution maps introduced in Sect. 6.2. Thus, any quantity depending upon states, controls and disturbances is now a random variable and, hence, when bounded or nonnegative, admits an integral with respect to probability $\mathbb{P}$.

Let $A : \Omega \to \mathbb{R}$ be a measurable function. In such a probabilistic context, we use the notation $\mathbb{E}_{w(\cdot)}[A\big(w(\cdot)\big)]$ for the expected value of the random variable $A\big(w(\cdot)\big)$, when integrable.

- For discrete probability laws (products of Bernoulli, binomial...), this means that:
$$\mathbb{E}_{w(\cdot)}[A\big(w(\cdot)\big)] = \sum_{w(\cdot)\in\Omega} A\big(w(\cdot)\big)\mathbb{P}\big(w(\cdot)\big) .$$
- For continuous probability laws (Gaussian, uniform, exponential, beta...) on $\mathbb{W} = \mathbb{R}^q$, this gives
$$\mathbb{E}_{w(\cdot)}[A\big(w(\cdot)\big)] = \int_\Omega A\big(w(\cdot)\big)q\big(w(\cdot)\big)dw(t_0)dw(t_0+1)\ldots, dw(T-1)dw(T) ,$$
with $q$ the density of $\mathbb{P}$ on $\Omega$.

We shall generally assume that the primitive random process $w(\cdot)$ is made of *independent and identically distributed (i.i.d.)* random variables $\big(w(t_0), w(t_0+1), \ldots, w(T-1), w(T)\big)$ under $\mathbb{P}$. In this configuration, all random variables $w(t)$ take values from the same domain $\mathbb{S}$, and the probability $\mathbb{P}$ on the domain of scenarios is $\Omega = \mathbb{S}^{T+1-t_0}$ is chosen as the product of $T - t_0 + 1$ copies of a probability $\mu$ over $\mathbb{S}$.

[5] For instance the usual Borelian $\sigma$-field $\mathcal{F} = \bigotimes_{t=t_0}^{T} \mathcal{B}(\mathbb{R}^q)$.

[6] Recall that a *random variable* is a measurable function on $(\Omega, \mathcal{F})$. Here, $w(\cdot)$ is identified with the identity mapping on $(\Omega, \mathcal{F})$.

[7] Thus, the sets $\mathbb{X} = \mathbb{R}^n$ and $\mathbb{U} = \mathbb{R}^p$ are assumed to be equipped with the Borel $\sigma$-fields $\mathcal{B}(\mathbb{R}^n)$ and $\mathcal{B}(\mathbb{R}^p)$ respectively, the dynamic $F$ is assumed to be measurable and, by feedback, we mean a measurable feedback.

[8] See the previous footnote 6.

## 6.4 Decision criteria under uncertainty

In this textbook, we focus on robust and expected criteria. However, other approaches for decision-making under uncertainty exist. To quote a few, we here expose optimistic, Hurwicz criteria and multi-prior approaches. We also mention the quadratic case which is connected to mean-variance analysis. Let us pick up a criterion $\pi$ among the discounted, maximin or Chichilnisky forms as in Sect. 6.1.

- **Robust or pessimistic.** Robust control sheds interesting light on decision-making under uncertainty by adopting a *pessimistic*, *worst case* or *totally risk-averse* point of view. It aims at maximizing the worst payoff

$$\sup_{\mathfrak{u}\in\mathfrak{U}^{ad}} \inf_{w(\cdot)\in\Omega} \pi\big(x(\cdot), u(\cdot), w(\cdot)\big) , \tag{6.14}$$

  where the last expression is abusively used, even if convenient and traditional, in which $x(\cdot)$ and $u(\cdot)$ need to be replaced by $x(t) = x_F[t_0, x_0, \mathfrak{u}, w(\cdot)](t)$ and $u(t) = \mathfrak{u}\big(t, x(t)\big)$, referring to solution state and control introduced in Sect. 6.2. Such a pessimistic approach does not require any probabilistic hypothesis on uncertainty $w(.)$ as only the set of possible scenarios $\Omega$ is involved.
- **Stochastic or expected.** The most usual approach to handle decision under uncertainty corresponds to optimizing the *expected payoff*

$$\sup_{\mathfrak{u}\in\mathfrak{U}^{ad}} \mathbb{E}\Big[\pi\big(x(\cdot), u(\cdot), w(\cdot)\big)\Big] , \tag{6.15}$$

  where again in the last expression $x(\cdot)$ and $u(\cdot)$ refer to solution state and control introduced in Sect. 6.2.
  Of course, this requires a probabilistic structure for the uncertainties. One weakness of such an approach is to promote means, hence to neglect rare events which may generate catastrophic paths. This situation may be especially critical for environmental concerns where irreversibility is important.
- **Optimistic.** Instead of maximizing the worst cost as in a robust approach, the optimistic perspective focuses on the most favorable payoff, namely:

$$\sup_{\mathfrak{u}\in\mathfrak{U}^{ad}} \sup_{w(\cdot)\in\Omega} \pi\big(x(\cdot), u(\cdot), w(\cdot)\big) .$$

- **Hurwicz criterion.** This approach adopts an intermediate attitude between optimistic and pessimistic approaches. A proportion $\alpha \in [0,1]$ graduates the level of prudence as follows:

$$\sup_{\mathfrak{u}\in\mathfrak{U}^{ad}} \Big\{\alpha \inf_{w(\cdot)\in\Omega} \pi\big(x(\cdot), u(\cdot), w(\cdot)\big) + (1-\alpha) \sup_{w(\cdot)\in\Omega} \pi\big(x(\cdot), u(\cdot), w(\cdot)\big)\Big\} .$$

- **Quadratic and variance approach.** We consider a quadratic function $L(z) = az - z^2$ to assess the performance of the payoff $\pi$ within a stochastic context:
$$\sup_{u \in \mathcal{U}^{ad}} \mathbb{E}_{w(\cdot)} \left[ L\Big( \pi\big(x(\cdot), u(\cdot), w(\cdot)\big)\Big)\right] .$$
The criterion is connected to mean-variance analysis since, denoting temporarily $\pi = \pi\big(x(\cdot), u(\cdot), w(\cdot)\big)$, we have:
$$\mathbb{E}[L(\pi)] = \mathsf{var}[\pi] + \mathbb{E}[\pi]^2 - a\mathbb{E}[\pi] .$$
Hence, if the expected payoff is a fixed $\mathbb{E}[\pi\big(x(\cdot), u(\cdot), w(\cdot)\big)] = \overline{\pi}$, the problem reads:
$$\inf_{u \in \mathcal{U}^{ad}} \mathsf{var}[\pi\big(x(\cdot), u(\cdot), w(\cdot)\big)] .$$
Since the variance is a well-known measure of dispersion and volatility, its minimization captures the idea of reduction of risks.
- **Multi-prior approach.** It is assumed that different probabilities $\mathbb{P}$, termed as beliefs or priors and belonging to a set $\mathcal{P}$ of admissible probabilities on $\Omega$, are relevant for the uncertain scenarios $w(\cdot) \in \Omega$. The multi-prior approach combines robust and expected criterion by taking the worst beliefs in terms of expected[9] payoff, namely [7]:
$$\sup_{u \in \mathcal{U}^{ad}} \inf_{\mathbb{P} \in \mathcal{P}} \mathbb{E}^{\mathbb{P}} \Big[ \pi\big(x(\cdot), u(\cdot), w(\cdot)\big)\Big] .$$

## 6.5 Management of multi-species harvests

For different species $i = 1, \ldots, n$, the population dynamic $g_i$ in Sect. 2.2 is submitted to random factors $w(\cdot)$ including fluctuations in recruitment and mortality parameters imposed by demographic or environmental changes

$$N_i(t+1) = g_i\big(N(t) - h(t), w(t)\big) ,$$

where $N(t) = \big(N_1(t), \ldots, N_n(t)\big)$ is the vector of abundances and $h(t) = \big(h_1(t), \ldots, h_n(t)\big)$ the vector of catches. All possible ecological scenarios[10] are described by $w(\cdot) \in \Omega$. For aggregated and compact models, instances of such population uncertainties can be depicted by a combination of the uncertain intrinsic growth rate $r_i(t)$ and the carrying capacity $k_i(t)$:

$$N_i(t+1) = g_i\big(N(t) - h(t), r_i(t), k_i(t)\big) .$$

[9] The following notation $\mathbb{E}^{\mathbb{P}}$ stresses the dependency of the mathematical expectation upon the probability $\mathbb{P}$.

[10] Other instances of uncertainty are of an anthropic feature. Uncontrollability of catches suggests uncertain efforts which impact the harvests.

In that case, the uncertain variables for species $i$ is $w_i(t) = (r_i(t), k_i(t))$ which takes its values from some given set $\mathbb{S}_i(t)$.

In this context, robust optimal criteria and worst case approach for discounted profit maximization are given by

$$\sup_{h(\cdot)} \inf_{w(\cdot)\in\Omega} \sum_{t=t_0}^{T-1} \rho^t \sum_{i=1}^{n} \Big(p_i h_i(t) - C_i\big(h_i(t), N_i(t)\big)\Big) ,$$

where $p = (p_1, \ldots, p_n)$ stands for unitary prices and $C_i(h_i(t), N_i(t))$ for costs, $i = 1, \ldots, n$. Similarly, whenever a probability $\mathbb{P}$ equips $\Omega$, optimal expected discounted profit is:

$$\sup_{h(\cdot)} \mathbb{E}_{w(\cdot)} \left[ \sum_{t=t_0}^{T-1} \rho^t \sum_{i=1}^{n} \Big(p_i h_i(t) - C_i\big(h_i(t), N_i(t)\big)\Big) \right] ,$$

Whether diversification of catches among the targeted species is more efficient than specialization in harvesting in one species constitutes a basic issue in such a context.

From a conservation and ecological viewpoint, the regulating agency may also aim at ensuring non extinction of the populations at final time $T$ with a confidence level $\beta$ in the sense that:

$$\mathbb{P}\Big( N_1(T) \geq N_1^\flat, \ldots, N_n(T) \geq N_n^\flat \Big) \geq \beta .$$

This last requirement corresponds to a stochastic viability problem. Chapter 7 gives insights into this issue.

## 6.6 Robust agricultural land-use and diversification

When an environment has unknown features or when it is constantly changing, diversification of production assets can be a way of dealing with uncertainty. Diversification means retaining assets currently thought to be of little value when it is known that circumstances may change and alter that valuation. This issue is well known in finance and portfolio theory and a basic concern for biodiversity and environmental conservation problems. In agriculture, where uncertainty and risk are pervasive, diversification is clearly an important topic.

Here is displayed a dynamic model dealing with wealth allocation strategies through land-use. The modeled system is made up of a finite number of agricultural uses where each use accounts for total household wealth. It takes into account biological processes: each use is characterized by its own return rate which fluctuates with climatic situations. Economic processes appear through decisions of land-use allocation in order to guarantee a final wealth.

The natural productivity of the land-uses described through their biomass $B_i(t)$, $i = 1,\ldots,n$, are represented by an intrinsic growth rate $R_i(w)$ depending on a climatic parameter $w$ as follows

$$B_i(t+1) = R_i\big(w(t)\big)B_i(t) ,$$

where we assume that the climatic parameter $w(t)$ can fluctuate along time within a given set $\mathbb{S}(t)$ (for instance between two extremal values).

If we introduce fixed prices (sales and purchases) $p_i$ for each resource $i$, the wealth of the farm is given by:

$$v(t) = \sum_{i=1}^{n} p_i B_i(t) .$$

The wealth evolution is then described by

$$v(t+1) = v(t)\left(\sum_{i=1}^{n} u_i(t) R_i\big(w(t)\big)\right) ,$$

where

$$u_i(t) := \frac{p_i B_i(t)}{v(t)}$$

stands for the proportion of wealth generated by use $i$ ($\sum_{i=1}^{n} u_i(t) = 1,\ u_i(t) \geq 0$). Consequently, the allocation $u = (u_1,\ldots,u_n) \in \mathcal{S}^n$, belonging to the simplex $\mathcal{S}^n$ of $\mathbb{R}^n$, among the different land-uses appears as a decision variable representing the land-use structure (the "portfolio").

In a viability approach, the farmer may aim at ensuring a minimal wealth at final time $T$ in the following sense:

$$v(T) \geq v^\flat .$$

Another way of formulating the problem is to consider the stochastic optimal problem

$$\max_{u(\cdot)} \mathbb{E}[L(v(t))] ,$$

where $L$ is some utility function linked to risk-aversion [8]. In fact, as explained in the general case, the maximization of utility of the final wealth occurs with respect to feedback strategies. Here again, such models raise the following question: How do mixed and diversified strategies, compared with strategies of specialization in one use, improve overall performance?

## 6.7 Mitigation policies for uncertain carbon dioxyde emissions

Following the stylized model described in Sect. 2.3, we consider a climate-economy system depicted by two variables: the aggregated economic production level, such as gross world product, GWP, denoted by $Q(t)$ and the atmospheric $CO_2$ concentration level, denoted by $M(t)$. The decision variable related to mitigation policy is the emission abatement rate denoted by $a(t)$.

The description of the carbon cycle is similar to [13], namely a highly simple dynamical model

$$M(t+1) = M(t) + \alpha E_{\text{BAU}}(t)\big(1 - a(t)\big) - \delta\big(M(t) - M_{-\infty}\big) , \tag{6.16}$$

where

- $M(t)$ is the $CO_2$ atmospheric concentration, measured in ppm, parts per million (379 ppm in 2005);
- $M_{-\infty}$ is the pre-industrial atmospheric concentration (about 280 ppm);
- $E_{\text{BAU}}(t)$ is the baseline for the $CO_2$ emissions, and is measured in GtC, Gigatonnes of carbon (about 7.2 GtC per year between 2000 and 2005);
- the abatement rate $a(t)$ corresponds to the applied reduction of $CO_2$ emissions level ($0 \leq a(t) \leq 1$);
- the parameter $\alpha$ is a conversion factor from emissions to concentration; $\alpha \approx 0.471$ ppm.GtC$^{-1}$ sums up highly complex physical mechanisms;
- the parameter $\delta$ stands for the natural rate of removal of atmospheric $CO_2$ to unspecified sinks ($\delta \approx 0.01$ year$^{-1}$).

The baseline $E_{\text{BAU}}(t)$ can be taken under the form $E_{\text{BAU}}(t) = \mathfrak{E}_{\text{BAU}}(Q(t))$, where the function $\mathfrak{E}_{\text{BAU}}$ stands for the emissions of $CO_2$ resulting from the economic production $Q$ in a "business as usual" (BAU) scenario and accumulating in the atmosphere.

The global economics dynamic is represented by an uncertain rate of growth $g\big(w(t)\big) \geq 0$ for the aggregated production level $Q(t)$ related to *gross world product*, GWP:

$$Q(t+1) = \Big(1 + g\big(w(t)\big)\Big)Q(t) . \tag{6.17}$$

We consider a physical or environmental requirement through the limitation of concentrations of $CO_2$ below a tolerable threshold $M^\sharp$ at a specified date $T > 0$:

$$M(T) \leq M^\sharp . \tag{6.18}$$

The cost effectiveness problem faced by the social planner is an optimization problem under constraints. It consists in minimizing the expected discounted intertemporal abatement cost $\mathbb{E}[\sum_{t=t_0}^{T-1} \rho^t C\big(a(t), Q(t)\big)]$ while reaching the concentration tolerable window $M(T) \leq M^\sharp$. The parameter $\rho \in [0,1]$ stands for a discount factor. Therefore, the problem can be written as

$$\inf_{a(t_0),\ldots,a(T-1)} \mathbb{E}\left[\sum_{t=t_0}^{T-1} \rho^t C\big(a(t), Q(t)\big)\right] , \tag{6.19}$$

under the dynamic constraints (6.16) and (6.17) and target constraint (6.18).

Some projections are displayed in Fig. 6.1 together with the ceiling target $M^\sharp = 550$ ppm. They are built from SCILAB code 13 with scenarios of economic growth $g\big(w(t)\big)$ following a uniform probability law on $[0\%, 6\%]$. The "business as usual" no abatement path $a_{\mathrm{BAU}}(t) = 0\%$ does not display satisfying concentrations since the ceiling target is exceeded at time $t = 2035$. Another abatement path corresponding here to stationary $a(t) = 90\%$ provides some viable or non viable paths. In the robust sense, this is not a viable reduction strategy.

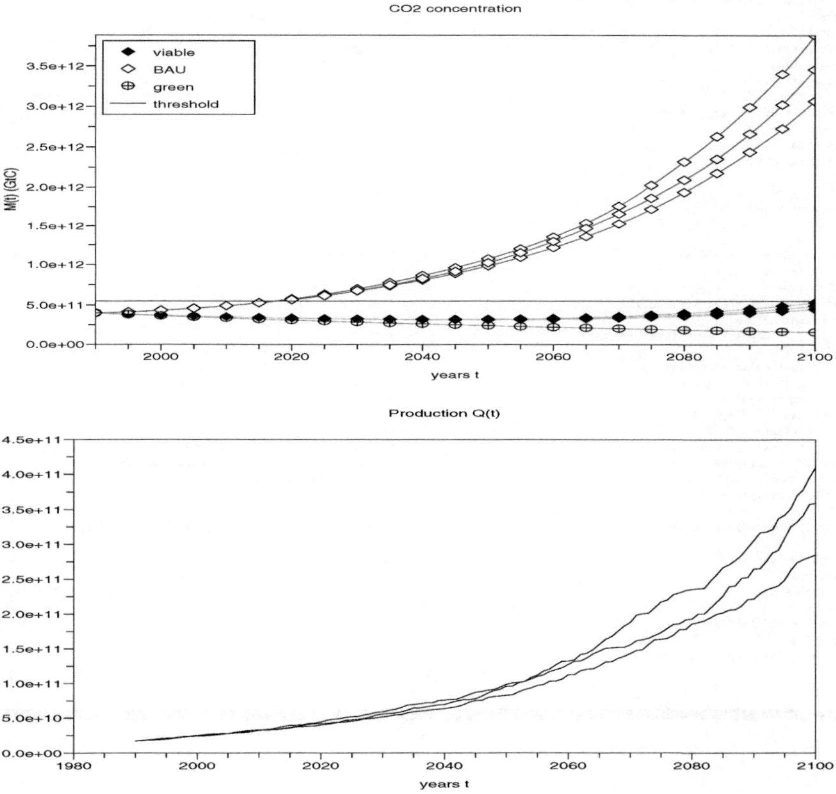


**Fig. 6.1.** Projections of $CO_2$ concentration $M(t)$ and economic production $Q(t)$ at horizon 2100 for different mitigation policies $a(t)$ together with ceiling target $M^\sharp = 550$ ppm in $\diamond$. In black, the non viable 'business as usual' path $a_{\mathrm{BAU}}(t) = 0\%$ and, in blue, a reduction $a(t) = 90\%$. The path in $\oplus$ relies on a total abatment $a(t) = 100\%$.

SCILAB CODE 13.

```
//
// PARAMETERS //

// initial time
t_0=1990;
// final Time
t_F=2100;
// Time step
delta_t=1;

mean_taux_Q=0.03;
sig_taux_Q=0.03;
// mean and standard deviation of growth rate
alphaa=0.64;
// dynamics parameter: atmospheric retention
// (uncertain +- 0.15)
sigma=0.519;
deltaa=1/120;

// concentration target (ppm)
M_sup=550;
M_inf=274;
// Initial conditions
t=t_0;
M0=354; //in (ppm)
M_bau=M0; M_g=M0;
Q0 = 20.9; // in (T US$)

function E=emissions(Q,a)
E = sigma * Q * (1-a);
endfunction
function Mnext=dynamics(t,M,Q,a)
E = emissions(Q,a) ;
Mnext =M  + alphaa* E -deltaa*(M-M_inf);
 endfunction
// Distinct abatment policies
a = 1*ones(1,t_F-t_0+1);
// Strong mitigation
a = 0*ones(1,t_F-t_0+1);
// No mitigation (BAU)
a = 0.9*ones(1,t_F-t_0+1);
// medium mitigation
//a = 1*rand(1,t_F-t_0+1);
// random mitigation

xbasc(1);xbasc(2);xbasc(4)

// System Dynamics
N_simu=3;
for ii=1:N_simu
  t=t_0;
  M=M0; M_bau=M; M_g=M;
  Q = Q0;
  //Initialisation
  L_t=[t_0];
  L_M=[M0]; L_bau=[M0]; L_g=[M0];
  L_Q=[];
  L_E=[]; L_Ebau=[];L_Eg=[];

  for (t=t_0:delta_t:t_F)
g_Q=sig_taux_Q*2*(rand(1,1)-0.5)+mean_taux_Q;
// random growth rate
E = emissions(Q,a(t-t_0+1));
M=dynamics(t,M,Q,a(t-t_0+1));
// Mitigation concentration
   E_bau= emissions(Q,0);
M_bau=dynamics(t,M_bau,Q,0);
//  Business as usual (BAU)
E_g = emissions(Q,1);
M_g=dynamics(t,M_g,Q,1);
// total abatement
Q=(1+g_Q)*Q;
L_Q=[L_Q Q];
L_t=[L_t t+1];
L_M=[L_M M];
L_E=[L_E E];
L_bau=[L_bau M_bau];
L_Ebau=[L_Ebau E_bau];
L_g=[L_g M_g];
L_Eg=[L_Eg E_g];
  end,

// Results printing

long=prod(size(L_t));
step=floor(long/20);
abcisse=1:step:long;
xset("window",1);
plot2d(L_t(abcisse),[L_E(abcisse)' L_Ebau(abcisse)' ...
   L_Eg(abcisse)'],style=-[4,5,3]) ;
legends(["Mitigation";"BAU";"green"],-[4,5,3],'ul');
xtitle('Emissions E(t)','t','E(t) (GtC)');

xset("window",2);
plot2d(L_t(abcisse),[L_M(abcisse)' L_bau(abcisse)' ...
L_g(abcisse)' ones(L_t(abcisse))'*M_sup],...
   style=-[4,5,3,-1], rect=[t_0,0,t_F,2000]) ;
legends(["Mitigation";"BAU";"green";"threshold"],...
     -[4,5,3,-1],'ul');
xtitle('Concentration CO2','t','M(t) (ppm)');

xset("window",4);
plot2d(L_t(abcisse),L_Q(abcisse),style=-[8]);
xtitle('Economie: Production Q(t)','t','Q(t) (T US$)');

end
//
```

## 6.8 Economic growth with an exhaustible natural resource

Let us expand to the uncertain context the stylized model introduced in Sect. (2.7) dealing with an economy exploiting an exhaustible natural resource

$$\begin{cases} S(t+1) = S(t) - r(t) \, , \\ K(t+1) = \Big(1 - \delta\big(w(t)\big)\Big)K(t) + Y\big(K(t), r(t), w(t)\big) - c(t) \, , \end{cases} \tag{6.20}$$

where $S(t)$ is the exhaustible resource stock, $r(t)$ stands for the extraction flow per discrete unit of time, $K(t)$ represents the accumulated capital, $c(t)$ stands for the consumption and the function $Y$ represents the technology of the economy. Parameter $\delta$ is the rate of capital depreciation. The last two parameters are now assumed to be uncertain in the sense that they both depend on some uncertain variable $w(t)$.

The controls of this economy are levels of consumption $c(t)$ and extraction $r(t)$ respectively.

Again state-control constraints can be taken into account. The extraction $r(t)$ is irreversible in the sense that:

$$0 \leq r(t) . \tag{6.21}$$

We take into account the scarcity of the resource by requiring:

$$0 \leq S(t) .$$

More generally, we can consider a stronger conservation constraint for the resource as follows:

$$S^\flat \leq S(t) . \tag{6.22}$$

The threshold $S^\flat > 0$ stands for some guaranteed resource target, referring to a strong sustainability concern whenever it has a strictly positive value.

We assume the investment in the reproducible capital $K$ to be irreversible in the sense that:

$$0 \leq Y\big(K(t), r(t), w(t)\big) - c(t) . \tag{6.23}$$

We also consider that the capital is non negative:

$$0 \leq K(t) . \tag{6.24}$$

A sustainability requirement can be imposed through some guaranteed consumption level $c^\flat$ along the generations:

$$0 < c^\flat \leq c(t) . \tag{6.25}$$

One stochastic optimality problem adapted from [5] is

$$\sup_{c(\cdot), r(\cdot)} \mathbb{E}\left[\sum_{t=t_0}^{+\infty} \rho^t L\big(S(t), c(t), w(t)\big)\right] ,$$

where $\rho \in [0, 1[$ is a discount factor. The uncertainty of the utility function $L(S, c, w)$ captures possible changes in environmental preferences.

A question that arises is whether the uncertainties $w(t)$ affect the consumption paths compared to the certain case.

# References


[1] D. P. Bertsekas. *Dynamic Programming and Optimal Control.* Athena Scientific, Belmont, Massachusets, second edition, 2000. Volumes 1 and 2.

[2] D. P. Bertsekas and S. E. Shreve. *Stochastic Optimal Control: The Discrete-Time Case.* Athena Scientific, Belmont, Massachusets, 1996.

[3] K. W. Byron, J. D. Nichols, and M. J. Conroy. *Analysis and Management of Animal Populations.* Academic Press, 2002.

[4] G. Chichilnisky, G. M. Heal, and A. Vercelli. *Sustainability: Dynamics and Uncertainty (Economics, Energy and Environment).* Springer, 1998.

[5] P. Dasgupta and G. Heal. The optimal depletion of exhaustible resources. *Review of Economic Studies*, 41:1–28, 1974. Symposium on the Economics of Exhaustible Resources.

[6] R. A. Freeman and P. V. Kokotowic. *Robust Nonlinear Control Design (State Space and Lyapunov Techniques).* Birkhäuser, Basel, 1996.

[7] I. Gilboa and D. Schmeidler. Maxmin expected utility with non-unique prior. *Journal of Mathematical Economics*, 18(2):141–153, April 1989.

[8] C. Gollier. *The Economics of Risk and Time.* MIT Press, Cambridge, 2001.

[9] F.H. Knight. *Risk, Uncertainty and Profit.* Houghton Mifflin Company, Boston, 1921.

[10] M. Kot. *Elements of Mathematical Ecology.* Cambridge University Press, Cambridge, 2001.

[11] J.-J. Laffont. *The Economics of Uncertainty and Information.* MIT Press, 1989.

[12] R. Lande, S. Engen, and B.-E. Saether. *Stochastic population dynamics in ecology and conservation.* Oxford series in ecology and evolution, 2003.

[13] W. D. Nordhaus. *Managing the Global Commons.* MIT Press, Cambridge, 1994.

[14] L. J. Olson and R. Santanu. Dynamic efficiency of conservation of renewable resources under uncertainty. *Journal of Economic Theory*, 95:186–214, 2000.

[15] P. Whittle. *Optimization over Time: Dynamic Programming and Stochastic Control*, volume 1. John Wiley & Sons, New York, 1982.

# 7

# Robust and stochastic viability

Vulnerability, risks, safety and precaution constitute major issues in the management of natural resources and sustainability concerns. Regarding these motivations, the role played by the acceptability constraints or targets (population extinction threshold, $CO_2$ ceiling...) is central but has to be articulated with uncertainty. The present chapter addresses the issue of constraints in the uncertain context and expands most of the concepts and mathematical and numerical tools examined in Chap. 4 about viable control.

In the uncertain framework, robust and stochastic approaches deserve to be distinguished although they are not disconnected. On the one hand, constraints "in the robust sense" rely on a worst case approach. The fundamental idea underlying such robust viability is to guarantee the satisfaction of the constraints whatever the uncertainties which may be related to a pessimistic, totally risk averse context [5, 11, 7]. On the other hand, constraints "in the probabilistic or stochastic sense" are basically related to risk assessment and management. Such stochastic viability includes, in the field of conservation biology, the problems and methods of *population viability analysis* (PVA) [2, 8, 9, 10]. The idea of stochastic viability is basically to require the respect of the constraints at a given confidence level (say 90%, 99%). It implicitly assumes that some extreme events render the robust approach irrelevant. The robust approach is closely related to the stochastic one with a confidence level of 100%.

Here, we adapt the notions of viability kernel and viable controls within the probability and robust frameworks. Some mathematical materials of stochastic viability can be found in [1, 3, 4] but they tend to focus on the continuous time case.

The chapter is organised as follows. After briefly introducing the uncertain viability problem in Sect. 7.1, it is detailed in the robust framework in Sect. 7.2, and the stochastic case is treated in Sect. 7.5. We present the dynamic programming method, and different examples illustrate the main statements.

## 7.1 The uncertain viability problem

The state equation introduced in (6.1) as the uncertain dynamic model is considered:

$$x(t+1) = F\big(t, x(t), u(t), w(t)\big) , \quad t = t_0, \ldots, T-1 \quad \text{with} \quad x(t_0) = x_0 . \quad (7.1)$$

Here again, $x(t) \in \mathbb{X} = \mathbb{R}^n$ represents the system state vector at time $t$, $x_0 \in \mathbb{X}$ is the initial condition at initial time $t_0$, $T > t_0$ is the horizon, $u(t) \in \mathbb{U} = \mathbb{R}^p$ represents the decision or control vector while $w(t) \in \mathbb{S}(t) \subset \mathbb{W} = \mathbb{R}^q$ stands for the uncertain variable, or disturbance, noise. Possible scenarios, or paths of uncertainties, $w(\cdot)$ are described by:

$$w(\cdot) \in \Omega := \mathbb{S}(t_0) \times \cdots \times \mathbb{S}(T) \subset \mathbb{W}^{T+1-t_0} .$$

As detailed in Sect. 6.1, the admissibility of decisions and states is restricted by the non empty subset $\mathbb{B}(t, x)$ of admissible controls in $\mathbb{U}$ for all $(t, x)$

$$u(t) \in \mathbb{B}\big(t, x(t)\big) \subset \mathbb{U} , \quad (7.2a)$$

together with a non empty subset $\mathbb{A}(t)$ of the state space $\mathbb{X}$ for all $t$

$$x(t) \in \mathbb{A}(t) \subset \mathbb{X} , \quad (7.2b)$$

and a target

$$x(T) \in \mathbb{A}(T) \subset \mathbb{X} . \quad (7.2c)$$

These control, state or target constraints may reduce the relevant paths of the system. Such a feasibility issue can be addressed in a robust or stochastic framework.

## 7.2 The robust viability problem

Here, we first deal with such a problem in a robust perspective. Namely we consider the admissible feedbacks $\mathfrak{u}$ in $\mathfrak{U}$ defined in (6.11) such that the control and state constraints (7.2a)-(7.2b)-(7.2c) hold true under the dynamics (7.1) *whatever the scenario* $w(\cdot) \in \Omega$. In the sequel, $\mathfrak{U}^{ad}$ is the set of admissible feedbacks as defined in (7.3). The *control constraints case* (7.2a) restricts feedbacks to *admissible feedbacks* as follows:

$$\mathfrak{U}^{ad} = \{\mathfrak{u} \in \mathfrak{U} \mid \mathfrak{u}(t, x) \in \mathbb{B}(t, x) , \quad \forall (t, x)\} . \quad (7.3)$$

### Robust viable controls and states

The viability kernel plays a basic role in viability analysis. In the robust case, it is the set of initial states $x_0$ such that the robust viability property holds true.

**Definition 7.1.** *The* robust viability kernel *at time $t_0$ is the set*

$$\mathbb{V}\mathrm{iab}_1(t_0) := \left\{ x_0 \in \mathbb{X} \;\middle|\; \begin{array}{l} \textit{there exists} \quad \mathfrak{u} \in \mathfrak{U}^{ad} \quad \textit{such that} \\ \textit{for all scenario} \quad w(\cdot) \in \Omega \\ x(t) \in \mathbb{A}(t) \quad \textit{for} \quad t = t_0, \ldots, T \end{array} \right\}, \tag{7.4}$$

*where $x(t)$ corresponds to the state map namely $x(t) = x_F[t_0, x_0, \mathfrak{u}, w(\cdot)](t)$ as defined in Sect. 6.2.*

Notice that the final viability kernel is the whole target set, namely

$$\mathbb{V}\mathrm{iab}_1(T) = \mathbb{A}(T) .$$

*Viable robust feedbacks* are feedbacks $\mathfrak{u} \in \mathfrak{U}^{ad}$ such that the robust viability property occurs.

**Definition 7.2.** *Viable robust feedbacks are defined by*

$$\mathfrak{U}_1^{viab}(t_0, x_0) := \left\{ \mathfrak{u} \in \mathfrak{U}^{ad} \;\middle|\; \begin{array}{c} \textit{for all scenario} \quad w(\cdot) \in \Omega \\ x(t) \in \mathbb{A}(t) \quad \textit{for} \quad t = t_0, \ldots, T \end{array} \right\}, \tag{7.5}$$

*where again $x(t)$ equals $x_F[t_0, x_0, \mathfrak{u}, w(\cdot)](t)$ as defined in Sect. 6.2.*

By definition, the robust iability kernel represents the initial states $x_0$ avoiding the emptiness of the set of robust viable feedbacks $\mathfrak{U}_1^{\mathrm{viab}}(t_0, x_0)$, *i.e.*:

$$x_0 \in \mathbb{V}\mathrm{iab}_1(t_0) \iff \mathfrak{U}_1^{\mathrm{viab}}(t_0, x_0) \neq \emptyset . \tag{7.6}$$

Thus, the viability problem consists in identifying the viability kernel $\mathbb{V}\mathrm{iab}_1(t_0)$ and the set of robust viable feedbacks $\mathfrak{U}_1^{\mathrm{viab}}(t_0, x_0)$.

## Robust dynamic programming equation

A characterization of robust viability in terms of dynamic programming can be exhibited. To achieve this, it is convenient to use the *indicator function*[1] $\mathbf{1}_{\mathbb{A}(t)}$ of the set $\mathbb{A}(t) \subset \mathbb{X}$.

**Definition 7.3.** *The* robust viability value function *or* Bellman function *$V(t, x)$, associated with dynamics (7.1), control constraints (7.2a) state constraints (7.2b) and target constraints (7.2c) is defined by the following backward induction, where $t$ runs from $T-1$ down to $t_0$:*

$$\begin{cases} V(T, x) := \mathbf{1}_{\mathbb{A}(T)}(x) , \\ V(t, x) := \mathbf{1}_{\mathbb{A}(t)}(x) \sup_{u \in \mathbb{B}(t,x)} \inf_{w \in \mathbb{S}(t)} V\big(t+1, F(t, x, u, w)\big) . \end{cases} \tag{7.7}$$

*This is the* robust viability dynamic programming equation *(or* Bellman equation*).*

[1] Recall that the *indicator function* of a set $A$ is defined by $\mathbf{1}_A(x) = 1$ if $x \in A$, and $\mathbf{1}_A(x) = 0$ if $x \notin A$.

Notice that $V(t,x) \in \{0,1\}$. It turns out that the robust viability value function $V(t,\cdot)$ at time $t$ is the indicator function of the robust viability kernel $\mathbb{V}\mathrm{iab}_1(t)$ (Proposition 7.5).

### Viable robust feedbacks

The backward equation of dynamic programming (7.7) makes it possible to define the value function $V(t,x)$ and reveals relevant viable robust feedbacks.

**Definition 7.4.** *For any time $t$ and state $x$, let us define* robust viable controls*:*

$$\mathbb{B}_1^{viab}(t,x) := \{u \in \mathbb{B}(t,x) \mid \forall w \in \mathbb{S}(t)\ ,\ F(t,x,u,w) \in \mathbb{V}\mathrm{iab}_1(t+1)\}\ . \quad (7.8)$$

The proof of the following Proposition 7.5 is given in the Appendix, Sect. A.5.

**Proposition 7.5.** *We have $V(t,x) = \mathbf{1}_{\mathbb{V}\mathrm{iab}_1(t)}(x)$, that is:*

$$V(t,x) = 1 \iff x \in \mathbb{V}\mathrm{iab}_1(t)\ . \quad (7.9)$$

*Robust viable controls exist at time $t$ if and only if the state $x$ belongs to the robust viability kernel at time $t$:*

$$\mathbb{B}_1^{viab}(t,x) \neq \emptyset \iff x \in \mathbb{V}\mathrm{iab}_1(t)\ . \quad (7.10)$$

*A solution to the viability problem is:*

$$\left.\begin{array}{r} x_0 \in \mathbb{V}\mathrm{iab}_1(t_0) \\ \mathfrak{u}(t,x) \in \mathbb{B}_1^{viab}(t,x)\ ,\ \forall t = t_0,\ldots,T-1\ ,\ \forall x \in \mathbb{V}\mathrm{iab}_1(t) \end{array}\right\} \Rightarrow \mathfrak{u} \in \mathfrak{U}_1^{viab}(t_0,x_0)\ .$$

Notice that the state constraints have now disappeared, being incorporated in the new control constraints $u(t) \in \mathbb{B}_1^{\mathrm{viab}}\big(t,x(t)\big)$.

The previous result also provides a geometrical formulation of robust viability dynamic programming since the robust viability kernels satisfy the backward induction, where $t$ runs from $T-1$ down to $t_0$:

$$\left\{\begin{array}{l} \mathbb{V}\mathrm{iab}_1(T) = \mathbb{A}(T)\ , \\ \mathbb{V}\mathrm{iab}_1(t) = \{x \in \mathbb{A}(t) \mid \exists u \in \mathbb{B}(t,x)\ ,\ \forall w \in \mathbb{S}(t)\ , \\ \qquad\qquad F(t,x,u,w) \in \mathbb{V}\mathrm{iab}_1(t+1)\}\ . \end{array}\right. \quad (7.11)$$

## 7.3 Robust agricultural land-use and diversification

Here we cope with the problem already introduced in Sect. 6.6 and inspired by [11]. The annual wealth evolution of the farm is described by

$$v(t+1) = v(t)\left(\sum_{i=1}^{n} u_i(t) R_i\big(w(t)\big)\right) = v(t)\langle u(t), R(w(t))\rangle \ ,$$

where

$$u_i(t) := \frac{p_i B_i(t)}{v(t)}$$

stands for the proportion of wealth generated by use $i$ ($\sum_{i=1}^{n} u_i(t) = 1,\ u_i(t) \geq 0$), and $w(t)$ corresponds to environmental uncertainties evolving in a given domain $\mathbb{S}$. The allocation $u = (u_1, \ldots, u_n) \in \mathcal{S}^n$, belonging to the simplex $\mathcal{S}^n$ of $\mathbb{R}^n$, among the different land-uses appears as a decision variable representing the land-use structure.

The farmer aims at ensuring a minimal wealth at final time $T$:

$$v(T) \geq v^\flat \ .$$

*Viability kernel*

**Result 7.6** *The robust viability kernel turns out to be the set*

$$\mathbb{V}\mathrm{iab}(t) = [v^\flat(t), +\infty[ \ ,$$

*with the viability threshold*

$$v^\flat(t) = \frac{v^\flat}{\Big(\sup_{u\in\mathcal{S}^n} \inf_{w\in\mathbb{S}} \langle u, R(w)\rangle\Big)^{T-t}} \ .$$

To prove such is the case, we reason backward using the dynamic programming method for indicator functions as in (7.7). The value function at final time $T$ is given by:

$$V(T, v) = \mathbf{1}_{[v^\flat, +\infty[}(v) \ .$$

Assume now that, at time $t+1$, the robust viability kernel is $\mathbb{V}\mathrm{iab}(t+1) = [v^\flat(t+1), +\infty[$ with:

$$v^\flat(t+1) = v^\flat (R^\sharp)^{t+1-T} \ , \quad R^\sharp = \sup_{u\in\mathcal{S}^n} \inf_{w\in\mathbb{S}} \langle u, R(w)\rangle \ .$$

Equivalently, the value function is $V(t+1, v) = \mathbf{1}_{[v^\flat(t+1), +\infty[}(v)$. Using the Bellman equation (7.7) for robust viability, we deduce that:

$$\begin{aligned} V(t,\upsilon) &= \sup_{u\in\mathcal{S}^n}\inf_{w\in\mathbb{S}} V\big(t+1,\upsilon(\langle u,R(w)\rangle)\big) \\ &= \sup_{u\in\mathcal{S}^n}\inf_{w\in\mathbb{S}} \mathbf{1}_{[\upsilon^\flat(t+1),+\infty[}\big(\upsilon(\langle u,R(w)\rangle)\big) \\ &= \mathbf{1}_{[\upsilon^\flat(t+1),+\infty[}\big(\upsilon(\sup_{u\in\mathcal{S}^n}\inf_{w\in\mathbb{S}}\langle u,R(w)\rangle)\big) \\ &= \mathbf{1}_{[\upsilon^\flat(t+1),+\infty[}\big(\upsilon R^\sharp\big) \\ &= \mathbf{1}_{[\upsilon^\flat(t+1)(R^\sharp)^{-1},+\infty[}\big(\upsilon\big) \\ &= \mathbf{1}_{[\upsilon^\flat(t),+\infty[}\big(\upsilon\big)\ . \end{aligned}$$

We conclude that $\mathbb{V}\mathrm{iab}(t) = [\upsilon^\flat(t),+\infty[$.

*Specialization versus diversification*

Now we aim at comparing specialized and diversified land-use. The specialized land-use in $i\in\{1,\dots,n\}$ corresponds to:

$$u_i(t)=1\ ,\quad u_j(t)=0\ ,\quad \forall j\neq i\ .$$

Hence we obtain the specialized robust viability kernel:

$$\mathbb{V}\mathrm{iab}^i(t) = [\upsilon_i^\flat(t),+\infty[\quad\text{with}\quad \upsilon_i^\flat(t) = \frac{\upsilon^\flat}{\big(\inf_{w\in\mathbb{S}} R_i(w)\big)^{T-t}}\ .$$

Of course, we have the inclusion $\bigcup_{i=1}^n \mathbb{V}\mathrm{iab}^i(t)\subset\mathbb{V}\mathrm{iab}(t)$. The equality holds true whenever no uncertainty occurs, namely with a fixed $w$. This means that the viability of specialized and diversified strategies coincide in this case. In the uncertain case, however, the equality does not hold in general. If the growth functions are such that

$$\max_{i=1,\dots,n}\inf_{w\in\mathbb{S}} R_i(w) < \sup_{u\in\mathcal{S}^n}\inf_{w\in\mathbb{S}}\langle R(w),u\rangle\ ,$$

then $\bigcup_{i=1}^n \mathbb{V}\mathrm{iab}^i \subsetneq \mathbb{V}\mathrm{iab}$. To capture the difference, it is enough to think of the following example:

$$\begin{cases} R_1(w) = \overline{R} - w\sigma, \\ R_2(w) = \overline{R} + w\sigma, \end{cases} \qquad w\in[-1,1]\ .$$

In this case, the worst growths are:

$$\begin{cases} \displaystyle\inf_{w\in\mathbb{S}} R_i(w) = \overline{R}-\sigma\ , \\ \displaystyle\sup_{u\in\mathcal{S}^n}\inf_{w\in\mathbb{S}}\langle u,R(w)\rangle = \overline{R}\ . \end{cases}$$

Note that the diversified viable land-use is given by $u^\star = (0.5;0.5)$. See Figs. 7.1 built with SCILAB code 14.

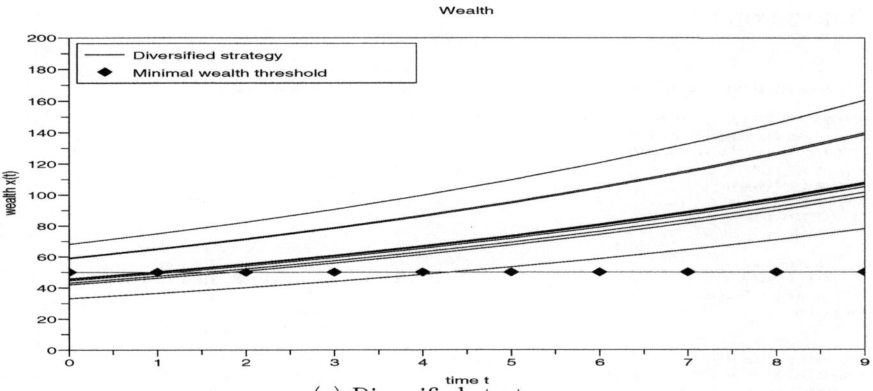


(a) Diversified strategy

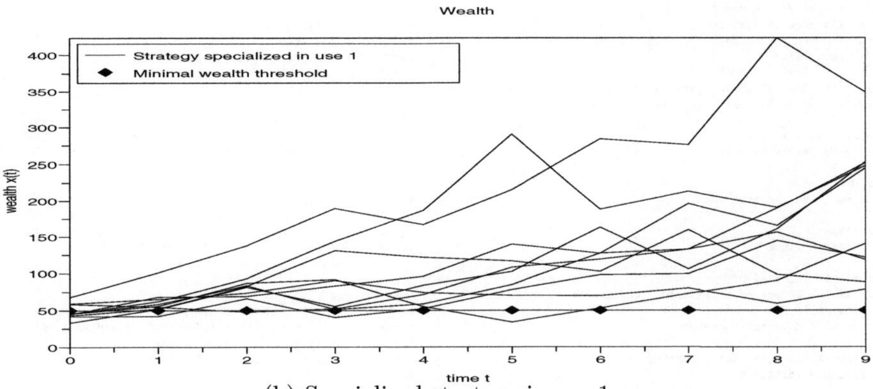


(b) Specialized strategy in use 1

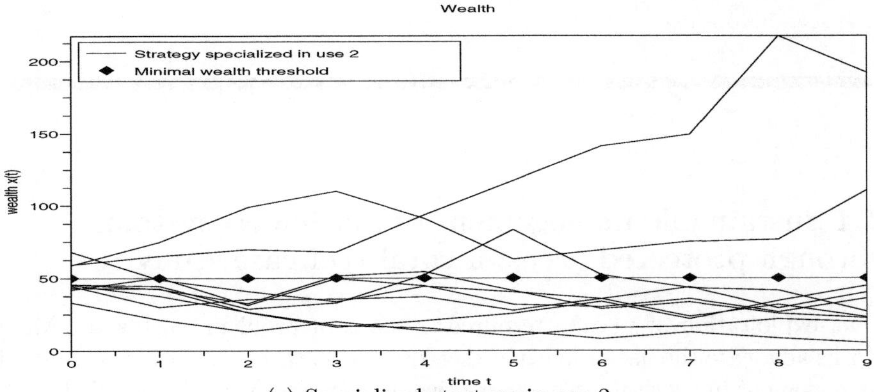


(c) Specialized strategy in use 2

**Fig. 7.1.** Wealth $v(t)$ for diversified $u^\star = (0.5; 0.5)$, specialized $u = (0, 1)$ and $u = (1, 0)$ for different environmental scenarios with time horizon $T = 9$ and $v^\flat = 50$. Many catastrophic scenarios $v(T) < v^\flat$ exist for specialized land-use, while diversified land-use ensures guaranteed wealth $v(T) \geq v^\flat$.

SCILAB CODE 14.

```
//
// exec robust_diversification.sce

////////////////////////////////////////////
//  Land-use and diversification
////////////////////////////////////////////

function [y]=f(x,u,w)
// dynamics
  y=x*(1+u'*retu(w));
endfunction

function r=retu(w)
// return of land-use
  r=[r_0+sig*w;r_0-sig*w];
endfunction


xset("window",0);xbasc();
xtitle("Wealth",'time t','wealth x(t)');
// diversified
xset("window",11);xbasc();
xtitle("Wealth",'time t','wealth x(t)');
// specialized in use 1
xset("window",21);xbasc();
xtitle("Wealth",'time t','wealth x(t)');
// specialized in use 2

xset("window",2);xbasc();xtitle("Allocation",'t','u(t)');
xset("window",3);xbasc();xtitle("Uncertainty",'t','w(t)');

w_min=-1; w_max=1;
// uncertainty margins

r_0=0.1;sig=0.5
// mean and variance return

Horizon=10;
// time horizon

x_min=50; x_max=100;
// wealth bounds // x_min safety constraint
r_star=r_0;
// Viable diversified return

x_prec=x_min/((1+r_star)^(Horizon-1));
// Viable initial state

u_viab=[0.5;0.5]*ones(1,Horizon-1);
// Diversified viable strategy

u_1=[1;0]*ones(1,Horizon-1);
u_2=[0;1]*ones(1,Horizon-1);
// Specialized strategies

N_simu=10;
// Number of simulations

for i=1:N_simu
  // Simulations
  x_viab=x_prec+rand(1,Horizon)*(x_max-x_min);
  x_1=x_viab;
  x_2=x_viab;
  //  precautionnary initial conditions
  // x(0)>= x_min/(1+r_star)^(Horizon-1)
  w_viab=w_min+(w_max-w_min)*rand(1,Horizon-1);
  // Random climate along time
  for (t=1:1:Horizon-1)
// Viable Trajectory x(.) u(.)
    x_viab(:,t+1)=f(x_viab(:,t),u_viab(:,t),w_viab(:,t));
    // Diversified wealth
    x_1(:,t+1)=f(x_1(:,t),u_1(:,t),w_viab(:,t));
    x_2(:,t+1)=f(x_2(:,t),u_2(:,t),w_viab(:,t));
// Specialized wealth
  end
  //
rect1=[0,0,Horizon-1,200];
rect2=[0,0,Horizon-2,1];
rect3=[0,w_min,Horizon-2,w_max];
xx=[0:1:Horizon-1];xu=[0:1:Horizon-2];
//
xset("window",0);
plot2d(xx,[x_viab' x_min+zeros(1,Horizon)' ],rect=rect1);
plot2d(xx,[x_viab' x_min+zeros(1,Horizon)' ],style=[1,-4])
legends(['Diversified strategy';'Minimal wealth threshold'],...
[1,-4],'ul')
//
xset("window",11);
plot2d(xx,[x_1' x_min+zeros(1,Horizon)' ],rect=rect1)
plot2d(xx,[x_1' x_min+zeros(1,Horizon)' ],style=[1,-4])
legends(['Strategy specialized in use 1';...
 'Minimal wealth threshold'],[1,-4],'ul')
//
xset("window",21);
plot2d(xx,[x_2' x_min+zeros(1,Horizon)' ],rect=rect1)
plot2d(xx,[x_2' x_min+zeros(1,Horizon)' ],style=[1,-4])
legends(['Strategy specialized in use 2';...
 'Minimal wealth threshold'],[1,-4],'ul')
//
  xset("window",2);plot2d(xu,u_viab',rect=rect2);
  xset("window",3);plot2d(xu,w_viab',rect=rect3);

end
//
```

## 7.4 Sustainable management of marine ecosystems through protected areas: a coral reef case study

Over-exploitation of marine resources remains a problem worldwide. Many works advocate the use of marine reserves as a central element of future stock management in a sustainable perspective. In the present model detailed in [6], the influence of protected areas upon the sustainability of fisheries within an ecosystemic framework is addressed through a dynamic bioeconomic model integrating a trophic web, catches and environmental uncertainties. The model is spatially implicit. It is inspired from data on the Aboré coral reef reserve in

New Caledonia. The evaluation of the ecosystem is designed through the respect along time of constraints of both conservation and guaranteed captures.

Hereafter, the time unit is assumed to be the day. Four trophic group densities ($g.m^{-2}$) and coral covers (percent) are considered in order to characterize the state of the ecosystem. Piscivores $N_1(t)$ are predators of fish and are often targeted by fishermen. Macrocarnivores $N_2(t)$ feed on macroinvertebrates and a few fish species. Herbivores $N_3(t)$ are represented by Scarus sp parrot fish. Other fish (small) $N_4(t)$ include sedentary and territorial organisms, microcarnivores (17 cm), coral feeders (16 cm) and zooplanktonophages (13 cm). Coral cover is denoted as $y_1(t)$.

*Trophodynamics of the ecosystem*

| | Piscivores (Pi) | Macro carnivores (MC) | Micro carnivores (mC) | Coral feeders (Co) | Herbivores (He) | Microalgae Detritivores (mAD) | Zooplankton feeders (Zoo) |
|---|---|---|---|---|---|---|---|
| GROUP FOR THE MODEL | $N_1$ | $N_2$ | $N_4$ | $N_4$ | $N_3$ | $N_3$ | $N_4$ |
| SPECIES RICHNESS | 46 | 112 | 50 | 26 | 10 | 73 | 54 |
| DIET COMPOSITION (%) | | | | | | | |
| - Nekton | 77 | 10 | 2 | 0 | 0 | 0.1 | 1 |
| - Macroinvertebrates | 21 | 82 | 20 | 2 | 0 | 2 | 1 |
| - Microinvertebrates | 0.3 | 6 | 67 | 11 | 3 | 5 | 6 |
| - Zooplankton | 1 | 0.4 | 3 | 2 | 0 | 3 | 79 |
| - Other plankton | 0 | 0 | 0 | 0 | 0 | 0 | 0.3 |
| - Macroalgae | 0 | 0.3 | 1 | 0 | 66 | 3 | 0.3 |
| - Microalgae | 0 | 1 | 5 | 7 | 28 | 80 | 11 |
| - Coral | 0 | 0.3 | 2 | 77 | 0 | 1 | 0.3 |
| - Detritus | 0 | 0.3 | 1 | 1 | 4 | 6 | 0.2 |
| MAXIMUM ADULT SIZE (CM) | 77 | 38 | 17 | 16 | 39 | 24 | 13 |

**Table 7.1.** Species richness, mean diet composition and adult size

The dynamics of the ecosystem rely on trophic interactions between groups $N_1(t), \ldots, N_4(t)$ and coral $y(t)$ evolutions. Based on diet composition in Table 7.1 and a Lotka-Volterra structure, the dynamic of the trophic groups $i = 1, \ldots, 4$ is summarized in matrix form by

$$N_i(t+1) = N_i(t)\Big(R + \exp\left(y_1^\sharp - y_1(t)\right) Sx(t)\Big)_i ,$$

where $R = (R_1, \ldots, R_4)$ and where the interaction matrix $S$ reads:

$$S = \begin{pmatrix} -0.093 & 0.013 & 0.013 & 0.013 \\ -0.106 & -0.012 & 0.002 & 0.002 \\ -0.076 & -0.01 & 0 & 0 \\ -0.53 & -0.069 & 0 & 0 \end{pmatrix} .$$

The predation intensity depends on coral cover $y_1(t)$ through a refuge effect

$$\exp\left(y_1^\sharp - y_1\right) S_{ij} ,$$

where the refuge parameter $y_1^\sharp$ corresponds to the maximal coral cover defined later in (7.12). The intrinsic growth rate $R_k$ includes mortality and recruitment of each trophic group $k$, independently of interactions with other trophic groups. Computed at the equilibrium, it reads:

$$R = \begin{pmatrix} 0.975 \\ 1.007 \\ 1.008 \\ 1.054 \end{pmatrix} .$$

*Habitat dynamics*

Coral evolution over time is described through the equation:

$$y_1(t+1) = y_1(t) \times \begin{cases} R_{\text{cor}} \left(1 - \dfrac{y_1(t)}{K_{\text{cor}}}\right) & \text{with probability } (1-p) \, , \\ 0.3 & \text{with probability } p \, . \end{cases}$$

- $p$ is the probability of a cyclonic event. In the model, cyclonic events occur randomly with probability $p$ at each time step and bring coral cover to 30% of its previous value. On average, a cyclone happens every 5 to 6 years and setting $p = 1/(6 \times 365)$ corresponds to the present cyclonic situation. We assume that the climatic change scenario corresponds to a rise of 50% in $p$ namely $p = 1/(4 \times 365)$.
- $R_{\text{cor}}$ is the intrinsic productivity at low cover levels. After a cyclonic event, the coral grows by 10% a year but not linearly: it takes 8 to 10 years to reach the initial cover. Simulations show that $R_{\text{cor}} = 1.002$ is a plausible value in this respect (recall that the time unit is assumed to be the day).
- $K_{\text{cor}}$ is related to the so-called *carrying capacity* $\bar{y}_1$ solution of:

$$1 = R_{\text{cor}} \left(1 - \frac{\bar{y}_1}{K_{\text{cor}}}\right) .$$

  We identify the maximal value of 80% with the carrying capacity:

$$y_1^\sharp = \frac{R_{\text{cor}} - 1}{R_{\text{cor}}} K_{\text{cor}} = 0.8 \, . \tag{7.12}$$

*Exploited dynamics with a protected area*

For the area under study, fishing is basically recreational and associated mainly with spear gun technology. It is assumed to affect only piscivores $N_1(t)$, carnivores $N_2(t)$ and (large) herbivores $N_3(t)$. Assuming a simple Gordon Schaefer production function where $e(t)$ is the fishing effort, we write

$$h_i(t) = q_i e(t) N_i(t)$$

with zero catchability $q_4 = 0$. Unfortunately, quantitative information on catches and effort in the area is not available. To overcome this difficulty, we impose some simplifications hereafter. We first assume that the effort rate in the overall area is targeted at some fixed level $e$:

$$e(t) = e\ , \quad t = t_0, \ldots, T-1\ .$$

We further assume that catchabilities are equal for each fished group in the sense:

$$q_1 = q_2 = q_3\ .$$

However we do not specify $e$ and we study the results for a range $e \in [0, 1]$.

Taking into account the protected area, it is assumed that only a part of the stock is available for fishing. In other words, catches are defined by

$$h_i(t) = q_i e(t)(1 - \text{MPA}) N_i(t)\ ,$$

where MPA is the proportion of the zone closed to fishing.

*A direct use value*

We assume that the ecosystem provides direct uses through harvests of predators $N_1$ and $N_2$ and herbivores $N_3$. The direct use $L$ is assumed to take the form of total catch in weight

$$L(h_1, h_2, h_3) = \upsilon_1 h_1 + \upsilon_2 h_2 + \upsilon_3 h_3\ , \tag{7.13}$$

where $\upsilon_i$ stands for the mean weight of group $i$. Weight values for each group are given by $\upsilon = (0.5 \quad 0.5 \quad 0.7 \quad 0.1)$ in kg. The direct use constraint reads

$$L\big(h_1(t), h_2(t), h_3(t)\big) \geq L^\flat\ , \tag{7.14}$$

where $L^\flat > 0$ stands for some guaranteed satisfaction level.

*A stronger conservation constraint*

We adopt a stronger conservation point of view and introduce a biodiversity constraint in the sense that trophic richness is guaranteed at a level $\mathcal{B}^\flat$:

$$\mathcal{B}(N(t)) = \sum_{i=1}^{4} \mathbf{1}_{\{N_i(t) > 0\}} \geq \mathcal{B}^\flat\ . \tag{7.15}$$

This guaranteed trophic threshold $\mathcal{B}^\flat$ which takes its values from $\{1, 2, 3, 4\}$ ensures a minimal number of non exhausted groups.

The indicator of robust viability (co-viability) is given by the robust viability kernel[2] $\mathbb{V}$iab defined by:

$$\mathbb{V}\text{iab}_{(A, L^\flat, \mathcal{B}^\flat)} = \left\{ (N(t_0), y(t_0)) \;\middle|\; \begin{array}{l} \sup_{e(\cdot)} \mathbb{P}\big(\ (N(t), h(t)) \text{ satisfies } (7.14), (7.15)\ , \\ \qquad\qquad t = t_0, \ldots, T-1\big) \geq 1 \end{array} \right\}.$$

A protection effect should capture processes through which both the conservation and catch services are enhanced by the existence of a reserve.

[2] See the following footnote 3 in this Chapter to explain the presence of $\mathbb{P}$ in the robust viability kernel.

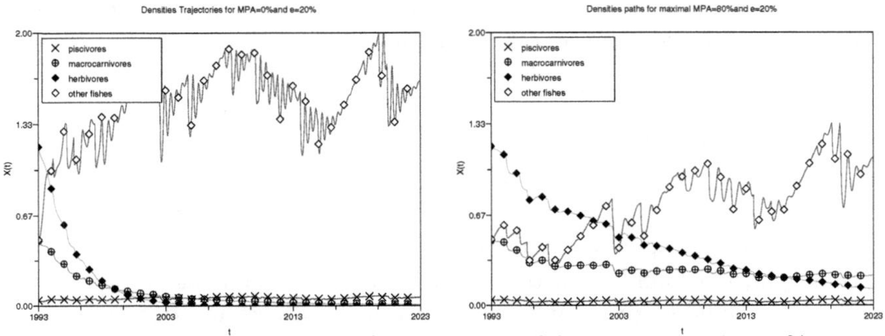


(a) Without reserve MPA= 0% (b) Optimal MPA= 80%

**Fig. 7.2.** Projections of state $(N(t), y(t))$ in the case of current cyclonic frequency $p = 1/(6 \times 365)$ for the moderate fishing effort $e = 20\%$. In (a) without reserve, *i.e.* MPA $= 0\%$, no guaranteed capture of carnivores is achieved while in (b) under the maximal reserve size MPA $= 80\%$, guaranteed utility of catch resulting from all trophic groups is exhibited.

*Results*

The results are based on simulations using the scientific software SCILAB. The initial time corresponds to year $t_0 = 1993$ and the time horizon is set to 30 years ahead, namely $T = t_0 + 31$. The initial state conditions derived from data of 1993 are:

$$N(t_0) = (0.04 \quad 0.48 \quad 1.17 \quad 0.49)\; g.m^{-2} \,.$$

We also set the initial habitat state $y_1(t_0)$ at equilibrium in the sense that:

$$y(t_0) = \bar{y}_1 = 0.8 = 80\% \,.$$

We assume that $p = 1/(6 \times 365)$ which is a current estimation of cyclonic probability by day. A catch reserve effect is obtained as displayed in Figs 7.2.

It turns out that a catch reserve effect is significant for low positive exploitation rates $e$ namely $10\% \leq e \leq 60\%$. In Fig. 7.2(a), without reserve (MPA $= 0\%$), carnivores and herbivores are depleted because of fishing. In other words, for some guaranteed capture level $L_0^\flat > 0$:

$$(N(t_0), y(t_0)) \in \mathbb{V}\mathrm{iab}_{(0, L_0^\flat, 2)} \,.$$

In Fig. 7.2(b) the maximal reserve size (MPA $= 80\%$) provides a larger guaranteed utility of captures $L^\flat > L_0^\flat$ resulting from every targeted trophic group including carnivores, piscivores and herbivores. In other words:

$$(N(t_0), y(t_0)) \in \mathbb{V}\mathrm{iab}_{(0.80, L^\flat, 4)} \,.$$

In this sense, a MPA catch effect is combined with a MPA biodiversity effect. Thus, in this case, catch reserve effects are compatible with biodiversity performance.

## 7.5 The stochastic viability problem

Here we address the issue of state constraints in the probabilistic sense. This is basically related to *risk assessment* which includes, in the field of conservation biology, the problems and methods of *population viability analysis* (PVA) and requires some specific tools inspired by the viability and invariance approach already exposed for the certain case in Chap. 4. In particular, within the probabilistic framework, we adapt the notions of viability kernel and viable controls. In the robust setting, the state constraints (7.2b)-(7.2c) were assumed to hold *whatever* the disturbances. In the probabilistic setting, one can relax the previous requirement by satisfying the state constraints along time with a given confidence level:

$$\mathbb{P}\Big(w(\cdot) \in \Omega \mid x(t) \in \mathbb{A}(t) \quad \text{for} \quad t = t_0, \dots, T\Big) \geq \beta \,.$$

### Stochastic viable controls and state

Probabilistic notations and assumptions are detailed in Sect. 6.3. *In the stochastic setting, all the objects considered will be implicitly equipped with appropriate measurability properties.* Thus, for instance, $\mathfrak{U}^{ad}$ as defined in (7.3) is now the set of measurable admissible feedbacks. The viability kernel plays a basic role in the viability analysis. In the stochastic case, it is the set of initial states $x_0$ such that the stochastic viability property holds true.

For sake of simplicity, the primitive random process $w(\cdot)$ is assumed to be a sequence of independent identically distributed (i.i.d.) random variables $\big(w(t_0), w(t_0+1), \dots, w(T-1), w(T)\big)$ under probability $\mathbb{P}$ on the domain of scenarios $\Omega = \mathbb{S}^{T+1-t_0}$.

**Definition 7.7.** *The* stochastic viability kernel at confidence level $\beta \in [0,1]$ *is*[3]

$$\mathbb{V}\mathrm{iab}_\beta(t_0) := \left\{ x_0 \in \mathbb{X} \;\middle|\; \begin{array}{l} \textit{there exists} \quad \mathfrak{u} \in \mathfrak{U}^{ad} \quad \textit{such that} \\ \mathbb{P}\Big(w(\cdot) \in \Omega \mid x(t) \in \mathbb{A}(t) \quad \textit{for} \quad t = t_0, \dots, T\Big) \geq \beta \end{array} \right\} \tag{7.16}$$

*where $x(t)$ corresponds to the solution map $x(t) = x_F[t_0, x_0, \mathfrak{u}, w(\cdot)](t)$ defined in Sect. 6.2.*

Stochastic viable feedbacks are feedback controls that allow the stochastic viability property to hold true.

---

[3] Notice that the notation $\mathbb{V}\mathrm{iab}_1(t_0)$ is consistent with that of the robust kernel in (7.4) when $\Omega$ is countable and that every scenario $w(\cdot)$ has strictly positive probability under $\mathbb{P}$.

**Definition 7.8.** Stochastic viable feedbacks *are those* $\mathfrak{u} \in \mathfrak{U}^{ad}$ *for which the above relations hold true*[4]

$$\mathfrak{U}_\beta^{viab}(t_0, x_0) := \left\{ \mathfrak{u} \in \mathfrak{U}^{ad} \,\middle|\, \mathbb{P}\Big( w(\cdot) \in \Omega \mid x(t) \in \mathbb{A}(t) \text{ for } t = t_0, \dots, T \Big) \geq \beta \right\}, \tag{7.17}$$

*where* $x(t)$ *corresponds to the solution map* $x(t) = x_F[t_0, x_0, \mathfrak{u}, w(\cdot)](t)$ *defined in Sect. 6.2.*

Similarly to the robust case, we have the following strong link between viable stochastic feedbacks and the viability kernel:

$$x_0 \in \mathbb{V}\mathrm{iab}_\beta(t_0) \iff \mathfrak{U}_\beta^{\mathrm{viab}}(t_0, x_0) \neq \emptyset \, .$$

### Stochastic dynamic programming equation

Consider state and target constraints as in (7.2b) and (7.2c). The stochastic viability value function is the maximal viability probability defined as follows.

**Definition 7.9.** *The* stochastic viability value function *or* Bellman function $V(t, x)$, *associated with dynamics (7.1), control constraints (7.2a) state constraints (7.2b) and target constraints (7.2c) is defined by the following backward induction*[5], *where* $t$ *runs from* $T-1$ *down to* $t_0$*:*

$$\begin{cases} V(T, x) := \mathbf{1}_{\mathbb{A}(T)}(x) \, , \\ V(t, x) := \mathbf{1}_{\mathbb{A}(t)}(x) \sup\limits_{u \in \mathbb{B}(t,x)} \mathbb{E}_{w(t)} \Big[ V\Big(t+1, F\big(t, x, u, w(t)\big)\Big) \Big] \, . \end{cases} \tag{7.18}$$

### Stochastic viable feedbacks

The backward equation of dynamic programming (7.18) makes it possible to define the value function $V(t, x)$. It turns out that the stochastic viability functions are related to the stochastic viability kernels, and that dynamic programming induction reveals relevant stochastic feedback controls. The proof of the following Proposition 7.10 is given in the Appendix, Sect. A.5.

**Proposition 7.10.** *Assume that the primitive random process* $w(\cdot)$ *is made of independent and identically distributed (i.i.d.) random variables* $\big(w(t_0), w(t_0+$

---

[4] See the previous footnote 3 in this Chapter for the case $\beta = 1$.

[5] All random variables $w(t)$ have the same distribution $\mu$ and take values from the same domain $\mathbb{S}$. Hence, we have the formula $\mathbb{E}_{w(t)}[V(t+1, F(t, x, u, w(t)))] = \int_{\mathbb{S}} V(t+1, F(t, x, u, w)) d\mu(w)$.

$1), \ldots, w(T-1), w(T)\big)$. *The viability kernel at confidence level* $\beta$ *is the section of level* $\beta$ *of the stochastic value function:*

$$V(t_0, x_0) \geq \beta \iff x_0 \in \mathbb{V}\mathrm{iab}_\beta(t_0) \, . \tag{7.19}$$

*For any time* $t$ *and state* $x$*, let us assume that*

$$\mathbb{B}^{viab}(t,x) := \operatorname*{arg\,max}_{u \in \mathbb{B}(t,x)} \Big( \mathbf{1}_{\mathbb{A}(t)}(x) \mathbb{E}_{w(t)} \Big[ V\Big(t+1, F\big(t,x,u,w(t)\big)\Big) \Big] \Big) \tag{7.20}$$

*is not empty. Then, any* $\mathfrak{u}^\star \in \mathfrak{U}$ *such that* $\mathfrak{u}^\star(t,x) \in \mathbb{B}^{viab}(t,x)$ *belongs to* $\mathfrak{U}^{viab}_\beta(t_0, x_0)$ *for* $x_0 \in \mathbb{V}\mathrm{iab}_\beta(t_0)$.

## 7.6 From PVA to CVA

Population viability analysis (PVA) is a process of identifying the threats faced by a species and evaluating the likelihood that it will persist for a given time into the future [2, 8, 9, 10]. Population viability analysis is often oriented towards the conservation and management of rare and threatened species, with the goal of applying the principles of population ecology to improve their chances of survival. Threatened species management has two broad objectives. The short term objective is to minimize the risk of extinction. The longer term objective is to promote conditions under which species retain their potential for evolutionary change without intensive management. Within this context, PVA may be used to address three aspects of threatened species management.

- Planning research and data collection. PVA may reveal that population viability is insensitive to particular parameters. Research may be guided by targeting factors that may have an important impact on extinction probabilities or on the rank order of management options.
- Assessing vulnerability. Together with cultural priorities, economic imperatives and taxonomic uniqueness, PVA may be used to set policies and priorities for allocating scarce conservation resources.
- Ranking management options. PVA may be used to predict the likely response of species to reintroduction, captive breeding, prescribed burning, weed control, habitat rehabilitation, or different designs for nature reserves or corridor networks.

We here advocate the use of a CVA (Co-Viability Analysis) approach combining PVA and viable control frameworks. We consider a model similar to that of agricultural land-use introduced in Sect. 6.6 and perform a stochastic analysis. A population abundance $N(t)$ evolves according to

$$N(t+1) = \Big(1 + r\big(N(t), w(t)\big)\Big)\Big(N(t) - h(t)\Big) \, ,$$

where $h(t)$ are catches of the resource and $r\big(N(t), w(t)\big)$ the uncertain growth rate of the population. A probabilistic structure on growth rate $r\big(N(t), w(t)\big)$ is assumed with $w(t)$ including both environmental and demographic stochasticity. Environmental stochasticity causes $r$ to fluctuate randomly in time with mean $\overline{r}$ and variance $\sigma_e^2$. Demographic (individual) stochasticity is characterized by the variance in individual fitness $\sigma_d^2$, so that the total variance in mean fitness or population growth $r\big(N(t), w(t)\big)$ is $\sigma^2(N(t)) = \sigma_e^2 + \frac{\sigma_d^2}{N(t)}$. We write this as

$$r(N(t), w(t)) = \overline{r} + \sqrt{\sigma_e^2 + \frac{\sigma_d^2}{N(t)}} w(t) ,$$

where $w(\cdot)$ is an i.i.d. sequence of random variables.

The regulating agency aims at ensuring a minimal catch $h^\flat$ at every time

$$h(t) \geq h^\flat ,$$

together with a precautionary stock $N^\flat$ at final time $T$ with a confidence level $\beta$ in the sense that:

$$\mathbb{P}\big(N(T) \geq N^\flat\big) \geq \beta .$$

Assume now that the uncertainty is specified by the following distribution:

$$w(t) = \begin{cases} -1 & \text{with probability} \quad p = 0.5 , \\ 1 & \text{with probability} \quad 1 - p = 0.5 . \end{cases}$$

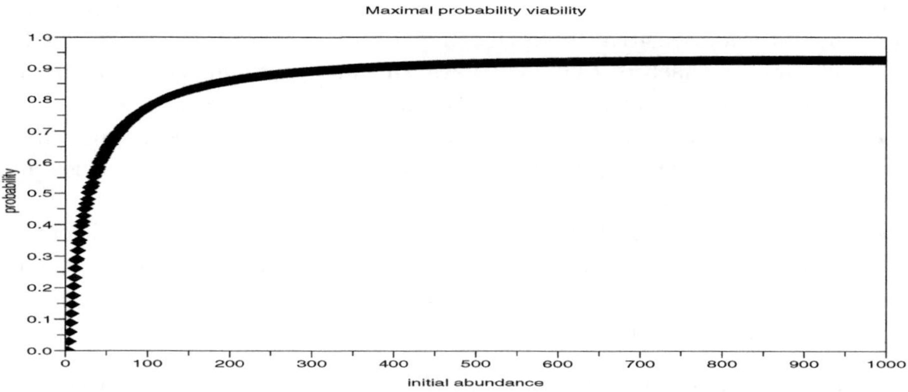


**Fig. 7.3.** Maximal viability probability $V(0, N_0)$ as a function of population $N_0$. The guaranteed catches and population are $h^\flat = 0.2$ and $N^\flat = 1$ respectively. The time horizon is set to $T = 250$. The stochastic viability kernel is $\mathbb{V}\mathrm{iab}_\beta(0) = \{N_0,\ V(0, N_0) \geq \beta\} = [N_\beta, +\infty[$.

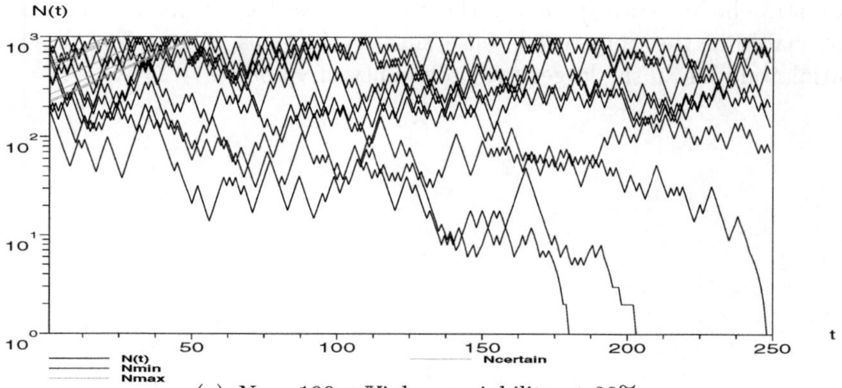


(a) $N_0 = 100 \in \mathbb{V}\mathrm{iab}_{80\%}$: viability at 80%.

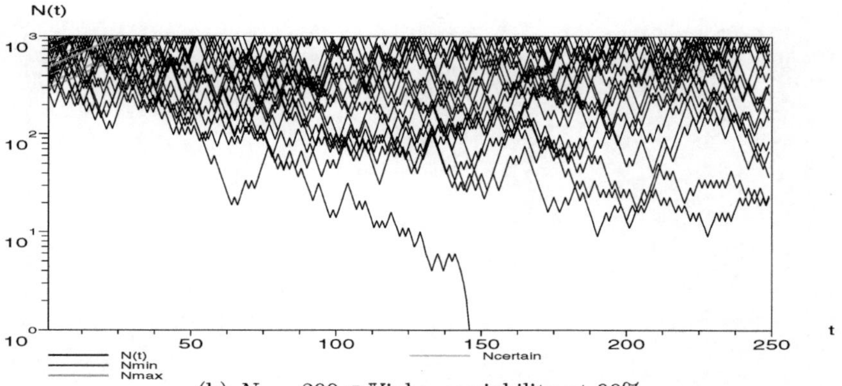


(b) $N_0 = 200 \in \mathbb{V}\mathrm{iab}_{90\%}$: viability at 90%

**Fig. 7.4.** Ten simulated trajectories $N(t)$ of a population with guaranteed catches $h^\flat = 0.2$ for two initial sizes of $N_0 \in \mathbb{V}\mathrm{iab}_{80\%}$ (a) and $N_0 \in \mathbb{V}\mathrm{iab}_{90\%}$ (b) individuals subject to stochastic and environmental stochasticity $w(t)$. The dynamic parameters are mean growth $\overline{r} = 3\%$, demographic $\sigma_d = 10\%$ and environmental $\sigma_e = 20\%$ standard deviations. The straight line corresponds to the certain trajectory. The Figures are generated with SCILAB code 15.

The Fig. 7.3, built with SCILAB code 15, plots the maximal co-viability probability $V(0, N_0)$ with guaranteed catch $h^\flat = 0.2$ as a function of population $N_0$. Not surprisingly, the viability probability $V(0, N_0)$ rises with confidence level $\beta$ while the stochastic viability kernel $\mathbb{V}\mathrm{iab}_\beta(0) = \{N_0,\ V(0, N_0) \geq \beta\} = [N_\beta, +\infty[$ decreases with it.

In Figs. 7.4, we exhibit simulated trajectories $N(t)$ of a population with an initial size of $N_0 = 100 \in \mathbb{V}\mathrm{iab}_{80\%}(0)$ (a) or $N_0 = 200 \in \mathbb{V}\mathrm{iab}_{90\%}(0)$ (b) individuals subject to stochastic and environmental stochasticity $w(\cdot)$ for guaranteed catch $h(t) \geq 0.2$. The dynamic parameters are mean growth $\overline{r} = 3\%$, demographic $\sigma_d = 10\%$ and environmental $\sigma_e = 20\%$ standard deviations.

The straight line corresponds to the certain trajectory. Some trajectories are not viable in the population sense since extinctions occur. The larger the initial abundance, the larger the probability of viability.

SCILAB CODE 15.

```
//
// exec proba_extinction.sce

// demographic stochasticity parameters
sig_e=0.2; sig_d=0.1; rbar=0.03;

// environmental stochasticity parameters
w_min=-1;w_max=1;p=0.5;q=1-p;

// Ricker dynamics parameters
Cap=0;
// carrying capacity

// constraints thresholds
N_min=1; N_max=10^3;
h_min=0.2;h_max=0.2;

// time horizon
Horizon=250 ;

function y=dynpop(x,w)
  y=zeros(x);
  z=x(x>=N_min);
  y(x>=N_min)= z.*(1+rbar+((sig_e^2+sig_d^2 ./z)^0.5).*
               w(x>=N_min));
// zero when x < N_min
endfunction

function y=dyna_exp(x,h,w)
y=dynpop(x-h,w);
endfunction

// Indicator function of state constraint
function y=Ind_x(t,x)
  y=bool2s(x>=N_min);
endfunction

// Characteristic function of control constraint (approximate)
function [y]=Phi_u(t,x,h)
  SGN = bool2s(h_min <= h) ;
  y=0*SGN + 1/%eps *(1-SGN) ;
endfunction

/////////////////////////////////////
// State and control Discretization
/////////////////////////////////////

// Grid state x
x_min=0; x_max=N_max;delta_x=1;
grille_x=x_min:delta_x:x_max;
S=size(grille_x);
NN=S(2);

// Grid control
u_min=0; u_max=h_max;delta_u=0.05;
u=u_min:delta_u:u_max;
R=size(u); MM=R(2);

function [z]=Projbis(x)
z=round(x./delta_x).*delta_x;
z=min(z,x_max);
z=max(z,x_min);
endfunction

function i=Indice(x)
i=int((x-x_min)./delta_x)+1;
endfunction

// Discretized dynamics
function [z]=dyna(x,h,w)
xsuiv=dynpop(x-h,w);
P=Projbis(xsuiv);
z=Indice(P);
endfunction

for (ii=1:NN)
Etat_x(ii)=x_min+(ii-1)*delta_x;
end;

for (ii=1:MM)
Control_u(ii)=u_min+(ii-1)*delta_u;
end;

/////////////////
// Graphics
//////////////////

xset("window",1);xbasc(1);
xtitle("Maximal viability probability",...
   'x','Max_{h}P(N(T)>=N_min)');

xset("window",2); xbasc(2);
xtitle("Population",'time t','N(t)');

xset("window",3); xbasc(3);
xtitle("Catches",'time t','h(t)');

//////////////////////
//   Dynamic programming
//////////////////////

x=grille_x;
W=zeros(Horizon,NN);

// Initialization at horizon T

for (i=1:NN)
W(Horizon,i)=Ind_x(Horizon,Etat_x(i));
end

// Bellman equation
for(t=Horizon-1:-1:1)
for (i=1:NN)
xx=Etat_x(i);
for (j=MM:-1:1)
uu=Control_u(j);
g_min(j)=-Phi_u(t,xx,uu)+W(t+1,dyna(xx,uu,w_min));
g_max(j)=-Phi_u(t,xx,uu)+W(t+1,dyna(xx,uu,w_max));
  g(j)=p*g_min(j)+(1-p)*g_max(j); // expected value
end,
[Vopt,jopt]=max(g) ;
W(t,i)=Vopt;
j_opt(t,i)=jopt;
end,
end,

// Viability kernel
Viab=W(1,:)';K=1-W(Horizon,:)';
xset("window",1);xbasc(1);
plot2d(Etat_x,Viab,-4,rect=[0,0,x_max,1]);
xtitle("Maximal probability viability","initial abundance",...
   "probability");
// legends("Maximal probability viability",-4,"lr");

//////////////////////////////////////////////////
// Simulations
//////////////////////////////////////////////////

// N0=200;
threshold_viab=0.8

if (max(W(1,:)) >= threshold_viab) then
for (k=1:10)
i(1)=int(rand(1)*(NN-1))+1;compt=0;
while(W(1,i(1))<threshold_viab )
then i(1)=int(rand(1)*(NN-1))+1;end,
// find viable state x(i(1))
traj_x(1)=Etat_x(i(1));
//traj_x(1)=N0; i(1)=Indice(N0);
for (t=1:Horizon-1)
if (rand(1)<p) then w(t)=w_min; else w(t)=w_max; end,
h_viab(t)=Control_u(j_opt(t,i(t)));
i(t+1)=dyna(traj_x(t),h_viab(t),w(t));
traj_x(t+1)=Etat_x(i(t+1));
end,
Tempx=(1:Horizon)-1;
Nmin=zeros(1,Horizon)'+N_min;
Nmax=zeros(1,Horizon)'+N_max;
NLin=traj_x(1)*(1+rbar).^Tempx';
//
xset("window",2);plot2d(Tempx,[traj_x Nmin Nmax NLin],...
[1,2,3,4],leg="N(t)@Nmin@Nmax@Ncertain",...
rect=[1,N_min,Horizon,N_max]);
xset("window",2);plot2d(Tempx,[traj_x Nmin Nmax NLin],...
rect=[1,N_min,Horizon,N_max]);
plot2d(Tempx,[traj_x Nmin Nmax NLin],style=-[1,2,3,4])
legends(["N(t)";"Nmin";"Nmax";"Ncertain"],-[1,2,3,4])
// STOP...
Temps=1:Horizon-1;
xset("window",3);plot2d(Temps,h_viab,...
rect=[1,u_min,Horizon,u_max]);

end,
else " beta kernel is empty "
end
//
```

# References


[1] J.-P Aubin and G. Da Prato. The viability theorem for stochastic differential inclusions. *Stochastic Analysis and Applications*, 16:1–15, 1998.

[2] S. R. Beissinger and M. I. Westphal. On the use of demographic models of population viability in endangered species management. *Journal of Wildlife Management*, 62(3):821–841, 1998.

[3] R. Buckdahn, M. Quincampoix, C. Rainer, and A. Rascanu. Stochastic control with exit time and contraints, application to small time attainability of sets. *Applied Mathematics and Optimization*, 49:99–112, 2004.

[4] L. Doyen. Guaranteed output feed-back control for nonlinear uncertain systems under state and control constraints. *Journal of Set-valued Analysis*, 8:149–162, 2000.

[5] L. Doyen and C. Béné. Sustainability of fisheries through marine reserves: a robust modeling analysis. *Journal of Environmental Management*, 69:1–13, 2003.

[6] L. Doyen, M. De Lara, J. Ferraris, and D. Pelletier. Sustainability of exploited marine ecosystems through protected areas: a viability model and a coral reef case study. *Ecological Modelling*, 208(2-4):353–366, November 2007.

[7] L. Doyen and J.-C. Pereau. The precautionary principle as a robust cost-effectiveness problem. *Environmental Modeling and Assessment*, revision.

[8] R. Ferrière, F. Sarrazin, S. Legendre, and J.-P. Baron. Matrix population models applied to viability analysis and conservation: theory and practice using the ULM software. *Acta Oecologica*, 17(6):629–656, 1996.

[9] R. Lande, S. Engen, and B.-E. Saether. *Stochastic population dynamics in ecology and conservation.* Oxford series in ecology and evolution, 2003.

[10] W. F. Morris and D. F. Doak. *Quantitative Conservation Biology: Theory and Practice of Population Viability Analysis.* Sinauer Associates, 2003.

[11] M. Tichit, B. Hubert, L. Doyen, and D. Genin. A viability model to assess the sustainability of mixed herd under climatic uncertainty. *Animal Research*, 53(5):405–417, 2004.

# 8

# Robust and stochastic optimization

In Chap. 7, viability issues were addressed in the uncertain context including both robust and stochastic cases. The present chapter focuses on the optimality process involving worst case or expected performance. For the uncertain case, material for the optimal management or conservation of natural resource and bioeconomic modeling can be found in [4] together with [3]. Stochastic optimality approaches to address the sustainability issues and, especially intergenerational equity and conservation issues, are proposed for instance in [5] including in particular the maximin, Green Golden and Chichilnisky approaches.

Again, it is worth noting that dynamic programming is a relevant method in the uncertain context in the sense that it is well-suited to both robust and stochastic problems. We have already seen in the certain case, that the so-called Bellman's principle expresses the fact that every "subpolicy" of an optimal policy remains optimal along the optimal trajectories. Such principle makes it possible to split a dynamic optimization problem into a sequence of static optimization problem interrelated by backward induction. We detail this process and these general ideas hereafter in the robust and stochastic cases, when the criterion is additive [1, 2, 7].

The chapter is organised as follows. After briefly introducing the criterion in the uncertain framework in Sect. 8.1, the robust optimality problem is presented in Sect. 8.2. The robust additive payoff case is treated with the dynamic programming method in Sect. 8.3, as well as the robust "maximin" approach in Sect. 8.5. The stochastic optimality problem is detailed in Sect. 8.6. Examples from natural resource modeling illustrate the abstract results and concepts. In particular, it is shown how some qualitative results of the certain case can be expanded to the uncertain framework using the certainty equivalent.

## 8.1 Dynamics, constraints, feedbacks and criteria

We briefly review some basic ingredients and conclude with the evaluation of the criterion along controlled trajectories.

### Dynamics

The state equation introduced in (6.1) as the uncertain dynamic model is considered

$$x(t+1) = F\big(t, x(t), u(t), w(t)\big)\ , \quad t = t_0, \ldots, T-1 \quad \text{with} \quad x(t_0) = x_0 \quad (8.1)$$

where again $x(t) \in \mathbb{X} = \mathbb{R}^n$ represents the system state vector at time $t \in \mathbb{N}$, $x_0 \in \mathbb{X}$ is the initial condition or initial state at initial time $t_0$, $T > t_0$ is the horizon, $u(t) \in \mathbb{U} = \mathbb{R}^p$ represents the decision or the control vector while $w(t) \in \mathbb{S}(t) \subset \mathbb{W} = \mathbb{R}^q$ stands for the uncertain variable, or disturbance, noise. Possible scenarios, or paths of uncertainties, $w(\cdot)$ are described by:

$$w(\cdot) \in \Omega := \mathbb{S}(t_0) \times \cdots \times \mathbb{S}(T) \subset \mathbb{W}^{T+1-t_0}\ .$$

### Constraints and feedbacks

The admissibility of decisions and states is restricted by non empty subset $\mathbb{B}(t, x)$ of admissible controls in $\mathbb{U}$ for all time $t$ and state $x$,

$$u(t) \in \mathbb{B}\big(t, x(t)\big) \subset \mathbb{U}\ , \tag{8.2a}$$

together with a non empty subset $\mathbb{A}(t)$ of the state space $\mathbb{X}$ for all $t = t_0, \ldots, T-1$,

$$x(t) \in \mathbb{A}(t) \subset \mathbb{X}\ , \tag{8.2b}$$

and a target

$$x(T) \in \mathbb{A}(T) \subset \mathbb{X}\ . \tag{8.2c}$$

These control, state or target constraints may reduce the relevant paths of the system. *Such a feasibility issue will only be addressed in a robust framework.*

For this purpose, let us consider the set $\mathfrak{U}^{ad}$ of admissible feedbacks as defined in (6.12)

$$\mathfrak{U}^{ad} = \{\mathfrak{u} \in \mathfrak{U} \mid \mathfrak{u}(t, x) \in \mathbb{B}(t, x)\ , \quad \forall (t, x)\}\ , \tag{8.3}$$

where $\mathfrak{U} = \{\mathfrak{u} : (t, x) \in \mathbb{N} \times \mathbb{X} \mapsto \mathfrak{u}(t, x) \in \mathbb{U}\}$ is defined in (6.11).

### Criterion

Let us pick up one of the finite horizon criteria $\pi$ defined previously in Sect. 6.1. The robust optimization results that we shall present in the sequel may be obtained with any of these criteria, while the stochastic optimization results may be obtained only with the finite horizon additive and multiplicative criteria.

For any $\mathfrak{u} \in \mathfrak{U}^{ad}$ (measurable in the stochastic context), and criterion $\pi : \mathbb{X}^{T+1-t_0} \times \mathbb{U}^{T-t_0} \times \mathbb{W}^{T+1-t_0} \to \mathbb{R}$, we put

$$\pi^{\mathfrak{u}}\big(t_0, x_0, w(\cdot)\big) := \pi\big(x_F[t_0, x_0, \mathfrak{u}, w(\cdot)](\cdot), u_F[t_0, x_0, \mathfrak{u}, w(\cdot)](\cdot), w(\cdot)\big) \tag{8.4}$$

where $t_0 \in \{0, \ldots, T-1\}$, $x_0 \in \mathbb{X}$, $w(\cdot) \in \Omega$ and $x_F$, $u_F$ are the solution maps introduced in Sect. 6.2. Thus, $\pi^{\mathfrak{u}}\big(t_0, x_0, w(\cdot)\big)$ is the evaluation of the criterion $\pi$ along the unique trajectory $x(\cdot) = x_F[t_0, x_0, \mathfrak{u}, w(\cdot)](\cdot)$, $u(\cdot) = u_F[t_0, x_0, \mathfrak{u}, w(\cdot)](\cdot)$, starting from $x(t_0) = x_0$, and generated by the dynamic (8.1), driven by feedback $u(t) = \mathfrak{u}\big(t, x(t)\big)$ and disturbance scenario $w(\cdot)$.

## 8.2 The robust optimality problem

In the robust optimality problem, we aggregate the scenarios $w(\cdot)$ in $\pi^{\mathfrak{u}}\big(t_0, x_0, w(\cdot)\big)$ by considering the worst case.

### Worst payoff

First, we fix an admissible feedback $\mathfrak{u}$. Then, we introduce the *worst performance* as in (6.14), namely the minimal payoff[1] with respect to the scenarios $w(\cdot) \in \Omega$:

$$\pi^{\mathfrak{u}}_{-}(t_0, x_0) := \inf_{w(\cdot) \in \Omega} \pi^{\mathfrak{u}}\big(t_0, x_0, w(\cdot)\big) \,. \tag{8.5}$$

Thus, the feedback $\mathfrak{u}$ being fixed, we let the scenario $w(\cdot)$ vary in $\Omega$ and evaluate the criterion by taking the lowest value.

### Maximal worst payoff

Second, we let the feedback $\mathfrak{u}$ vary, and aim at maximizing this worst payoff (8.5) by solving the optimization problem

$$\begin{aligned} \pi^{\star}_{-}(t_0, x_0) &:= \sup_{\mathfrak{u} \in \mathfrak{U}^{ad}} \pi^{\mathfrak{u}}_{-}(t_0, x_0) \\ &= \sup_{u(\cdot)} \inf_{w(\cdot) \in \Omega} \pi\big(x(\cdot), u(\cdot), w(\cdot)\big) \,, \end{aligned} \tag{8.6}$$

[1] Recall that $\pi$ measures payoffs. To address minimization problems, one should simply change the sign and consider $-\pi$.

where the last expression is abusively used, but practical and traditional, in which $x(\cdot)$ and $u(\cdot)$ need to be replaced by $x(t) = x_F[t_0, x_0, \mathfrak{u}, w(\cdot)](t)$ and $u(t) = \mathfrak{u}(t, x(t))$, referring to state and control solution maps introduced in Sect. 6.2.

**Definition 8.1.** *Given an initial condition $x_0 \in \mathbb{X}$ at time $t_0$, the optimal value $\pi_-^\star(t_0, x_0)$ in (8.6) is called the* maximal worst payoff *and any $\mathfrak{u}^\star \in \mathfrak{U}^{ad}$ such that*

$$\pi_-^\star(t_0, x_0) = \max_{\mathfrak{u} \in \mathfrak{U}^{ad}} \pi_-^{\mathfrak{u}}(t_0, x_0) = \pi_-^{\mathfrak{u}^\star}(t_0, x_0) \tag{8.7}$$

*is an* optimal feedback.

In the viability case, the infimum in (8.6) is taken for $\mathfrak{u} \in \mathfrak{U}_1^{\text{viab}}(t_0, x_0)$ defined in (7.5), giving

$$\mathfrak{U}_1^{\text{viab}}(t_0, x_0) := \left\{ \mathfrak{u} \in \mathfrak{U}^{ad} \;\middle|\; \begin{array}{c} \text{for all scenario} \quad w(\cdot) \in \Omega \\ x_F[t_0, x_0, \mathfrak{u}, w(\cdot)](t) \in \mathbb{A}(t) \\ \text{for} \quad t = t_0, \ldots, T \end{array} \right\}, \tag{8.8}$$

instead of $\mathfrak{U}^{ad}$ as defined in (8.3).

**Definition 8.2.** *Given an initial condition $x_0 \in \mathbb{X}$, the* maximal viable worst payoff *is*

$$\pi_-^\star(t_0, x_0) := \sup_{\mathfrak{u} \in \mathfrak{U}_1^{viab}(t_0, x_0)} \pi_-^{\mathfrak{u}}(t_0, x_0) \tag{8.9}$$

*and any $\mathfrak{u}^\star \in \mathfrak{U}_1^{viab}(t_0, x_0)$ such that*

$$\pi_-^\star(t_0, x_0) = \max_{\mathfrak{u} \in \mathfrak{U}_1^{viab}(t_0, x_0)} \pi_-^{\mathfrak{u}}(t_0, x_0) = \pi_-^{\mathfrak{u}^\star}(t_0, x_0) \tag{8.10}$$

*is an* optimal viable feedback strategy.

## 8.3 The robust additive payoff case

In the robust additive payoff case, we specify the criterion and consider the finite horizon additive criterion:

$$\pi\big(x(\cdot), u(\cdot), w(\cdot)\big) = \sum_{t=t_0}^{T-1} L\big(t, x(t), u(t), w(t)\big) + M\big(T, x(T), w(T)\big) . \tag{8.11}$$

### Robust additive dynamic programming without state constraints

We have already seen in the certain case that the so-called Bellman's principle expresses the fact that every "subpolicy" of an optimal policy remains optimal along the optimal trajectories. Such principle facilitaties splitting one optimization problem over time (dynamic) into a sequence of static optimization problem interrelated by backward induction. We detail this process and these general ideas hereafter in the robust case, when the criterion $\pi$ is additive, that is, given by (8.11). Here, we restrict the study to the case without state constraints, namely $\mathbb{A}(t) = \mathbb{X}$. The value function is defined by backward induction as follows.

**Definition 8.3.** *Assume no state constraints, that is* $\mathbb{A}(t) = \mathbb{X}$ *for* $t = t_0$, $\dots$, $T$. *The* value function *or* Bellman function$V(t,x)$, *associated with the additive criterion (8.11), the dynamics (8.1) and control constraints (8.2a), is defined by the following backward induction, where* $t$ *runs from* $T-1$ *down to* $t_0$*:*

$$\begin{cases} V(T,x) := \inf_{w \in \mathbb{S}(T)} M(T,x,w) \,, \\ V(t,x) := \sup_{u \in \mathbb{B}(t,x)} \inf_{w \in \mathbb{S}(t)} \Big[ L(t,x,u,w) + V\big(t+1, F(t,x,u,w)\big) \Big] \,. \end{cases} \tag{8.12}$$

## Optimal robust feedbacks

The backward equation of dynamic programming (8.12) makes it possible to define the value function $V(t,x)$. It turns out that the value $V(t_0,x_0)$ coincides with the maximal worst payoff $\pi_-^\star(t_0,x_0)$. In fact, we obtain a stronger result since dynamic programming induction maximization reveals relevant robust feedback controls.

The proof of the following Proposition 8.4 is given in the Appendix, Sect. A.6.

**Proposition 8.4.** *Assume that* $\mathbb{A}(t) = \mathbb{X}$ *for* $t = t_0, \dots, T$. *For any time* $t$ *and state* $x$, *assume the existence of the following feedback decision*

$$\mathfrak{u}^\star(t,x) \in \arg\max_{u \in \mathbb{B}(t,x)} \inf_{w \in \mathbb{S}(t)} \Big[ L(t,x,u,w) + V\big(t+1, F(t,x,u,w)\big) \Big] \,. \tag{8.13}$$

*Then* $\mathfrak{u}^\star \in \mathfrak{U}$ *is an optimal feedback of the robust problem (8.6), where* $\pi$ *is given by (8.11), and, for any initial state* $x_0$, *the maximal worst payoff is given by*

$$V(t_0,x_0) = \pi_-^\star(t_0,x_0) = \pi_-^{\mathfrak{u}^\star}(t_0,x_0) \,. \tag{8.14}$$

## Robust dynamic programming in the viability case

When state constraints restrict the choice of relevant paths, we have to adapt the dynamic programming equation using viability tools and especially the robust viability kernel $\mathbb{V}\mathrm{iab}_1$ introduced in Chap. 7. The value function is defined by backward induction as follows.

**Definition 8.5.** *The* value function *or* Bellman function$V(t,x)$, *associated with the additive criterion (8.11), the dynamics (8.1), control constraints (8.2a), state constraints (8.2b) and target constraints (8.2c), is defined by the following backward induction, where t runs from* $T-1$ *down to* $t_0$,

$$\begin{cases} V(T,x) := M(T,x) \ , \quad \forall x \in \mathbb{V}\mathrm{iab}_1(T) = \mathbb{A}(T) \ , \\ V(t,x) := \sup_{u \in \mathbb{B}_1^{viab}(t,x)} \inf_{w \in \mathbb{S}(t)} \Big[ L(t,x,u,w) + V\big(t+1, F(t,x,u,w)\big) \Big] \ , \\ \qquad \forall x \in \mathbb{V}\mathrm{iab}_1(t) \ , \end{cases} \tag{8.15}$$

*where* $\mathbb{V}\mathrm{iab}_1(t)$ *is given by the backward induction (7.11)*

$$\begin{cases} \mathbb{V}\mathrm{iab}_1(T) = \mathbb{A}(T) \ , \\ \mathbb{V}\mathrm{iab}_1(t) = \{ x \in \mathbb{A}(t) \mid \exists u \in \mathbb{B}(t,x) \ , \ \forall w \in \mathbb{S}(t) \ , \\ \qquad F(t,x,u,w) \in \mathbb{V}\mathrm{iab}_1(t+1) \} \ , \end{cases} \tag{8.16}$$

*and where the supremum in (8.15) is over viable controls in* $\mathbb{B}_1^{viab}(t,x)$ *given by*

$$\mathbb{B}_1^{viab}(t,x) = \{ u \in \mathbb{B}(t,x) \mid \forall w \in \mathbb{S}(t) \ , \ F(t,x,u,w) \in \mathbb{V}\mathrm{iab}_1(t+1) \} \ . \tag{8.17}$$

## Optimal robust viable feedbacks

The backward equation of dynamic programming (8.15) makes it possible to define the value function $V(t,x)$. It turns out that the value $V(t_0,x_0)$ coincides with the maximal worst payoff $\pi_-^\star(t_0,x_0)$. In fact, we obtain a stronger result since dynamic programming induction maximization reveals relevant feedbacks.

The proof[2] of the following Proposition 8.6 is given in the Appendix, Sect. A.6.

**Proposition 8.6.** *For any time* $t = t_0, \ldots, T-1$ *and state* $x \in \mathbb{V}\mathrm{iab}_1(t)$, *assume the existence of the following feedback decision*

$$\mathfrak{u}^\star(t,x) \in \arg\max_{u \in \mathbb{B}_1^{viab}(t,x)} \inf_{w \in \mathbb{S}(t)} \Big[ L(t,x,u,w) + V\big(t+1, F(t,x,u,w)\big) \Big] \ . \tag{8.18}$$

*Then* $\mathfrak{u}^\star \in \mathfrak{U}$ *is an optimal feedback of the viable robust problem (8.9), where* $\pi$ *is given by (8.11), and, for any initial state* $x_0$, *the maximal viable worst payoff is given by*

$$V(t_0,x_0) = \pi_-^\star(t_0,x_0) = \pi_-^{\mathfrak{u}^\star}(t_0,x_0) \ . \tag{8.19}$$

[2] For the proof, we require additional technical assumptions: $\inf_{x,u,w} L\big(t,x,u,w\big) > -\infty$ and $\inf_{x,w} M\big(T,x,w\big) > -\infty$.

Hence, there is no theoretical problem in coupling viability and optimality requirements in the robust framework. This will no longer be the situation in the stochastic case.

## 8.4 Robust harvest of a renewable resource over two periods

We consider a model of the management of a renewable resource over two periods $T = 2$ and we perform a robust analysis. The uncertain resource productivity $R(t)$ is supposed to vary within an interval $\mathbb{S} = [R^\flat, R^\sharp] \subset \mathbb{W} = \mathbb{R}$, with $R^\flat < R^\sharp$. We aim at maximizing the worst benefit namely the minimal sum of the discounted successive harvesting revenues

$$\sup_{0 \leq h(0) \leq B(0),\ 0 \leq h(1) \leq B(1)} \inf_{R(1), R(2)} \Big[ ph(0) + \rho p h(1) \Big] \, ,$$

where the resource dynamics corresponds to

$$B(1) = R(1)\big(B(0) - h(0)\big) \, , \quad B(2) = R(2)\big(B(1) - h(1)\big) \, .$$

- Final time $t = T = 2$. The robust value function is $V(2, B) = 0$.
- Time $t = 1$. By virtue of dynamic programming equation (8.12), one has

$$V(1, B) = \sup_{0 \leq h \leq B} \inf_{R(2) \in [R^\flat, R^\sharp]} \big[\rho p h + V\big(2, R(2)(B - h)\big)\big] = \sup_{0 \leq h \leq B} \{\rho p h\} \, .$$

The optimality problem over $h$ is linear and we obtain similarly the optimal feedback $\mathfrak{u}^\star(1, B) = B$ while the value function is

$$V(1, B) = \rho p B \, .$$

- Time $t = 0$. By virtue of dynamic programming equation (8.12), one has

$$\begin{aligned} V(0, B) &= \sup_{0 \leq h \leq B} \inf_{R(1) \in [R^\flat, R^\sharp]} \big[ph + V\big(1, R(1)(B - h)\big)\big] \\ &= \sup_{0 \leq h \leq B} \Big\{ ph + \inf_{R(1) \in [R^\flat, R^\sharp]} [\rho p R(1)(B - h)] \Big\} \\ &= p \sup_{0 \leq h \leq B} \Big\{ h(1 - \rho R^\flat) + \rho R^\flat B \Big\} \end{aligned}$$

where $R^\flat = \inf_{R(1) \in [R^\flat, R^\sharp]} R(1)$. The optimality problem over $h$ is linear and we obtain

$$\mathfrak{u}^\star(0, B) = \begin{cases} B & \text{if} \quad \rho R^\flat < 1 \\ 0 & \text{if} \quad \rho R^\flat > 1 \end{cases}$$

and the value function is

$$V(0, B) = \begin{cases} pB & \text{if} \quad \rho R^\flat < 1 \\ \rho p R^\flat B & \text{if} \quad \rho R^\flat > 1 \, . \end{cases}$$

Thus, the results are similar to the certain case using the worst equivalent $R^\flat$. We deduce that sustainability is even more difficult to achieve in such a robust framework.

## 8.5 The robust "maximin" approach

Here, we consider the finite horizon maximin criterion

$$\pi\big(x(\cdot), u(\cdot), w(\cdot)\big) = \inf_{t=t_0,\ldots,T-1} L\big(t, x(t), u(t), w(t)\big) . \tag{8.20}$$

We shall also consider the final payoff case with

$$\pi\big(x(\cdot), u(\cdot), w(\cdot)\big) = \min\left(\min_{t=t_0,\ldots,T-1} L\big(t, x(t), u(t), w(t)\big), M\big(T, x(T), w(T)\big)\right) .$$

However, by changing $T$ in $T+1$ and defining $L(T,x,u,w) = M\big(T,x,w\big)$, the minimax with final payoff may be interpreted as one without on a longer time horizon.

### Robust dynamic programming equation

We state the results without state constraints, though robust ones would not pose problems. The value function is defined by backward induction as follows.

**Definition 8.7.** *Assume no state constraints* $\mathbb{A}(t) = \mathbb{X}$ *for* $t \in \{t_0, \ldots, T\}$. *The* value function *or* Bellman function$V(t,x)$, *associated with the maximin criterion (8.20), the dynamics (8.1), control constraints (8.2a), is defined by the following backward induction, where* $t$ *runs from* $T-1$ *down to* $t_0$*:*

$$\begin{cases} V(T,x) := \inf_{w \in \mathbb{S}(T)} M(T,x,w) , \\ V(t,x) := \sup_{u \in \mathbb{B}(t,x)} \inf_{w \in \mathbb{S}(t)} \min\Big(L(t,x,u,w), V\big(t+1, F(t,x,u,w)\big)\Big) . \end{cases} \tag{8.21}$$

### Optimal robust feedbacks

The backward equation of dynamic programming (8.21) makes it possible to define the value function $V(t,x)$. It turns out that the value $V(t_0,x_0)$ coincides with the maximal worst payoff $\pi^\star_-(t_0,x_0)$. In fact, we obtain a stronger result since dynamic programming induction maximization reveals relevant feedbacks.

The proof of the following Proposition 8.8 is given in the Appendix, Sect. A.6.

**Proposition 8.8.** *Assume that* $\mathbb{A}(t) = \mathbb{X}$ *for* $t = t_0, \ldots, T$. *For any time* $t$ *and state* $x$, *assume the existence of the following feedback decision*

$$\mathfrak{u}^\star(t,x) \in \arg\max_{u \in \mathbb{B}(t,x)} \inf_{w \in \mathbb{S}(t)} \min\Big(L(t,x,u,w), V\big(t+1, F(t,x,u,w)\big)\Big) . \tag{8.22}$$

*Then* $\mathfrak{u}^\star \in \mathfrak{U}$ *is an optimal feedback of the robust problem (8.6), where* $\pi$ *is given by (8.20), and, for any initial state* $x_0$, *the maximal worst payoff is given by*

$$V(t_0,x_0) = \pi_-^\star(t_0,x_0) = \pi_-^{\mathfrak{u}^\star}(t_0,x_0) . \tag{8.23}$$

## 8.6 The stochastic optimality problem

Suppose that $L$ and $M$ in (8.11) are *measurable* and either bounded or non-negative so as to be integrable when composed with measurable state and decision maps. Suppose that the assumptions in Sect. 6.3 are satisfied.

In all that follows, , the primitive random process $w(\cdot)$ is assumed to be a sequence of independent identically distributed (i.i.d.) random variables $\big(w(t_0), w(t_0+1), \ldots, w(T-1), w(T)\big)$ under probability $\mathbb{P}$ on the domain of scenarios $\Omega = \mathbb{S}^{T+1-t_0}$.

Let $\mathbb{B}(t,x) \subset \mathbb{U}$ be a non empty subset of the control space $\mathbb{U}$ for all $(t,x)$, and $\mathfrak{U}^{ad}$ be the set of *measurable* admissible feedbacks as defined in (8.3).

### Mean payoff

We aim at computing the optimal expected value of the criterion $\pi$. As long as the criterion $\pi$ represents payoffs, we aim at identifying the maximal mean payoffs and the associated feedbacks or pure Markovian strategies.

For any admissible feedback strategy $\mathfrak{u} \in \mathfrak{U}^{ad}$ and initial condition $x_0 \in \mathbb{X}$ at time $t_0$, let us consider the or *mean payoff* as in (6.15):

$$\overline{\pi}^{\mathfrak{u}}(t_0,x_0) := \mathbb{E}\Big[\pi^{\mathfrak{u}}\big(t_0,x_0,w(\cdot)\big)\Big] . \tag{8.24}$$

### Maximal mean payoff

The stochastic optimization problem is

$$\begin{aligned} \overline{\pi}^\star(t_0,x_0) &:= \sup_{\mathfrak{u} \in \mathfrak{U}^{ad}} \mathbb{E}\Big[\pi^{\mathfrak{u}}\big(t_0,x_0,w(\cdot)\big)\Big] \\ &= \sup_{u(\cdot)} \mathbb{E}\Big[\pi\big(x(\cdot),u(\cdot),w(\cdot)\big)\Big] , \end{aligned} \tag{8.25}$$

where the last expression is abusively used, but practical and traditional, in which $x(\cdot)$ and $u(\cdot)$ need to be replaced by $x(t) = x_F[t_0, x_0, \mathfrak{u}, w(\cdot)](t)$ and $u(t) = \mathfrak{u}\big(t, x(t)\big)$, referring to state and control solution maps introduced in Sect. 6.2.

**Definition 8.9.** *Consider any initial condition* $x_0 \in \mathbb{X}$. *The optimal value* $\overline{\pi}^\star(t_0, x_0)$ *in (8.25) is called the* maximal expected payoff *or* maximal mean payoff *and any* $\mathfrak{u}^\star \in \mathfrak{U}^{ad}$ *such that*

$$\overline{\pi}^\star(t_0, x_0) = \max_{\mathfrak{u} \in \mathfrak{U}^{ad}} \overline{\pi}^{\mathfrak{u}}(t_0, x_0) = \overline{\pi}^{\mathfrak{u}^\star}(t_0, x_0) \tag{8.26}$$

*is an* optimal feedback.

In the viability case where state constraints reduce the admissibility of decisions and feedbacks, let us emphasize that the stochastic optimality problem with stochastic state constraints is not easy to cope with and that substantial mathematical difficulties arise. This is why we restrict the viability problem to the robust approach. Hence, the maximum in (8.25) is taken for robust feedbacks $\mathfrak{u} \in \mathfrak{U}_1^{\mathrm{viab}}(t_0, x_0)$ defined as follows[3]

$$\mathfrak{U}_1^{\mathrm{viab}}(t_0, x_0) := \left\{ \mathfrak{u} \in \mathfrak{U}^{ad} \;\middle|\; \mathbb{P}\Big( w(\cdot) \in \Omega \mid x(t) \in \mathbb{A}(t) \;\text{ for }\; t = t_0, \ldots, T \Big) = 1 \right\}, \tag{8.27}$$

where $x(t)$ corresponds to the solution map $x(t) = x_F[t_0, x_0, \mathfrak{u}, w(\cdot)](t)$ defined in Sect. 6.2, instead of $\mathfrak{U}^{ad}$ as defined in (8.3).

**Definition 8.10.** *Given an initial condition* $x_0 \in \mathbb{X}$, *the* maximal viable mean payoff *is*

$$\pi_-^\star(t_0, x_0) := \sup_{\mathfrak{u} \in \mathfrak{U}_1^{viab}(t_0, x_0)} \pi_-^{\mathfrak{u}}(t_0, x_0) \tag{8.28}$$

*and any* $\mathfrak{u}^\star \in \mathfrak{U}_1^{viab}(t_0, x_0)$ *such that*

$$\pi_-^\star(t_0, x_0) = \max_{\mathfrak{u} \in \mathfrak{U}_1^{viab}(t_0, x_0)} \pi_-^{\mathfrak{u}}(t_0, x_0) = \pi_-^{\mathfrak{u}^\star}(t_0, x_0) \tag{8.29}$$

*is an* optimal viable feedback.

### Stochastic dynamic programming without state constraints

We first restrict the study to the case without state constraints, namely $\mathbb{A}(t) = \mathbb{X}$ for $t = t_0, \ldots, T$.

Again a backward inductive equation defines the value function $V(t, x)$ and we can split up the maximization operation into two parts for the following reasons: the criterion $\pi$ in (8.11) is additive, the expectation operator is linear, the dynamic is a first order induction equation and, finally, constraints at time $t$ depend only on time $t$ and state $x$.

[3] See the footnotes 3 and 4 in the previous Chap. 7.

**Definition 8.11.** *The* value function *or* Bellman function$V(t,x)$, *associated with the additive criterion (8.11), the dynamics (8.1), control constraints (8.2a) and no state constraints ($\mathbb{A}(t) = \mathbb{X}$ for $t = t_0, \ldots, T$), is defined by the following backward induction*[4], *where $t$ runs from $T-1$ down to $t_0$:*

$$\begin{cases} V(T,x) := \mathbb{E}_{w(T)}\Big[M\big(T,x,w(T)\big)\Big] \ , \\ V(t,x) := \sup\limits_{u \in \mathbb{B}(t,x)} \mathbb{E}_{w(t)}\Big[L\big(t,x,u,w(t)\big) + V\big(t+1, F(t,x,u,w)\big)\Big] \ . \end{cases} \tag{8.30}$$

### Stochastic optimal feedback

The backward equation of dynamic programming (8.30) makes it possible to define the value function $V(t,x)$. It turns out that the value $V(t_0,x_0)$ at time $t_0$ coincides with the optimal mean payoff $\overline{\pi}^\star(t_0,x_0)$. Moreover, dynamic programming induction maximization reveals relevant feedback controls or pure Markovian strategies. Indeed, assuming the additional hypothesis that the infimum is achieved in (8.30) for at least one decision, if we denote by $\mathfrak{u}^\star(t,x)$ a value $u \in \mathbb{B}(t,x)$ which achieves the maximum in equation (8.30), then $\mathfrak{u}^\star(t,x)$ is an optimal feedback for the optimal control problem in the following sense.

The proof of the following Proposition 8.12 is given in the Appendix, Sect. A.6.

**Proposition 8.12.** *Assume that $\mathbb{A}(t) = \mathbb{X}$ for $t = t_0, \ldots, T$. For any time $t$ and state $x$, assume the existence of the following feedback decision:*

$$\mathfrak{u}^\star(t,x) \in \arg\max_{u \in \mathbb{B}(t,x)} \mathbb{E}_{w(t)}\Big[L\big(t,x,u,w(t)\big) + V\Big(t+1, F\big(t,x,u,w(t)\big)\Big)\Big] \ . \tag{8.31}$$

*If $\mathfrak{u}^\star : (t,x) \to \mathfrak{u}^\star(t,x)$ is measurable, we thus have an optimal strategy of the maximization problem (8.25), where $\pi$ is given by (8.11) and, for any initial state $x_0$, the optimal expected payoff is given by:*

$$V(t_0,x_0) = \overline{\pi}^\star(t_0,x_0) = \overline{\pi}^{\mathfrak{u}^\star}(t_0,x_0) \ . \tag{8.32}$$

### Stochastic dynamic programming with state constraints

If we introduce state constraints and deal with them in a robust sense, the hereabove results hold true with $\mathbb{B}(t,x)$ replaced by $\mathbb{B}_1^{\mathrm{viab}}(t,x)$.

[4] See the footnote 5 in Definition 7.9.

**Definition 8.13.** *The* value function *or* Bellman function$V(t,x)$, *associated with the additive criterion (8.11), the dynamics (8.1), control constraints (8.2a), state constraints (8.2b) and target constraints (8.2c), is defined by the following backward induction, where $t$ runs from $T-1$ down to $t_0$,*

$$\begin{cases} V(T,x) := \mathbb{E}_{w(T)}\Big[M\big(T,x,w(T)\big)\Big] , & \forall x \in \mathbb{V}\mathrm{iab}_1(T) = \mathbb{A}(T) , \\ V(t,x) := \sup_{u\in\mathbb{B}(t,x)} \mathbb{E}_{w(t)}\Big[L\big(t,x,u,w(t)\big) + V\Big(t+1, F\big(t,x,u,w(t)\big)\Big)\Big] , \\ \qquad \forall x \in \mathbb{V}\mathrm{iab}_1(t) , \end{cases} \tag{8.33}$$

*where* $\mathbb{V}\mathrm{iab}_1(t)$ *is given by the backward induction (7.11)*

$$\begin{cases} \mathbb{V}\mathrm{iab}_1(T) = \mathbb{A}(T) , \\ \mathbb{V}\mathrm{iab}_1(t) = \{x \in \mathbb{A}(t) \mid \exists u \in \mathbb{B}(t,x) , \forall w \in \mathbb{S}(t) , \\ \qquad F(t,x,u,w) \in \mathbb{V}\mathrm{iab}_1(t+1)\} , \end{cases} \tag{8.34}$$

*and where the supremum in (8.33) is over viable controls in* $\mathbb{B}_1^{viab}(t,x)$ *given by*

$$\mathbb{B}_1^{viab}(t,x) = \{u \in \mathbb{B}(t,x) \mid \forall w \in \mathbb{S}(t) , F(t,x,u,w) \in \mathbb{V}\mathrm{iab}_1(t+1)\} . \tag{8.35}$$

The proof[5] of the following Proposition 8.14 is given in the Appendix, Sect. A.6.

**Proposition 8.14.** *For any time $t$ and state $x$, assume the existence of the following feedback decision*

$$\mathfrak{u}^\star(t,x) \in \operatorname*{arg\,max}_{u\in\mathbb{B}_1^{viab}(t,x)} \mathbb{E}_{w(t)}\Big[L\big(t,x,u,w(t)\big) + V\Big(t+1, F\big(t,x,u,w(t)\big)\Big)\Big] . \tag{8.36}$$

*If $\mathfrak{u}^\star : (t,x) \to \mathfrak{u}^\star(t,x)$ is measurable, we thus have an optimal strategy of the maximization problem (8.28), where $\pi$ is given by (8.11), and, for any initial state $x_0$, the optimal expected payoff is given by:*

$$V(t_0,x_0) = \overline{\pi}^\star(t_0,x_0) = \overline{\pi}^{\mathfrak{u}^\star}(t_0,x_0) . \tag{8.37}$$

## Robust is more stringent than stochastic

Not surprisingly, maximal payoffs are smaller in a robust perspective than in a mean approach since the inequality

[5] For the proof, we require the additional technical assumptions that $L$ and $M$ are bounded.

$$\mathbb{E}_w[A(w)] \geq \inf_{w \in \mathbb{S}} A(w)$$

holds. Thus, for any initial state $x_0$, the maximal expected payoff is larger than the maximal worst payoff:

$$\overline{\pi}^\star(t_0, x_0) \geq \pi_-^\star(t_0, x_0) \; .$$

Moreover, the robust and stochastic maximization problems coincide whenever the uncertainty disappears, which corresponds to the deterministic case. In the certain case where $\mathbb{S} = \{\overline{w}\}$, the maximal expected payoff and the maximal worst payoff coincide:

$$\mathbb{S} = \{\overline{w}\} \Longrightarrow \pi_-^\star(t_0, x_0) = \overline{\pi}^\star(t_0, x_0) \; .$$

## 8.7 Stochastic management of a renewable resource

### Over two periods

We consider the biomass linear model over two periods $T = 2$,

$$B(1) = R(1)\big(B(0) - h(0)\big) \; , \quad B(2) = R(2)\big(B(1) - h(1)\big) \; ,$$

for which we aim at maximizing the expectation of the sum of the discounted successive harvesting revenues

$$V(0, B(0)) = \sup_{0 \leq h(0) \leq B(0),\; 0 \leq h(1) \leq B(1)} \mathbb{E}_{R(1),R(2)} \Big[ ph(0) + \rho ph(1) \Big] \; ,$$

where $R(1)$ and $R(2)$ are two independent random variables.

- Final time $t = T = 2$. The value function is $V(2, B) = 0$.
- Time $t = 1$. By virtue of dynamic programming equation (8.30), one has:

$$V(1, B) = \sup_{0 \leq h \leq B} \mathbb{E}_{R(2)} \left[\rho ph + V\big(2, R(2)(B - h)\big)\right] = \sup_{0 \leq h \leq B} \{\rho ph\} \; .$$

  The optimality problem over $h$ is linear and we obtain the optimal feedback harvesting $\mathfrak{u}^\star(1, B) = B$ while the value function is:

$$V(1, B) = \rho p B \; .$$

- Time $t = 0$. By (8.30), one has

$$\begin{aligned} V(0, B) &= \sup_{0 \leq h \leq B} \mathbb{E}_{R(1)} \left[ph + V\big(1, R(1)(B - h)\big)\right] \\ &= \sup_{0 \leq h \leq B} \left\{ph + \mathbb{E}_{R(1)} \left[\rho p R(1)(B - h)\right]\right\} \\ &= p \sup_{0 \leq h \leq B} \left\{h(1 - \rho \overline{R}) + \rho \overline{R} B\right\} \end{aligned}$$

where $\overline{R} = \mathbb{E}[R(1)]$. The optimality problem over $h$ is linear and we obtain:

$$\mathfrak{u}^\star(0,B) = \begin{cases} B & \text{if} \quad \rho\overline{R} < 1 \\ 0 & \text{if} \quad \rho\overline{R} > 1 \,. \end{cases}$$

The value function is

$$V(0,B) = \begin{cases} pB & \text{if} \quad \rho\overline{R} < 1 \\ p\rho\overline{R}B & \text{if} \quad \rho\overline{R} > 1 \,. \end{cases}$$

Thus the results are similar to the certain case using the certainty equivalent $\overline{R}$. Consequently, we deduce that:

- the resource $B$ becomes extinct in at least in two periods;
- conservation problems are stronger if $\overline{R}$ is weaker since the resource is completely harvested at the first period in this case;
- no intergenerational equity occurs since either the whole resource is captured in the first or second period.

### Over $T$ periods

The dynamic model still is

$$B(t+1) = R(t)\big(B(t) - h(t)\big)\,, \quad 0 \leq h(t) \leq B(t)\,,$$

where $R(t_0)$, ..., $R(T-1)$ are independent and identically distributed positive random variables. We consider expected intertemporal discounted utility maximization

$$\sup_{h(t_0),\dots,h(T-1)} \mathbb{E}\left[\sum_{t=t_0}^{T-1} \rho^t L\big(h(t)\big) + \rho^T L\big(B(t)\big)\right] .$$

We restrict the study to the isoelastic case where

$$L(h) = h^\eta \quad \text{with} \quad 0 < \eta < 1\,.$$

For $t = t_0, \dots, T-1$, dynamic programming equation (8.30) gives

$$V(t,B) = \sup_{h\in[0,B]} \Big(\rho^t L(h) + \mathbb{E}_R\big[V\big(t+1, R(B-h)\big)\big]\Big)\,,$$

where $R$ is a random variable standing for the uncertain growth of the resource and having the same distribution as any of the random variables $R(t_0)$, ..., $R(T-1)$.

Using a backward induction starting from $V(T,B) = \rho^T L(B)$, it turns out that the optimal strategies of harvesting $\mathfrak{h}^\star(t,B)$ along with the value function $V(t,B)$ are characterized by

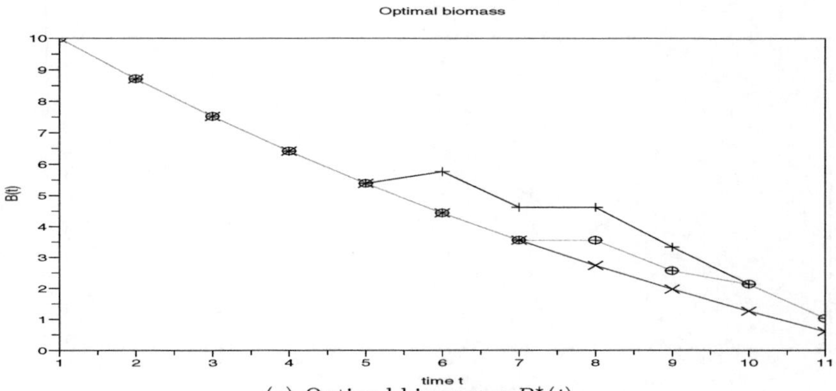

(a) Optimal biomasses $B^\star(t)$

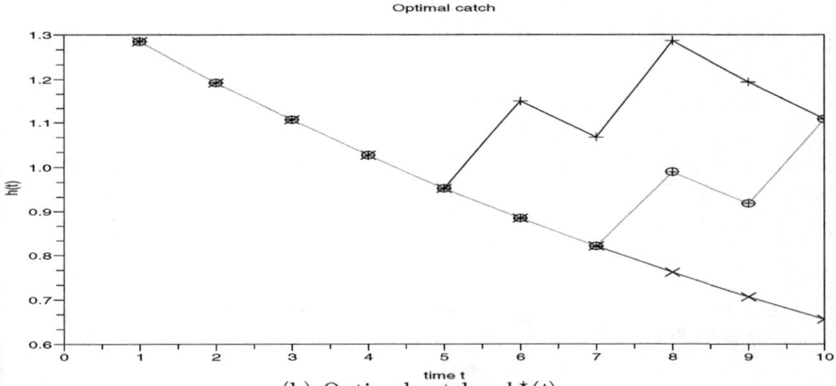

(b) Optimal catches $h^\star(t)$

**Fig. 8.1.** Preference for present: optimal stochastic paths $B^\star(t)$ and $h^\star(t)$ over time horizon $T = 10$ with catch utility function $L(h) = h^{0.5}$, discount factor $\rho = 0.95$ while initial biomass is $B_0 = 10$. Resource growth parameters $R_\sharp = 1.3$, $R_\flat = 1$ and $p = \mathbb{P}(R = 1.3) = 0.1$. Consequently, certainty equivalent is $\widehat{R} \approx 1.028$ and sustainability problems occurs as catches are strong for first periods and resource is exhausted.

$$V(t, B) = \rho^t b(t)^{\eta-1} B^\eta \quad \text{and} \quad \mathfrak{h}^\star(t, B) = b(t)B \, .$$

The recursive relation revealing $b(t)$ is given by

$$b(t) = \frac{ab(t+1)}{1 + ab(t+1)} \, , \quad b(T) = 1 \, ,$$

depending on the term

$$a = (\rho \widehat{R}^\eta)^{\frac{1}{\eta-1}} \, ,$$

where the certainty equivalent $\widehat{R}$ is defined by the implicit equation (the utility function $L(h) = h^\eta$ is strictly increasing)

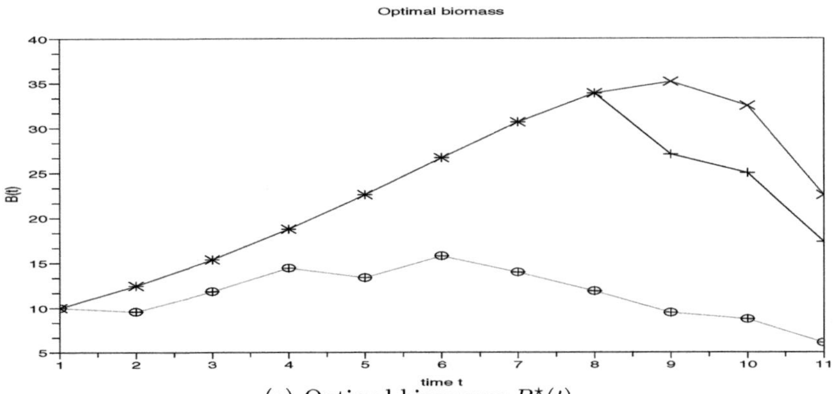


(a) Optimal biomasses $B^\star(t)$

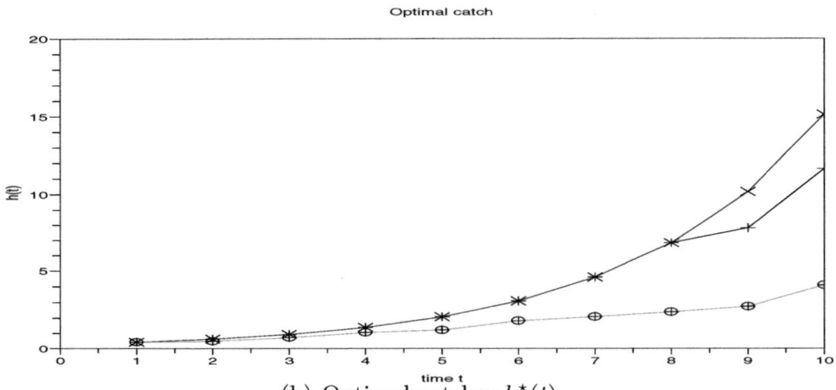


(b) Optimal catches $h^\star(t)$

**Fig. 8.2.** Preference for future: optimal stochastic paths $B^\star(t)$ and $h^\star(t)$ over time horizon $T = 10$ with catch utility function $L(h) = h^{0.5}$, discount factor $\rho = 0.95$ while initial biomass is $B_0 = 10$. Resource parameters are $R_\sharp = 1.3$; $R_\flat = 1$ and $p = \mathbb{P}(R = 1.3) = 0.9$. Thus resource certainty equivalent is $\widehat{R} = 1.26$. Sustainability problems occurs as catches are weak for first periods.

$$L(\widehat{R}) = \mathbb{E}_R[L(R)] \,. \tag{8.38}$$

Optimal stochastic biomasses $B^\star(t)$ and catches $h^\star(t)$ are defined by

$$B^\star(t+1) = R(t)\big(B^\star(t) - h^\star(t)\big) \,, \quad h^\star(t) = \mathfrak{h}^\star\big(t, B^\star(t)\big) = b(t)B^\star(t) \,,$$

which gives

$$h^\star(t+1) = b(t+1)B^\star(t+1) = b(t+1)R(t)B^\star(t)(1-b(t)) = \frac{R(t)b(t)}{a}B^\star(t) = \frac{R(t)}{a}h^\star(t).$$

Since $L(h) = h^\eta$, we obtain that

$$\mathbb{E}_R\left[\frac{L(h^\star(t+1))}{L(h^\star(t))}\right] = \frac{\mathbb{E}_R[L(R)]}{L(a)} = \frac{L(\widehat{R})}{L(a)} = \frac{\widehat{R}^\eta}{(\rho\widehat{R}^\eta)^{\frac{\eta}{\eta-1}}} = \left(\rho\widehat{R}\right)^{\frac{\eta}{1-\eta}} .$$

It turns out that the sustainability issues depend critically upon the product

$$\underbrace{\rho}_{\text{economic discount factor}} \times \underbrace{\widehat{R}}_{\text{biological growth factor}} \tag{8.39}$$

which mixes economic and biological characteristics of the problem.

**Result 8.15** *The relative variation of the utility of optimal stochastic catches depends on the index $\rho\widehat{R}$ as follows.*

1. *If $\rho\widehat{R} > 1$, then the utility of optimal stochastic catches increases along time in the mean following sense:*

$$\mathbb{E}\left[\frac{L\big(h^\star(t+1)\big)}{L\big(h^\star(t)\big)}\right] > 1 .$$

2. *If $\rho\widehat{R} < 1$, then the utility of optimal stochastic catches decreases along time in the mean following sense:*

$$\mathbb{E}\left[\frac{L\big(h^\star(t+1)\big)}{L\big(h^\star(t)\big)}\right] < 1 .$$

3. *If $\rho\widehat{R} = 1$, then the utility of optimal stochastic catches remains stationary in the mean following sense:*

$$\mathbb{E}\left[\frac{L\big(h^\star(t+1)\big)}{L\big(h^\star(t)\big)}\right] = 1 .$$

Figs. 8.1 and 8.2 show that sustainability problems occur with such optimal stochastic catches. It is assumed that resource productivity $R$ is a random variable taking values $R_\sharp = 1.3$ and $R_\flat = 1$.

In the first case depicted by Fig. 8.1 where the uncertain resource natural growth features are not favorable, we have $\mathbb{P}(R = 1.3) = 0.1$ and thus $\widehat{R} = 1.028$, so that the resource stocks $B^\star(t)$ quickly decrease. Similarly, the catches $h^\star(t)$ decrease and tend to collapse. In such a case, there is a preference for the present which condemns the conservation of both the resource and exploitation. As a consequence, intergenerational equity is not guaranteed.

In the second case displayed by Fig. 8.2, resource natural growth features are more favorable and $\mathbb{P}(R = 1.3) = 0.9$ yields the certainty equivalent $\widehat{R} = 1.26$: conservation of the resource is achieved. However a problem of intergenerational equity appears since the catches $h^\star(t)$ are very weak (almost zero) for the first periods before they quickly increase at the end of the concerned period. There is a clear preference for the future.

SCILAB CODE 16.

```
//
// exec opti_uncertain_resource.sce


//parameters
puis=0.5;
Horizon=10;
rho=0.95 ;
B0=10;

p=0.1; q=1-p;
p=0.9; q=1-p;

R#=1.3;Rb=1;

// isoelastic utility
function [u]=util(x)
  u=x^puis
endfunction

// certainty equivalent
Rchap=(p*util(R#)+q*util(Rb))^(1/puis) ;

a=(rho * util(Rchap))^(1/(puis-1)) ;

// proportion of consumption  b(t)
b=[];
b(Horizon+1)=1;
for t=Horizon:-1:1
  b(t)=a*b(t+1)/(1+a*b(t+1)) ;
end

// optimal catches and resource
function [h]=PREL_OPT(t,B)
  h =b(t)*B
endfunction

function [x]=DYN_OPT(t,B,R)
  x=R*(B-PREL_OPT(t,B))
endfunction

xset("window",1) ; xbasc()
xset("window",2) ; xbasc()


N_simu=3;
for i=1:N_simu

// simulation of random productivity R(t)
z=rand(1,Horizon,'uniform');
for t=1:Horizon
  if (z(t) <= p) then
Rsimu(t)=R#;
  else Rsimu(t)=Rb;
  end
end

// computation of optimal trajectories
Bopt(1)=B0;J=0;
for t=1: Horizon
  hopt(t)=PREL_OPT(t,Bopt(t));
  Bopt(t+1)=DYN_OPT(t,Bopt(t),Rsimu(t));
end

// graphics
xset("window",1) ;
plot2d(1:Horizon+1,Bopt,style=i) ;
plot2d(1:Horizon+1,Bopt,style=-i) ;
xtitle("Optimal biomass","time t","B(t)")

xset("window",2) ;
plot2d(1:Horizon,hopt,style=i) ;
plot2d(1:Horizon,hopt,style=-i) ;
xtitle("Optimal catch","time t","h(t)")

end


//
```

## 8.8 Optimal expected land-use and specialization

We now cope with the problem introduced in Sect. 6.6. The annual wealth evolution of the farm was described by

$$v(t+1) = v(t)\left(\sum_{i=1}^{n} u_i(t) R_i\big(w(t)\big)\right) = v(t)\langle u_i(t), R\big(w(t)\big)\rangle ,$$

where $R(w) = \big(R_1(w), \dots, R_n(w)\big)$ and $\langle , \rangle$ denotes scalar product on $\mathbb{R}^n$,

$$u_i(t) := \frac{p_i B_i(t)}{v(t)}$$

stood for the proportion of wealth generated by use $i$ ($\sum_{i=1}^{n} u_i(t) = 1,\ u_i(t) \geq 0$), and $w(t)$ corresponded to environmental uncertainties varying in a given domain $\mathbb{S} \subset \mathbb{W}$. The allocation $u = (u_1, \dots, u_n) \in \mathcal{S}^n$, belonging to the simplex $\mathcal{S}^n$ of $\mathbb{R}^n$, to the different land-uses appeared as a decision variable representing the land-use structure.

Now the farmer aims at optimizing the expected discounted final wealth at time $T$:

$$\sup_u \mathbb{E}\left[\rho^T v(T)\right] .$$

Let us prove that the optimal allocation is a specialized one defined by use $i^\star \in \arg\max_{i=1,\dots,n} \mathbb{E}_w[R_i(w)]$ in the sense that

$$\begin{cases} u_i^\star(t,v) = \begin{cases} 0 & \text{if} \quad i \neq i^\star \\ 1 & \text{if} \quad i = i^\star \end{cases} \\ V(t,v) = v\rho^T \mathbb{E}_w[R_{i^\star}(w)]^{T-t} . \end{cases} \tag{8.40}$$

This result suggests that without risk aversion (utility is linear), specialization in one land-use is optimal when the land-use remains constant. The chosen use is the one which maximizes the expected growth $\mathbb{E}_w[R_i(w)]$.

**Result 8.16** *Without risk aversion (utility L is linear), specialization in one land-use is optimal in the stochastic sense. The chosen use is the* $i^\star \in \arg\max_{i=1,\dots,n} \mathbb{E}_w[R_i(w)]$ *which maximizes the expected growth* $\mathbb{E}_w[R_i(w)]$. *The optimal allocation and value are given by:*

$$\begin{cases} u_i^\star(t,v) = \begin{cases} 0 & \textit{if} \quad i \neq i^\star \\ 1 & \textit{if} \quad i = i^\star \end{cases} \\ V(t,v) = v\rho^T \mathbb{E}_w[R_{i^\star}(w)]^{T-t} . \end{cases} \tag{8.41}$$

To prove this, we reason backward. Clearly the relation holds at final time since we have $V(T,v) = v\rho^T$. Assume the relation (8.41) at time $t+1$. From dynamic programming equation (8.12), we deduce that:

$$\begin{aligned} V(t,v) &= \sup_{u \in \mathcal{S}^n} \mathbb{E}_w \left[ V\big(t+1, F(v,u,w)\big) \right] \\ &= \sup_{u \in \mathcal{S}^n} \mathbb{E}_w \left[ v(\langle R(w), u\rangle) \rho^T \mathbb{E}_w[R_{i^\star}(w)]^{T-(t+1)} \right] \\ &= v\rho^T \mathbb{E}_w[R_{i^\star}(w)]^{T-(t+1)} \sup_{u \in \mathcal{S}^n} \mathbb{E}_w[\langle R(w), u\rangle] \\ &= v\rho^T \mathbb{E}_w[R_{i^\star}(w)]^{T-(t+1)} \sup_{u \in \mathcal{S}^n} \langle \mathbb{E}_w[R(w)], u\rangle \end{aligned}$$

.

The optimization problem $\sup_{u \in \mathcal{S}^n} \langle \mathbb{E}_w[R(w)], u\rangle$ is a linear programming problem which admits $u_i^\star(t,v)$ as solution. Therefore, we obtain the desired result:

$$V(t,v) = v\rho^T \mathbb{E}_w[R_{i^\star}(w)]^{T-t} .$$

## 8.9 Cost-effectiveness of grazing and bird community management in farmland

Farmland biodiversity has undergone severe and widespread decline in recent decades. Such is particularly true for bird species and agricultural intensification is designated as a major cause. Targeting agricultural practices may therefore help to restore habitat quality and enhance biodiversity. Hence, livestock raising and grazing should be considered as effective tools for the conservation of bird biodiversity, especially for wader populations foraging and nesting in wet grasslands. Hereafter, following [6], we present a bio-economic model of habitat - biodiversity interactions to provide intensity and timing of grazing for the sustainability of both wader populations and farmer practices. The wader community here includes Lapwings, Redshanks and Godwit species. The model integrates sward and bird population dynamics, grazing strategies and uncertain climatic impacts on a discrete monthly basis at a local scale. Major constraints are based on specific sward heights which promote bird viability. The evaluation is conducted in terms of cost-effectiveness as the grazing profile is chosen from among the robust viable strategies which minimize the expected economic cost of indoor livestock feeding. The numerical method relies on stochastic dynamic programming under the context of robust constraints. Data used for model calibration come from the wet grasslands of the Marais Poitevin (France). The whole set of variables and parameters are summarized in Tabs. 8.1, 8.2 and 8.3.

*Dynamics*

The first component of the model represents a grass sward grazed by suckling cattle on a monthly basis. The sward state consists of a biomass (organic matter) including live $B_L(t)$ and standing dead $B_D(t)$ grass in $(g.m^{-2})$. The live grass increases via new growth and senesces to become dead grass. Both live and dead grass are lost through grazing, with dead grass also vanishing through decay. The management decision is represented by grazing intensity $u(t)$ expressed in livestock unit ($\mathrm{LU.m}^{-2}$). The grass dynamics $B(t) = \big(B_L(t), B_D(t)\big)$ controlled by grazing $u(t)$ may be summarized in the following compact form:

$$B(t+1) = \mathcal{A}\big(t, B(t), w(t)\big)B(t) - \mathcal{G}\big(u(t), B(t)\big)\ . \tag{8.42}$$

Here we comment the different terms.

- The matrix $\mathcal{A}$ reads:

$$\mathcal{A}\big(t, B, w(t)\big) = \begin{pmatrix} \exp\big(-r_S(t)\big) + r_G\big(t, B, w(t)\big) & 0 \\ 1 - \exp\big(-r_S(t)\big) & \exp\big(-r_D(t)\big) \end{pmatrix}.$$

- $r_S(t)$ and $r_D(t)$ stand respectively for the senescence and decay rate coefficients which are time dependent.

- The uncertain growth rate $r_G\big(t,B,w(t)\big)$ is the product of a potential growth rate $r_P\big(t,w(t)\big)$ which varies according to the year and the relative light interception by live mass based on Beer's law:

$$r_G\big(t,B,w(t)\big) = r_P\big(t,w(t)\big)\frac{1-\exp\big(-\beta\mu(B_L+B_D)\big)}{B_L+B_D} ,$$

with $\beta$ a coefficient attenuation related to sun angle and $\mu$ a specific leaf area.
- Stochastic uncertainty $w(t) \in \{-1,1\}$ is assumed to be a sequence of i.i.d. random variables under probability $\mathbb{P}$ with common law $\mathbb{P}(w(t)=1) = p \in ]0,1[$ and $\mathbb{P}(w(t)=-1) = 1-p$, giving a random potential growth rate $r_P(t,w)$ as follows:

$$r_P(t,w) = \overline{r_P}(t)\begin{cases}(1+\sigma) & \text{with probability} \quad p ,\\ (1-\sigma) & \text{with probability} \quad 1-p .\end{cases}$$

Hence the potential growth fluctuates around its mean value $\overline{r_P}(t)$ with a dispersion level of $\sigma$.
- It is here assumed that cow grazing exhibits a preference for live biomass. To meet this requirement, cattle first consume all available live grass $\mathcal{G}_L$ and then all available dead grass $\mathcal{G}_D$ (in $g.m^{-2}$) as follows

$$\begin{cases}\mathcal{G}_L(u,B) &= \min(qu,B_L)\\ \mathcal{G}_D(u,B) &= qu - \mathcal{G}_L(u,B) ,\end{cases}$$

with $q$ the amount of grass required per month.

The second component of the model describes a community of three wader species breeding in the grass sward. The life history of each species $i=1,2,3$ is modeled by a life-cycle graph with two age-classes. The first class $N_{i1}(t)$ consists of sub-adults (first-year individuals) and the second class $N_{i2}(t)$ of adults (second-year or older). Only females are considered. Assuming a pre-breeding census, the monthly dynamics of the species $i$ corresponds to

$$N_i(t+1) = \mathcal{M}_i\Big(t,N_i(t),h\big(B(t)\big)\Big)N_i(t) \tag{8.43}$$

where $h\big(B(t)\big)$ is the height of grass sward $B(t)$. The matrix $\mathcal{M}_i$ is defined by:

$$\mathcal{M}_i(t,N,h) = \begin{cases}\begin{pmatrix} f_{i1}(N,h) & f_{i2}(N,h)\\ s_{i1} & s_{i2}\end{pmatrix} & \text{if} \quad t = t_i^\star \quad [\text{modulo } 12] ,\\ \begin{pmatrix}1 & 0\\ 0 & 1\end{pmatrix} & \text{if} \quad t \neq t_i^\star \quad [\text{modulo } 12] .\end{cases} \tag{8.44}$$

We detail the parameters.

- $s_{ij}$ is the survival of class $j$ in species $i$.
- $t_i^*$ is the month of chick rearing.
- $f_{ij}(N, h)$ is the breeding success of individuals of class $j$ in species $i$. Such breeding success is the product of the proportion of breeding females $\gamma_i$, clutch size $f_{ij}$, primary sex-ratio (i.e. the proportion of females at birth) $\upsilon_i$ and chick survival. This survival depends on two factors that affect reproduction: grass height $h$ as depicted by Fig. 8.3 and density of breeders $N_i$. Specifically, we use a Beverton-Holt-like density-dependence. The breeding success of individuals of class $j$ in species $i$ at time $t$ is thus

$$f_{ij}(N, h) = \gamma_i \upsilon_i f_{ij} \frac{s_{i0}(h)}{1 + b_i(N_{i1} + N_{i2})} \tag{8.45}$$

where $s_{i0}(h)$ is depicted by Fig. 8.3.

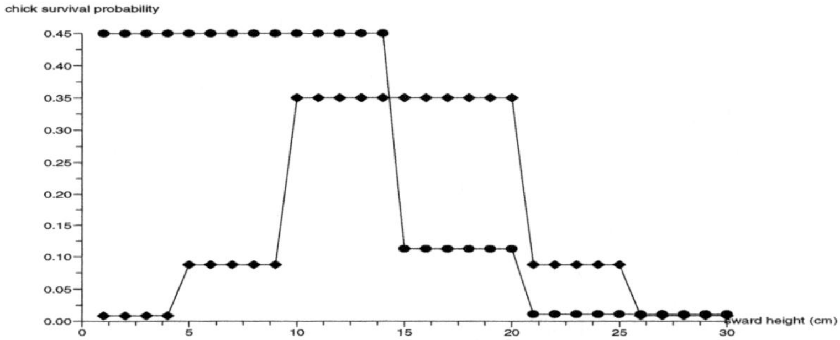


**Fig. 8.3.** Chick survival rates $s_{i0}(h)$ depending on sward height $h$ (cm) for two breeding wader species, lapwings (circle) and redhsanks (diamond).

- We rely on field data from the Marais Poitevin to determine a linear relationship between grass height and biomass:

$$h = h(B) = a_1(B_L + B_D) + a_0 \, . \tag{8.46}$$

*Biodiversity and production constraints*

We first consider an ecological constraint that requires sward states suitable for the chick rearing period of the different bird species. In this context, for each bird species $i$, appropriate sward quality consists of both minimal $h_i^\flat(t)$ and maximal $h_i^\sharp(t)$ grass heights. Typically, a maximal survival for each species can be required at the appropriate rearing period $t_i^*$ through the height corridor:

$$[h_i^\flat(t_i^*), h_i^\sharp(t_i^*)] = \arg\max_h s_{i0}(h) \, .$$

The combination of these thresholds for every species $i$ yields a global height corridor summarized by the following constraints:

$$h^\flat(t) = \max_i h_i^\flat(t) \leq h(t) \leq \min_i h_i^\sharp(t) = h^\sharp(t) . \tag{8.47}$$

Another important requirement we impose is related to the satisfaction of cattle feeding requirements along the months. This production constraint assumes that the demand of sward mass for grazing cannot exceed the available biomass:

$$qu(t) \leq B_L(t) + B_D(t) . \tag{8.48}$$

Moreover the size of the herd is assumed to be set at $u^\sharp$ which implies the following control constraint:

$$u(t) \leq u^\sharp . \tag{8.49}$$

*Cost-effectiveness*

We focus on the economic grazing strategy that consists in minimizing the expected discounted cost related to indoor cattle feeding:

$$\min_{u \in \mathfrak{U}_1^{\mathrm{viab}}(t_0, B_0)} \mathbb{E}_{w(\cdot)} \left[ \sum_{t=0}^{T} \rho^t c\big(u^\sharp - u(t)\big) \right] ,$$

where $\rho \in [0, 1]$ stands for the discount factor and $\mathfrak{U}_1^{\mathrm{viab}}(t_0, B_0)$ means that control $u(t)$ is a viable feedback in the robust sense. A linear relation is here assumed for costs through the relation $c(u^\sharp - u)$ with $c$ standing for the indoor feeding cost of one livestock unit.

*Numerical results*

Here dynamic programming and simulations are performed over a period of $T = 48$ months with neiter present nor future preference $\rho = 1$. The whole set of used variables and parameters detailed in [6] are expounded in Tab. 8.1 for birds (Lapwings, Redshanks, Godwit), in Tab. 8.2 for the sward and in Tab. 8.3 for the agronomic and economic data. The environmental uncertainty $w(t)$ is characterized by a dispersion level $\sigma = 5\%$ of potential growth rate $r_P(t, w(t))$. Expected cost-effective grazing strategies $\mathrm{u}^\star(t, B)$ give grazing decisions $u^\star(t) = \mathrm{u}^\star(t, B(t))$ (Fig. 8.4b) which are characterized by dense grazing intensity in spring followed by autumn grazing. Spring grazing is the major determinant for sward height requirements (Fig. 8.4a). Autumn grazing is the key to increasing the economic efficiency of this strategy. During all periods, productive constraints related to feeding requirements are satisfied. The bird population dynamics $N_i(t)$ resulting from the sward requirements are viable in the robust sense (Fig. 8.4c) despite the uncertain climatic scenarios.

**Table 8.1.** Bird demographic parameters.

| | **Lapwings** | **Redshanks** | **Godwit** |
|---|---|---|---|
| Sub-adult survival $s_{i1}$ | 0.6 | 0.7 | 0.7 |
| Adult survival $s_{i2}$ | 0.7 | 0.8 | 0.8 |
| Adult clutch-size $f_{i2}$ | 4.2 | 4.2 | 4.2 |
| Sub-adult cluth size $f_{i1}$ | 3.7 | 3.7 | 3.7 |
| Proportion of breeding females $\gamma_i$ | 0.75 | 0.75 | 0.75 |
| Sex ratio $\upsilon_i$ | 0.5 | 0.5 | 0.5 |
| Maximal chick survival $s_{i0}$ | 0.45 | 0.35 | 0.35 |
| Upper desirable sward height $h_i^\flat$(t) (cm) | 14 if $t_1^*$=May<br>$+\infty$ otherwise | 20 if $t_2^*$=June<br>$+\infty$ otherwise | $+\infty$ |
| Lower desirable sward height $h_i^\flat$(t) (cm) | 0 | 10 if $t_2^*$=June<br>0 otherwise | 0 |
| Capacity charge parameter $b_i$ | 0.0077 | 0.0077 | 0.0077 |

**Table 8.2.** Sward parameters

| | |
|---|---|
| Senescence of green biomass $r_S$(t) | $(0, 0, 0, 0.9, 0.9, 0.9, 1.35, 0.45, 0.45, 0.45, 0, 0)$ |
| Decay rate of death biomass $r_D$ (t) | $(0, 0, 0, 0, 0, 0, 0.52, 0.52, 1.05, 1.05, 0.52, 0.52)$ |
| Mean potential growth rate $\overline{r_P}(t)$ | $(3, 3, 60, 330, 390, 150, 150, 150, 150, 150, 0, 0)$ |
| Probability $p$ of high growth | 0.5 |
| Maximal deviation $\sigma$ of high growth | 0.1 |
| Leaf area $\mu$ $(m^2 g^{-1})$ | 0.01 |
| Attenuation $\beta$ | 0.5 |
| Slope $a_1$ and intercept $a_0$ relation height - biomass | $(0.07, 4.05)$ |

**Table 8.3.** Agronomic and economic parameters

| | |
|---|---|
| Livestock demand $q$ (g.LU$^{-1}$.month$^{-1}$) | 412.10$^3$ |
| Feeding cost $c$ (euros.LU$^{-1}$.month$^{-1}$) | 30 |
| Maximal livestock $u^\sharp$ (LU.ha$^{-1}$) | 5 |

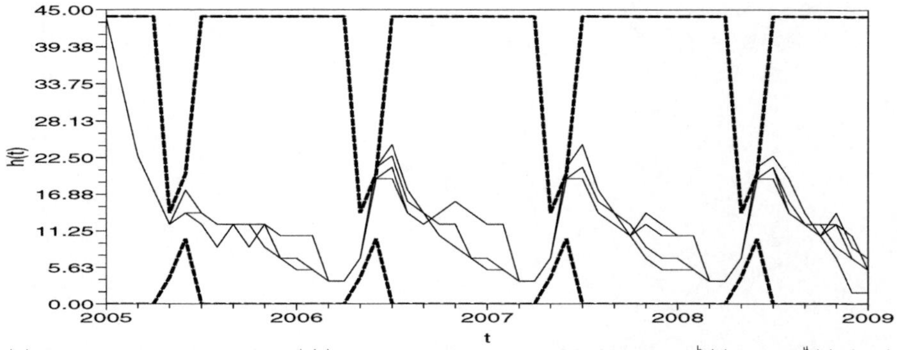


(a) Optimal sward height $h^\star(t)$; upper and lower viable heights $h^\flat(t)$ and $h^\sharp(t)$ (cm)

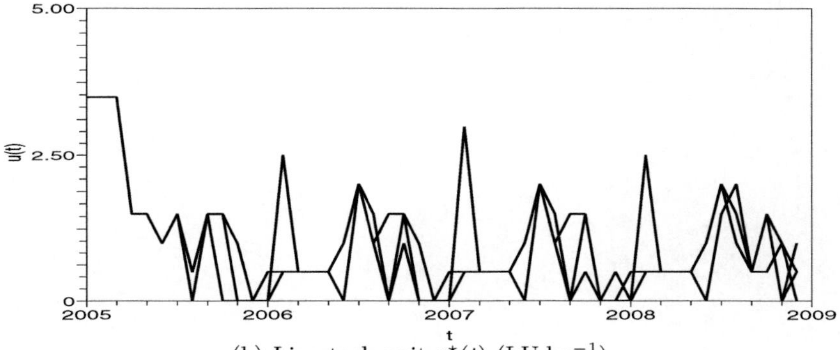


(b) Livestock unit $\mathfrak{u}^\star(t)$ (LU.ha$^{-1}$)

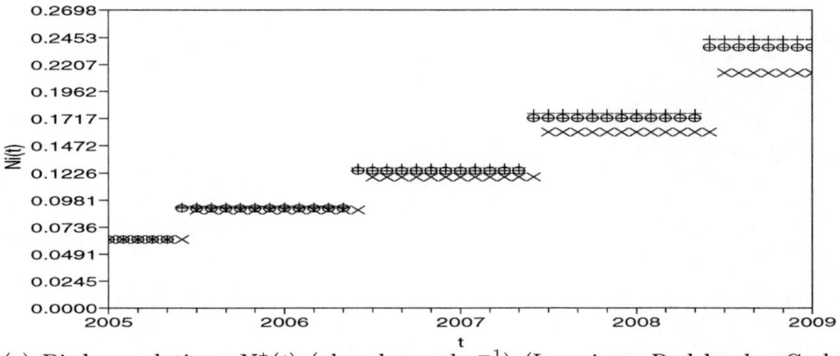


(c) Bird populations $N^\star(t)$ (abundances.ha$^{-1}$) (Lapwings, Redshanks, Godwit)

**Fig. 8.4.** Expected cost-effectiveness paths over time over 4 years from viable initial sward states $B_0 = (475, 150)$ (in g.$m^{-2}$) and bird states $N_{10} = (0.02, 0.04)$, $N_{20} = (0.01, 0.05)$ and $N_{30} = (0.01, 0.05)$ (abundances.ha$^{-1}$). Optimal viable paths are plotted including grazing $\mathfrak{u}^\star(t)$, sward height $h^\star(t)$, and population densities $N_i^\star(t)$ for different climatic scenarios $w(t)$. Every bird population (Lapwings, Redshanks, Godwit) is maintained at a sustainable level through robust grazing strategies $\mathfrak{u}^\star(t, B)$ despite climatic uncertainties $w(t)$.

# References

[1] D. P. Bertsekas. *Dynamic Programming and Optimal Control.* Athena Scientific, Belmont, Massachusets, second edition, 2000. Volumes 1 and 2.

[2] D. P. Bertsekas and S. E. Shreve. *Stochastic Optimal Control: The Discrete-Time Case.* Athena Scientific, Belmont, Massachusets, 1996.

[3] K. W. Byron, J. D. Nichols, and M. J. Conroy. *Analysis and Management of Animal Populations.* Academic Press, 2002.

[4] C. W. Clark. *Mathematical Bioeconomics.* Wiley, New York, second edition, 1990.

[5] G. Heal. *Valuing the Future, Economic Theory and Sustainability.* Columbia University Press, New York, 1998.

[6] M. Tichit, L. Doyen, J.Y. Lemel, and O. Renault. A co-viability model of grazing and bird community management in farmland. *Ecological Modelling*, 206(3-4):277–293, August 2007.

[7] P. Whittle. *Optimization over Time: Dynamic Programming and Stochastic Control*, volume 1. John Wiley & Sons, New York, 1982.

# 9

# Sequential decision under imperfect information

Up to now, it has been postulated that decision makers, regulating agencies or planners know the whole state of the system at each time step, even in the uncertain framework, and use this knowledge for appropriate feedback controls. This situation refers to the case of *perfect information.* Unfortunately, in many problems, only a partial component of the state is known. Likewise, the observation of the state may be corrupted by noise. For instance, a fishery only provides data on targeted species within an ecosystem whose global state remains unknown. In a renewable resource management problem, data on catch effort may be quite uncertain because of non compliance of agents exploiting the natural stock. These situations correspond to *imperfect information* problems [4, 14]. Of course, such a context may deeply alter the quality of the decisions and change the global controlled dynamics and trajectories. This context where information is at stake is of particular relevance for environmental problems since it is related to notions such as *value of information, flexibility* or *learning effect* and *precautionary issues.*

The chapter is organised as follows. The intertemporal decision problem with imperfect observation is detailed in Sect. 9.1. The concept of *value of information* is introduced in Sect. 9.2, and illustrated by precautionary catches and climate change mitigation. The so-called *precautionary effect* is presented in Sect. 9.5 and related to monotone variation of the value of information.

## 9.1 Intertemporal decision problem with imperfect observation

Now, the state is not available in totality, being partially known and/or corrupted by uncertainty. First, we must specify the dynamics/observation couple. Second, admissible feedbacks have to be redefined as now being a function of observation only.

### 9.1.1 Dynamics and observation

The control dynamical system of previous chapters is now assumed not to provide the exact state, but only a partial and/or corrupted function of it. It is described by

$$\begin{cases} x(t+1) = F\big(t, x(t), u(t), w(t)\big) \ , & t = t_0, \dots, T-1 \\ x(t_0) = x_0 & \\ y(t) = H\big(t, x(t), w(t)\big) \ , & t = t_0, \dots, T-1 \\ y(t_0) = y_0 \ , & \end{cases} \tag{9.1}$$

where again $x(t) \in \mathbb{X} = \mathbb{R}^n$ is the state, $u(t) \in \mathbb{U} = \mathbb{R}^p$ stands for the control or decision and $T \in \mathbb{N}^*$ corresponds to the time horizon.

Vector $y(t) \in \mathbb{Y} = \mathbb{R}^m$ represents the system's *observation* or *output* at time $t$ which implies that the decisions strategies will now be feedbacks of the outputs. The *initial observation* is $y_0 \in \mathbb{Y}$.

Compared to the perfect information case, $w(t)$ is now an *extended* disturbance $w(t) = \big(w_{\mathbb{X}}(t), w_{\mathbb{Y}}(t)\big)$ evolving in some space $\mathbb{W}_{\mathbb{X}} \times \mathbb{W}_{\mathbb{Y}}$, which includes both uncertainties $w_{\mathbb{X}}(t) \in \mathbb{W}_{\mathbb{X}} = \mathbb{R}^{q_{\mathbb{X}}}$ affecting the state *via* the dynamics $F$ and uncertainties $w_{\mathbb{Y}}(t) \in \mathbb{W}_{\mathbb{Y}} = \mathbb{R}^{q_{\mathbb{Y}}}$ affecting the observation *via* the *output function*[1] $H$ in (9.1). The initial state $x_0$ is also uncertain now, only partially known through the relation $y_0 = H\big(t_0, x_0, w(t_0)\big)$. The set of possible scenarios for the uncertainties is again represented by $\Omega \subset (\mathbb{W}_{\mathbb{X}} \times \mathbb{W}_{\mathbb{Y}})^{T-t_0}$ in the sense that:

$$w(\cdot) \in \Omega \subset (\mathbb{W}_{\mathbb{X}} \times \mathbb{W}_{\mathbb{Y}})^{T-t_0} \ .$$

The *observation function* $H : \mathbb{N} \times \mathbb{X} \times \mathbb{W} \to \mathbb{Y}$ represents the system's output which may depend on state $x$ and on uncertainty $w$. Here we find important instances of output functions.

- The case of *perfect information*, treated in the previous chapters, refers to:

$$H(t, x, w) = (t, x) \ .$$

- Deterministic systems can be associated with the case:

$$H(t, x, w) = (t, x, w) \ .$$

- The case of observations corrupted by additive noise is depicted by the situation where $y(t) = x(t) + w_{\mathbb{X}}(t)$:

$$H(t, x, w) = H(t, x, w_{\mathbb{X}}, w_{\mathbb{Y}}) = x + w_{\mathbb{X}} \ .$$

[1] The output function $H$ depends on time $t$, state $x$ and disturbance $w$. It might also depend on the control $u$. However, we shall not treat this case.

- The case of *learning* may correspond to information on the state after a given time $t^* < T$, namely $y(t) = h\big(x(t)\big)$ for $t = t^*, \ldots, T-1$ and $y(t) = 0$ for $t = t_0, \ldots, t^* - 1$. The corresponding observation function is:

$$H(t, x, w) = \begin{cases} 0 & \text{if} \quad t = t_0, \ldots, t^* - 1 \,, \\ h(x) & \text{if} \quad t = t^*, \ldots, T-1 \,. \end{cases}$$

### 9.1.2 Decisions, solution map and admissible feedback strategies

As in the certain case, we may require state and decision constraints to be satisfied:

$$\begin{cases} u(t) \in \mathbb{B}\big(t, y(t)\big) \subset \mathbb{U} \,, \\ x(t) \in \mathbb{A}(t) \subset \mathbb{X} \,, \\ x(T) \in \mathbb{A}(T) \,. \end{cases} \tag{9.2}$$

However, since the state $x(t)$ is only partially known by the decision maker, the question arises as to whether state constraints may be achieved by decisions depending only upon partial and corrupted knowledge of the state. Notice also that the control constraints are written as $u(t) \in \mathbb{B}\big(t, y(t)\big)$ and not $u(t) \in \mathbb{B}\big(t, x(t)\big)$.

Decision issues are more complicated than in the perfect observation case in the sense that decisions $u(t) = \mathfrak{u}\big(t, y(t)\big)$ now depend on the output $y(t)$ which stands for the information available to the decision maker. We thus now define *feedbacks* as the set of all mappings from $\mathbb{N} \times \mathbb{Y}$ to $\mathbb{U}$ (and no longer from $\mathbb{N} \times \mathbb{X}$ to $\mathbb{U}$):

$$\mathfrak{U} = \{\mathfrak{u} : \mathbb{N} \times \mathbb{Y} \to \mathbb{U}\} \,. \tag{9.3}$$

Furthermore, given a feedback $\mathfrak{u} \in \mathfrak{U}$, a scenario $w(\cdot) \in \Omega$ and an initial state $x_0$ at time $t_0 \in \{0, \ldots, T-1\}$, the *state map* $x_H[t_0, x_0, \mathfrak{u}, w(\cdot), w(\cdot)]$, *control map* $u_H[t_0, x_0, \mathfrak{u}, w(\cdot), w(\cdot)]$ and *observation map* $y_H[t_0, x_0, \mathfrak{u}, w(\cdot)]$ are now defined by $x_H[t_0, x_0, \mathfrak{u}, w(\cdot), w(\cdot)](t) = x(t)$, $y_H[t_0, x_0, \mathfrak{u}, w(\cdot)] = y(t)$ and $u_H[t_0, x_0, \mathfrak{u}, w(\cdot), w(\cdot)](t) = \mathfrak{u}\big(t, y(t)\big)$ respectively, where $x(\cdot)$ satisfies the dynamic

$$x(t+1) = F\Big(t, x(t), \mathfrak{u}\big(t, y(t)\big), w(t)\Big) \,, \quad t = t_0, \ldots, T-1 \,,$$

with initial condition $x(t_0) = x_0$, and where

$$y(t) = H\big(t, x(t), w(t)\big) \,, \quad t = t_0, \ldots, T-1 \,.$$

In this context, we extend the notion of *admissible feedbacks* $\mathfrak{u} \in \mathfrak{U}_H^{ad}$ as follows treating state constraints in the robust sense:

$$\mathcal{U}_H^{ad}(t_0, y_0) := \left\{ \mathfrak{u} \in \mathfrak{U} \;\middle|\; \begin{array}{l} \text{for all scenario} \quad w(\cdot) \in \Omega \\ \text{for all} \quad x_0 \in \mathbb{X} \quad \text{such that} \quad H(t_0, x_0, w_0) = y_0 \\ x(t) = x_H[t_0, x_0, \mathfrak{u}, w(\cdot)](t) \quad \text{and} \\ u(t) = u_H[t_0, x_0, \mathfrak{u}, w(\cdot)](t) \\ \text{satisfy (9.2).} \end{array} \right\} .$$

Notice how the initial state $x_0$ has now been treated as an uncertain variable as the scenario $w(\cdot)$.

### 9.1.3 Criterion to optimize

The *criterion* $\pi$ is introduced as in Sect. 6.1. The usual case is the separable or additive one:

$$\pi\big(x(\cdot), u(\cdot), w(\cdot)\big) = \sum_{t=t_0}^{T-1} L\big(t, x(t), u(t), w(t)\big) + M\big(T, x(T), w(T)\big) .$$

We shall focus on expected[2] criterion. For this purpose, we assume that a probability $\mathbb{P}$ is given on the space $\mathbb{X} \times \Omega$ (recall that the initial state $x_0$ is an uncertain variable), and that all measurability and integrability assumptions are satisfied for the following expressions to hold true.

The maximal value criterion $V_H(t_0, y_0)$ of the criterion $\pi$ with respect to acceptable feedbacks depends on the observation function $H$ and is now assessed with respect to the initial observation[3] $y_0$

$$V_H(t_0, y_0) := \sup_{\mathfrak{u} \in \mathcal{U}_H^{ad}(t_0, y_0)} \mathbb{E}_{(x_0, w(\cdot))} \Big[ \pi\big(t_0, x(\cdot), u(\cdot), w(\cdot)\big) \mid H(t_0, x_0, w(t_0)) = y_0 \Big] , \tag{9.4}$$

with states $x(t) = x_F[t_0, x_0, \mathfrak{u}, w(\cdot)](t)$ and controls $u(t) = \mathfrak{u}\big(t, y(t)\big)$ the solution maps introduced hereabove. The maximal value criterion is an expectation with respect to probability $\mathbb{P}$ on $\mathbb{X} \times \Omega$, and is *conditional*[4] to those $\big(x_0, w(t_0)\big)$ compatible with the output $y_0$ at time $t_0$.

---

[2] For the robust case, a parallel between probability and cost measures may be found in [1, 3].

[3] In the sequel, the reader will also see an evaluation $V_H(t, x)$ with respect to the state depending on the context. See the following footnote 4 for the mathematical proper dependence with respect to a new state.

[4] The maximal value criterion could be assessed with respect to the conditional law of the state $x_0$ knowing the output $y_0$. More generally, an appropriate new state for indexing the maximal value criterion is the conditional law of the state knowing past observations. We shall not develop this point in this monograph.

## 9.2 Value of information

Consider two information structures $H_\beta$ and $H_\gamma$ (observation functions) taking values from the same space $\mathbb{Y}$. We define the value of information between $H_\beta$ and $H_\gamma$ as the difference between the two associated optimal criteria:

$$\Delta(H_\gamma, H_\beta) := V_{H_\gamma}(t_0, y_0) - V_{H_\beta}(t_0, y_0) \,. \tag{9.5}$$

Whenever the value $\Delta(H_\gamma, H_\beta)$ is strictly positive, an information effect occurs which means that the information structure $H_\gamma$ improves the performance[5] through more adequate decisions compared to the information structure $H_\beta$.

## 9.3 Precautionary catches

We consider the management of an animal population focusing on the trade-off between conservation and invasion issues. According to [12], the population, described by its biomass $B(t)$, varies in time according to uncertain dynamics $g$ and catches $h(t)$ as follows

$$B(t+1) = g\big(R(t), B(t) - h(t)\big) \,,$$

where $R(t)$ plays the role of uncertainty. Assuming a no density-dependent and linear function $g(R, B) = RB$ with $r = R - 1$ the natural resource growth rate, we obtain

$$B(t+1) = R(t)B(t)\big(1 - e(t)\big) \,, \tag{9.6}$$

where $e(t) = h(t)/B(t)$ stands for the catch effort decision constrained by

$$e(t) \in \mathbb{B} = [0, 1] \subset \mathbb{U} = \mathbb{R} \,.$$

Conservation intensity corresponds to $1 - e(t)$. Productivity $R(t)$ of the population is uncertain between two values $0 < R^\flat \leq R^\sharp$:

$$R(t) \in \mathbb{S} = \{R^\flat, R^\sharp\} \subset \mathbb{W} = \mathbb{R} \,.$$

We assume that the sequence $R(0)$, $R(1)\ldots$ consists of independent and identically distributed random variables, and that $\mathbb{P}(R(t) = R^\flat) > 0$ and $\mathbb{P}(R(t) = R^\sharp) > 0$.

We shall consider a two-periods problem: $t_0 = 0$ and $T = 2$. It is postulated that the policy goal is to constrain the biomass level of the population within an ecological window, namely between minimal (survival, conservation) and

---

[5] In the following examples, it may happen that the performance is measured by cost minimization. In this case, $\Delta(H_\beta, H_\gamma) = V_{H_\beta}(t_0, y_0) - V_{H_\gamma}(t_0, y_0)$ is the value of information between $H_\beta$ and $H_\gamma$.

maximal (invasive threshold) safety values $0 < B^\flat < B^\sharp$ at time horizon $T = 2$:

$$B(2) \in \mathbb{A}(2) = [B^\flat, B^\sharp] \subset \mathbb{X} = \mathbb{R} . \tag{9.7}$$

We focus on cost-effective policies which minimize[6] the expected intertemporal costs of conservation $C\big(1 - e(t)\big)$, where the cost function $C(\cdot)$ is assumed to be linear

$$C(1 - e) = c \times (1 - e) ,$$

and which ensure that the target (9.7) is achieved for all scenario.

First, it is only assumed information on state $B(t)$ at each time $t$ and we aim at computing

$$V_{H_\beta}(0, B_0) = \inf_{e(0)\in[0,1]} \mathbb{E}_{R(0)} \left[ \inf_{e(1)\in[0,1]} \mathbb{E}_{R(1)} \left[ C\big(1 - e(0)\big) + \rho C\big(1 - e(1)\big) \right] \right] ,$$

where $\rho \in [0, 1[$ is a discount factor, and under the dynamics constraints (9.6) and safety target (9.7). The observation function is thus:

$$H_\beta(t, B, R) = B \quad \text{for} \quad t = 0, 1.$$

Now assume that learning of the population growth $R(1)$ occurs at the second period ($t^* = 1$). The decision maker faces a more informative structure of information:

$$H_\gamma(1, B, R) = (B, R) , \quad H_\gamma(0, B, R) = B .$$

Focusing on cost-effective policies, we aim at computing:

$$V_{H_\gamma}(0, B_0) = \inf_{e(0)\in[0,1]} \mathbb{E}_{R(0)} \left[ \mathbb{E}_{R(1)} \left[ \inf_{e(1)\in[0,1]} \left[ C\big(1 - e(0)\big) + \rho C\big(1 - e(1)\big) \right] \right] \right] .$$

The difference between the expressions of $V_{H_\beta}(0, B_0)$ and $V_{H_\gamma}(0, B_0)$ is the following: without information one evaluates $\inf_{e(1)\in[0,1]} \mathbb{E}_{R(1)}$ in $V_{H_\beta}(0, B_0)$, while the infimum can go deeper into the expectation in $V_{H_\gamma}(0, B_0)$ which reads $\mathbb{E}_{R(1)} \inf_{e(1)\in[0,1]}$ when information is available. Hence, we have:

$$V_{H_\beta}(0, \widetilde{M_0}) \geq V_{H_\gamma}(0, \widetilde{M_0}) .$$

A question that arises is whether a value of information occurs, *i.e.* $V_{H_\beta}(0, B_0) > V_{H_\gamma}(0, B_0)$ (recall that we minimize costs and see the footnote 6). As examined in [6], the answer depends on the relative size of both the conservation window and the growth uncertainty. The proof is exposed in the Appendix, Sect. A.7.

[6] This differs from the general utility maximization approach followed thus far in the book.

**Result 9.1** *An information effect occurs if the safety ratio $B^\sharp/B^\flat$ is strictly smaller that the growth ratio $R^\sharp/R^\flat$. In this case, the value of information $\Delta(H_\beta, H_\gamma)$ is strictly positive for $B_0 \in \left[\frac{B^\flat}{(R^\flat)^2}, +\infty\right[$ since $V_{H_\beta}(0, B_0) = +\infty$ and $V_{H_\gamma}(0, B_0) < +\infty$.*
*Moreover, for any $B \geq \frac{B^\flat}{(R^\flat)^2}$, the optimal catch feedback (with learning) at first period is given by*

$$\mathfrak{e}^\star(0, B) = 1 - \sqrt{\frac{\rho B^\flat}{\widehat{R}^2 B}} ,$$

*with the harmonic mean $\widehat{R}^{-1} = \mathbb{E}_R[R^{-1}]$.*

This result shows that, when the uncertainty concerning the growth rate of the resource is too large and when no information is revealed to the decision maker, no decisions exist which can ensure a safe final state in the robust sense. However, the resolution of the uncertainty allows for cost-effective management with initial biomass higher than a precautionary threshold given by $B^\flat/(R^\flat)^2$. For invasive species management and conservation, this suggests that a precautionary biomass lower bound makes sense. In the linear case, it is worth pointing out that the "precautionary" decision (with learning) at first period $\mathfrak{e}^\star\big(0, B(0)\big)$ combines distinct population features as it involves both viability biomass $B^\flat$ and certainty equivalent productivity $\widehat{R}$. Furthermore note that this initial decision $\mathfrak{e}^\star\big(0, B(0)\big)$ is not zero. Hence it is the opposite of indecision before the arrival of the information.

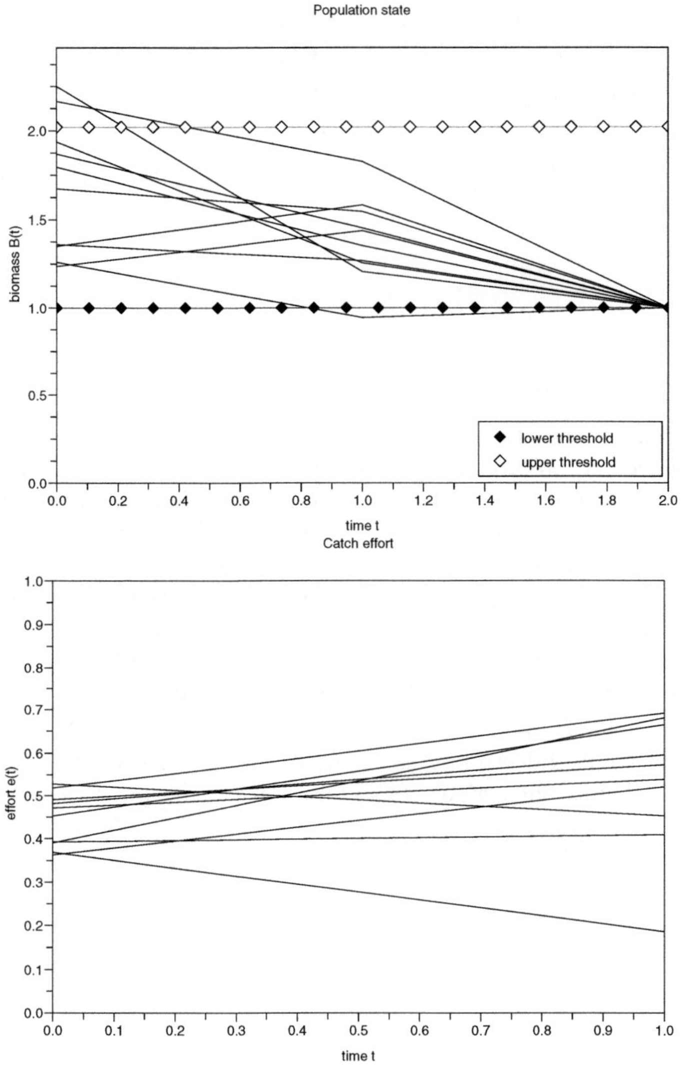


**Fig. 9.1.** Biomasses $(B(0), B(1), B(2))$ and cost-effective (with learning) efforts $(e(0), e(1))$ obtained with the SCILAB code 17. Parameters are $R^\flat = 0.9$, $R^\sharp = 2$, $B^\flat = 1$, $B^\flat = 2$ while $\rho = 0.9$.

SCILAB CODE 17.

```
//
// exec precaution.sce

///////////////////////////////////////////////////////
//                 Cost-effectiveness
// for renewable resource management with learning
///////////////////////////////////////////////////////
//

function [y]=f(B,w)
// linear population dynamics
  y=w*B;
endfunction

function z=dyna(B,e,w)
// controlled uncertain dynamics
  z=f(B*(1-e),w);
endfunction

function e=cost_effective(t,y)
// cost-effective effort
  w_r=w_hat;
  if t==1 then e=1-sqrt(rho*B_min./((w_r^2).*y(1)));end
  if t==2 then e=1-B_min./(y(2).*y(1));end
  e=max(0,e);
endfunction

function y=Obser(t,B,w)
// observation with learning
  y=zeros(1,2);
  if t==1 then y=[B,0];end
  if t==2 then y=[B,w];end
endfunction

xset("window",1);xbasc();
xtitle("Population state",'time t','biomass B(t)');

xset("window",2);xbasc();
xtitle("Catch effort",'time t','effort e(t)');

xset("window",3);xbasc();
xtitle("Uncertainty",'time t','uncertainty w(t)');

w_min=0.9; w_max=2;
// parameters dynamics
w_hat=(w_max-w_min)/(log(w_max)-log(w_min));

// Certainty equivalent  of 1/w
Horizon=3;
// two period problem
B_min=1; B_max=B_min*w_max/(1.1*w_min);
// Safety Constraint
B_prec=B_min/(w_min^(Horizon-1));
// Precautionnary state
rho=1/1.05;
//discount factor
B_simu=10;
// Simulation number

for i=1:B_simu
// Simulations
B=B_prec+rand(1,Horizon)*(B_max-B_min);
//  precautionnary initial conditions
// B(0)>= B_min/(w_min^(Horizon-1))
  w=w_min+(w_max-w_min)*rand(1,Horizon-1);
  // Random or growth along time
for t=1:Horizon-1
// precautionary Trajectory B(.) e(.)
    y=Obser(t,B(t),w(t));
    e(t)=cost_effective(t,y);
    B(t+1)=dyna(B(t),e(t),w(t));
end,

  rect1=[0,0,Horizon-1,(Horizon-1)*B_prec];
  rect2=[0,0,Horizon-2,1];
  rect3=[0,w_min,Horizon-2,w_max];
  xB=[0:1:Horizon-1];xe=[0:1:Horizon-2];
  //
  xset("window",1);
  plot2d(xB,[B' B_min+zeros(1,Horizon)' ...
 B_max+zeros(1,Horizon)'],rect=rect1);
  // drawing diamonds, crosses, etc. to identify the curves
  // plot2d(xB,B',style=1);
  abcisse=linspace(0,Horizon-1,20);
  plot2d(abcisse,[B_min*ones(abcisse)' B_max*ones(abcisse)'],..
 style=-[4,5]);
 legends(["lower threshold";"upper threshold"],-[4,5],'lr');
 //
  xset("window",2);plot2d(xe,e,rect=rect2);
  xset("window",3);plot2d(xe,w,rect=rect3);
end

//
```

## 9.4 Information effect in climate change mitigation

We first present an example developed in [8] (see also [9]). Consider a version of mitigation policies for carbon dioxyde emissions as exposed in Sect. 2.3. Introducing the following difference between $CO_2$ concentrations

$$\widetilde{M}(t) := M(t) - M_{-\infty} \tag{9.8}$$

the carbon cycle (2.20) may be written as

$$\widetilde{M}(t+1) = (1-\delta)\,\widetilde{M}(t) + \alpha E_{\text{BAU}}\big(1 - a(t)\big)\,, \quad t = 0, 1 \tag{9.9}$$

where we have assumed stationary emissions $E_{\text{BAU}}$, with abatement rate $a(t) \in [0,1]$. Here, only two periods are considered, say $t = 0$ being today, $t = 1$ for 25 years ahead and $t = 2$ for 50 years.

The decision problem is one of cost minimization, inspired by [5]. Abatement costs are $C\big(a(0)\big)$ and $C\big(a(1)\big)$, with the convex functional form

$$C(a) = ca^2 \tag{9.10}$$

taken for simplicity. There is a final cost $\theta D(\widetilde{M}(2))$ corresponding to damages due to climate change, with here also a quadratic expression

$$D(\widetilde{M}) = \widetilde{M}^2 \tag{9.11}$$

and $\theta \geq 0$ an unknown parameter which measures the sensibility of damages to the $\mathrm{CO}_2$ concentration level. Two successive reduction decisions $a(0)$ and $a(1)$ have to be taken in order to minimize[7] the whole discounted cost

$$\pi(\widetilde{M}(\cdot), \theta(\cdot), a(\cdot)) = C\big(a(0)\big) + \rho C\big(a(1)\big) + \theta(2) D\big(\widetilde{M}(2)\big) .$$

We have introduced a new state variable $\theta(t)$ with state equation

$$\theta(t+1) = \theta(t) \,, \quad t = 0,1 \quad \text{and} \quad \theta(0) = \theta_0 \,. \tag{9.12}$$

Thus $\theta(2) = \theta(1) = \theta_0$ is the value of the unknown parameter.

In the perfect information case, the whole state $\big(\widetilde{M}(t), \theta(t)\big)$ is observed at each time $t$. This corresponds to perfect knowledge of $\theta$. We shall not consider this case.

In absence of information about the parameter $\theta$ (now become component of the state $(\widetilde{M}, \theta)$), only part $\widetilde{M}(t)$ of the whole state is observed. This corresponds to an output function

$$H_\beta(t, \widetilde{M}, \theta) = \widetilde{M} \,, \quad t = 0,1 \,. \tag{9.13}$$

Now, scientific information may be revealed at mid-term $t^\star = 1$, after the first decision, but before the climatic damages. Such a case of *learning* about the parameter $\theta$ is depicted by

$$H_\gamma(0, \widetilde{M}, \theta) = (\widetilde{M}, 0) \quad \text{and} \quad H_\gamma(1, \widetilde{M}, \theta) = (\widetilde{M}, \theta) \,. \tag{9.14}$$

Notice that no additional disturbance set $\mathbb{W}$ is needed, since no disturbances affect the dynamics, the observations[8] or the costs.

In the stochastic framework, a probability distribution $\mathbb{P}_0$ on $\theta$ is supposed to be given, while $\widetilde{M}_0 = \widetilde{M}(0)$ is supposed to be known. In both cases of information structure, the initial decision $a(0)$ will only depend upon $\widetilde{M}(0)$.

In absence of information, $a(1)$ will only depend upon $\widetilde{M}(1)$ and we obtain[9]

---

[7] See the previous footnote 6 in this Chapter.

[8] Noisy observations would require an uncertainty $w$ and $H(1, \widetilde{M}, \theta, w) = (\widetilde{M}, \theta{+}w)$ for instance.

[9] The expression $\mathbb{E}_{\theta_0}^{\mathbb{P}_0}$ denotes an expectation with respect to the probability distribution $\mathbb{P}_0$ of the random variable $\theta_0 = \theta$.

$$V_{H_\beta}(0,\widetilde{M}_0) = \inf_{0\leq a(0)\leq 1}\left[C\big(a(0)\big) + \inf_{0\leq a(1)\leq 1}\mathbb{E}_{\theta_0}^{\mathbb{P}_0}\left[\rho C\big(a(1)\big) + \theta(2)D(\widetilde{M}(2))\right]\right] .$$

When scientific information may be revealed at mid-term, the value of $\theta$ is revealed at $t^* = 1$, so that $a(1)$ will depend upon both $\widetilde{M}(1)$ and $\theta$, giving thus

$$V_{H_\gamma}(0,\widetilde{M}_0) = \inf_{0\leq a(0)\leq 1}\left[C\big(a(0)\big) + \mathbb{E}_{\theta_0}^{\mathbb{P}_0}\left[\inf_{0\leq a(1)\leq 1}\Big(\rho C\big(a(1)\big) + \theta(2)D(\widetilde{M}(2))\Big)\right]\right] .$$

The difference between the expressions of $V_{H_\beta}(0,\widetilde{M}_0)$ and $V_{H_\gamma}(0,\widetilde{M}_0)$ is the following: without information one evaluates $\inf_{0\leq a(1)\leq 1}\mathbb{E}_{\theta_0}^{\mathbb{P}_0}$ in $V_{H_\beta}(0,\widetilde{M}_0)$, while the infimum can go deeper into the expectation in $V_{H_\gamma}(0,\widetilde{M}_0)$ which reads $\mathbb{E}_{\theta_0}^{\mathbb{P}_0}\inf_{0\leq a(1)\leq 1}$ when information is available. Hence, we have:

$$V_{H_\beta}(0,\widetilde{M}_0) \geq V_{H_\gamma}(0,\widetilde{M}_0) .$$

It turns out that an information effect generally holds for such a mitigation problem. This suggests that costs of $CO_2$ mitigation are reduced when learning of uncertainties occurs. The proof is given in the Appendix, Sect. A.7.

**Result 9.2** *Assume that the unknown parameter $\theta$ follows a probability distribution $\mathbb{P}_0$ having support not reduced to a single point and included in $[\theta^\flat, \theta^\sharp]$ such that*

$$0 \leq \theta^\flat < \theta^\sharp \leq \frac{\rho c}{\alpha E_{\text{BAU}}(1-\delta)\big((1-\delta)\widetilde{M}_0 + \alpha E_{\text{BAU}}\big)} . \tag{9.15}$$

*Then, there is an information effect in the sense that*

$$\Delta(H_\beta, H_\gamma) = V_{H_\beta}(0,\widetilde{M}_0) - V_{H_\gamma}(0,\widetilde{M}_0) > 0 .$$

The question whether first optimal abatement $a^\star(0)$ is also reduced with such a learning mechanism is examined in Sect. 9.5.

## 9.5 Monotone variation of the value of information and precautionary effect

In the theoretical literature on environmental irreversibility and uncertainty [2, 10, 11], an important issue is the so called “irreversibility effect”. This effect states roughly that, when there is a source of irreversibility in the system we control, then "the learning effect is precautionary" in the following sense. By learning effect we refer to how the first-period optimal decision is modified when the decision maker considers that information will arrive in the future. By precautionary, we mean that the first-period optimal decision is lower (less emissions, less consumption, more cautious) with than without information prospect. Such question is also treated in [7, 13]. We present here a general framework and an analysis which focuses upon the value of information as a function of any possible initial decision.

Consider a two-periods control system where, for simplicity, the dynamics is not affected by uncertainty as in

$$\begin{cases} x(t+1) = F\big(t, x(t), u(t)\big) \ , & t = 0, 1 \\ x(t_0) = x_0 & \\ y(t) = H\big(t, x(t), w(t)\big) \ , & t = 0, 1 \\ y(t_0) = y_0 \ . & \end{cases} \tag{9.16}$$

Suppose that the control variable $u$ is scalar:

$$u \in \mathbb{B}(0) \subset \mathbb{U} = \mathbb{R} \ .$$

Irreversibility may be captured by the property that the second decision $u(1)$ belongs to $\mathbb{B}(1, x(1))$, where the set $\mathbb{B}(1, x)$ indeed depends upon state $x$.

The optimization problem is (without discount factor)

$$\sup_{\begin{cases} u(0) \in \mathbb{B}(0) \\ u(1) \in \mathbb{B}(1, x(1)) \end{cases}} \mathbb{E}\Big[L\big(0, x(0), u(0), w(0)\big) + L\big(1, x(1), u(1), w(1)\big) + M\big(2, x(2), w(2)\big)\Big] \ .$$

Suppose that the initial decision $u(0)$ is taken without information, that is $H(0, x_0, w_0) = 0$. For the second time, let us consider $H_\beta$ and $H_\gamma$, two information structures (not necessarily comparable in the sense that one is finer than the other).

Define the value of information at time $t = 1$ starting from state $x_1$ by:

$$V_{H_\beta}(1, x_1) := \sup_{u(1) \in \mathbb{B}(t, x_1)} \mathbb{E}\Big[L\big(1, x_1, u(1), w(1)\big) + M\big(2, x(2), w(2)\big)\Big] \ . \tag{9.17}$$

The *value of substituting information structure* $H_\beta$ *for* $H_\gamma$ at time $t = 1$ is the following function of the state $x_1$ at time $t = 1$:

$$\Delta^{(1)}(H_\gamma, H_\beta)(x_1) := V_{H_\gamma}(1, x_1) - V_{H_\beta}(1, x_1) \ . \tag{9.18}$$

**Proposition 9.3.** *Assuming their existence and uniqueness, let* $u^\star_{H_\beta}(0)$ *be the optimal initial decision with information structure* $H_\beta$*, and the same for* $u^\star_{H_\gamma}(0)$*. If the function* $u_0 \in \mathbb{B}(0) \mapsto \Delta^{(1)}(H_\gamma, H_\beta)\big(F(0, x_0, u_0)\big)$ *is decreasing,* *then precautionary effect holds in the sense that the initial optimal decisions are ranked as follows:*

$$u^\star_{H_\gamma}(0) \leq u^\star_{H_\beta}(0) .$$

The proof, exposed in [8] and given in the Appendix, Sect. A.7, relies on the mathematical property that two functions $f_\beta$ and $f_\gamma$ having the *increasing differences property* ($u \mapsto f_\beta(u) - f_\gamma(u)$ is increasing) verify $u_\gamma \leq u_\beta$, where $u_\beta := \arg\max_u f_\beta(u)$ and $u_\gamma := \arg\max_u f_\gamma(u)$ are supposed to exist and to be unique.

## 9.6 Precautionary effect in climate change mitigation

Returning to the climate change mitigation problem exposed in Sect. 9.4, let us see whether precautionary effect holds true.

Without information, we have:

$$V_\beta(1, \widetilde{M}_1) = \inf_{0 \leq a(1) \leq 1} \mathbb{E}^{\mathbb{P}_0}_{\theta_0} \Big[ \rho C\big(a(1)\big) + \theta(2) D\Big((1-\delta)\widetilde{M}_1 + \alpha E_{\text{BAU}}\big(1 - a(1)\big)\Big)\Big] .$$

Similarly, in the case of learning about damage intensity $\vartheta = \theta_0$, we compute

$$V_\gamma(1, \widetilde{M}_1) = \mathbb{E}^{\mathbb{P}_0}_{\theta_0} \left[ \inf_{0 \leq a(1) \leq 1} \Big( \rho C\big(a(1)\big) + \theta(2) D\big((1-\delta)\,\widetilde{M}_1 + \alpha E_{\text{BAU}}(1 - a(1))\big)\Big)\right],$$

Under the assumptions of Result 9.2, it is proved in the Appendix, Sect. A.7, that

$$V_\gamma(1, \widetilde{M}_1) - V_\beta(1, \widetilde{M}_1) =$$
$$\left( \mathbb{E}^{\mathbb{P}_0}_{\theta_0} \left[ \frac{\theta \rho c}{\rho c + \theta \alpha^2 E^2_{\text{BAU}}} \right] - \frac{\mathbb{E}^{\mathbb{P}_0}_{\theta_0}[\theta]\, \rho c}{\rho c + \mathbb{E}^{\mathbb{P}_0}_{\theta_0}[\theta]\, \alpha^2 E^2_{\text{BAU}}} \right) \left((1-\delta)\widetilde{M}_1 + \alpha E_{\text{BAU}}\right)^2 .$$

Thus, the value of substituting information structure $H_\beta$ for $H_\gamma$ at time $t = 1$ (recall that we minimize costs and see the footnote 6 in this Chapter) is decreasing with the state $\widetilde{M}_1$. Indeed, we have

$$\frac{d}{d\widetilde{M}_1} \big( V_{H_\gamma}(1, \widetilde{M}_1) - V_{H_\beta}(1, \widetilde{M}_1) \big) =$$
$$\left( \mathbb{E}^{\mathbb{P}_0}_{\theta_0} \left[ \frac{\theta \rho c}{\rho c + \theta \alpha^2 E^2_{\text{BAU}}} \right] - \frac{\mathbb{E}^{\mathbb{P}_0}_{\theta_0}[\theta]\, \rho c}{\rho c + \mathbb{E}^{\mathbb{P}_0}_{\theta_0}[\theta]\, \alpha^2 E^2_{\text{BAU}}} \right) (1-\delta)\big((1-\delta)\widetilde{M}_1 + \alpha E_{\text{BAU}}\big) < 0.$$

This follows from the strict concavity of $\theta \mapsto \dfrac{\theta \rho c}{\rho c + \theta \alpha^2 E^2_{\text{BAU}}}$ and from Jensen inequality.

Now since the function $a(0) \mapsto \widetilde{M}(1) = (1-\delta)\,\widetilde{M}_0 + \alpha E_{\text{BAU}}\big(1 - a(0)\big)$ is strictly decreasing, we can apply Proposition 9.3. This gives the following Result.

**Result 9.4** *Under the assumptions of Result 9.2, there is a precautionary effect in the sense that the initial optimal abatement is lower with information than without:*

$$a^{\star}_{H_{\gamma}}(0) \leq a^{\star}_{H_{\beta}}(0) .$$

# References


[1] M. Akian, J.-P. Quadrat, and M. Viot. Duality between probability and optimization. In J. Gunawardena, editor, *Idempotency*. Cambridge University Press, 1998.

[2] K. J. Arrow and A. C. Fisher. Environmental preservation, uncertainty, and irreversibity. *Quarterly Journal of Economics*, 88:312–319, 1974.

[3] P. Bernhard. A separation theorem for expected value and feared value discrete time control. Technical report, INRIA, Projet Miaou, Sophia Antipolis, Décembre 1995.

[4] D. P. Bertsekas and S. E. Shreve. *Stochastic Optimal Control: The Discrete-Time Case*. Athena Scientific, Belmont, Massachusets, 1996.

[5] A. Dixit and R. Pindick. *Investment under Uncertainty*. Princeton University Press, 1994.

[6] L. Doyen and J.-C. Pereau. The precautionary principle as a robust cost-effectiveness problem. *Environmental Modeling and Assessment*, revision.

[7] L. G. Epstein. Decision making and temporal resolution of uncertainty. *International Economic Review*, 21:269–283, 1980.

[8] L. Gilotte. *Incertitude, inertie et choix optimal. Modèles de contrôle optimal appliqués au choix de politiques de réduction des émissions de gaz à effet de serre.* PhD thesis, École des ponts, ParisTech, Université Paris-Est, 2004.

[9] C. Gollier, B. Jullien, and N. Treich. Scientific progress and irreversibility: an economic interpretation of the "precautionary principle". *Journal of Public Economics*, 75:229–253, 2000.

[10] C. Henry. Investment decisions under uncertainty: The "irreversibility effect". *American Economic Review*, 64(6):1006–1012, 1974.

[11] C. Henry. Option values in the economics of irreplaceable assets. *Review of Economic Studies*, 41:89–104, 1974.

[12] L. J. Olson and S. Roy. Dynamic efficiency of conservation of renewable resources under uncertainty. *Journal of Economic Theory*, 95(2):186–214, December 2000.

[13] A. Ulph and D. Ulph. Global warming, irreversibility and learning. *The Economic Journal*, 107(442):636–650, 1997.

[14] P. Whittle. *Optimization over Time: Dynamic Programming and Stochastic Control*, volume 1. John Wiley & Sons, New York, 1982.

# A

# Appendix. Mathematical Proofs

## A.1 Mathematical proofs of Chap. 3

### Proof of Theorem 3.3

*Proof.* This proof is inspired by [5]. Recall that the solution of equation $x(t+1) = Ax(t)$ is given by:

$$x(t) = A^t x(0) .$$

Moreover, for any matrix $A$, there exists a nonsingular matrix $P$ (possibly complex) that transforms $A$ into its Jordan form, that is

$$P^{-1}AP = J = \mathrm{Diag}[J_1, J_2, \ldots, J_r] ,$$

where $r$ is the number of distinct eigenvalues of $A$ and $J_i$ is the Jordan block associated with the eigenvalue $\lambda_i$ of $A$ ($\mathrm{spec}(A) = \{\lambda_1, \ldots, \lambda_r\}$). The Jordan Block $J_i$ takes the form of

$$J_i = \begin{pmatrix} \lambda_i & 1 & 0 & \ldots\ldots & & 0 \\ 0 & \lambda_i & 1 & 0 & \ldots & 0 \\ \vdots & & \ddots & & & \vdots \\ \vdots & & & \ddots & & 0 \\ \vdots & & & & \ddots & 1 \\ 0 & & \ldots\ldots & & 0 & \lambda_i \end{pmatrix}_{\nu_i \times \nu_i} ,$$

where $\nu_i = \nu(\lambda_i)$ is the order of multiplicity of eigenvalue $\lambda_i$ of $A$. Therefore

$$A^t = PJ^tP^{-1} = P\mathrm{Diag}[J_1^t, J_2^t, \ldots, J_r^t]P^{-1} \sum_{i=1}^{r} \sum_{k=1}^{\nu_i} t^{k-1} \lambda_i^t R_{ik} ,$$

where the matrix with general term $R_{ik}$ depends upon $P$ and the upper part of the $J_i$. The state $x(t)$ converges toward 0 if and only if $\lim_{t \to +\infty} A^t = 0$. Equivalently, we have

$$\lim_{t \to +\infty} t^{k-1} \lambda_i^t = 0 \,, \quad \forall i = 1, \ldots, r \,,$$

which holds true if and only if $\sup_{i=1,\ldots,r} |\lambda_i| < 1$.

## Proof of Result 3.6

*Proof.* From (3.23a)-(3.23b), $(N_{\mathrm{E}}, R_{\mathrm{E}})$ is an equilibrium if, and only if,

$$0 = N_{i,\mathrm{E}}(f_i R_{\mathrm{E}} - d_i) \,, \quad i = 1, \ldots, n \,,$$
$$0 = S_{\mathrm{E}} - a R_{\mathrm{E}} - \sum_{i=1}^{n} w_i f_i R_{\mathrm{E}} N_{i,\mathrm{E}} \,.$$

Excluding the exceptional case where $d_i/f_i = d_j/f_j$ for at least one pair $i \neq j$, a non-zero equilibrium $N_{\mathrm{E}}$ has only one non-zero component. This is the *exclusion principle.*

We thus have at most $n$ possible equilibria. Let us focus on one of them, characterized as follows with $S_{\mathrm{E}}$ large enough. For the sake of simplicity, assume temporarily that $i_{\mathrm{E}} = n$ in the sense that:

$$\begin{cases} R_{\mathrm{E}} = \dfrac{d_n}{f_n} \,, \\ N_{i,\mathrm{E}} = \begin{cases} \dfrac{S_{\mathrm{E}} - R_{\mathrm{E}} a}{R_{\mathrm{E}} w_n f_n} > 0 & \text{if} \quad i = n \,, \\ 0 & \text{if} \quad i \neq n \,. \end{cases} \end{cases}$$

Putting $x = (N, R)$, the Jacobian matrix of dynamic $F$ given by (3.23a)-(3.23b) at $x_{\mathrm{E}} = (N_{\mathrm{E}}, R_{\mathrm{E}})$ is:

$$\frac{\partial F}{\partial x}(N_{\mathrm{E}}, R_{\mathrm{E}}) = \begin{pmatrix} 1 + \Delta_t(f_1 R_{\mathrm{E}} - d_1) & 0 & 0 \ \ldots & 0 & 0 \\ 0 & 1 + \Delta_t(f_2 R_{\mathrm{E}} - d_2) & 0 & 0 & 0 \\ 0 & 0 & \ddots \ \ddots & \vdots & \vdots \\ \vdots & \ddots & \ddots & 0 & \vdots \\ 0 & \ldots & \ldots \ 0 & 1 & \Delta_t N_{n,\mathrm{E}} f_n \\ -\Delta_t w_1 f_1 R_{\mathrm{E}} & -\Delta_t w_2 f_2 R_{\mathrm{E}} & \ldots \ \ldots & -\Delta_t w_n f_n R_{\mathrm{E}}; & 1 - \Delta_t(a + w_n f_n N_{n,\mathrm{E}}) \end{pmatrix} .$$

The eigenvalues $\lambda_i$ are solutions of $\mathrm{Det}(\frac{\partial F}{\partial x}(N_{\mathrm{E}}, R_{\mathrm{E}}) - \lambda I) = 0$, that is:

$$0 = \prod_{i=1}^{n-1} (1 + \Delta_t(f_i R_{\mathrm{E}} - d_i) - \lambda) \begin{vmatrix} 1 - \lambda & ; & \Delta_t N_{n,\mathrm{E}} f_n \\ -\Delta_t w_n f_n R_{\mathrm{E}} & ; & 1 - \Delta_t(a + w_n f_n N_{n,\mathrm{E}}) - \lambda \end{vmatrix} .$$

We can see at once that the $n-1$ first eigenvalues are $\lambda_i = 1 + \Delta_t(f_i R_{\mathrm{E}} - d_i)$. The last two $\lambda_n$ and $\lambda_{n+1}$ are solution of $P(\lambda) = 0$ where

$$P(\lambda) = (1 - \lambda)(1 - \Delta_t(a + w_n f_n N_{n,\mathrm{E}}) - \lambda) + \Delta_t^2 N_{n,\mathrm{E}} f_n^2 w_n R_{\mathrm{E}} \,.$$

For a small enough $\Delta_t$, it appears that $|\lambda_n| < 1$ and $|\lambda_{n+1}| < 1$ because the necessary and sufficient conditions $P(1) > 0$, $P(-1) > 0$ and $P(0) < 1$ (valid for a second-order polynomial $P$) of the so-called Jury test [4, 1] are satisfied. Indeed,

$$\begin{aligned} P(1) &= \Delta_t^2 N_{n,\mathrm{E}} f_n^2 w_n R_{\mathrm{E}} \; , \\ P(-1) &= 2\big(2 - \Delta_t(a + w_n f_n N_{n,\mathrm{E}})\big) + \Delta_t^2 N_{n,\mathrm{E}} f_n^2 w_n R_{\mathrm{E}} \; , \\ P(0) &= 1 - \Delta_t(a + w_n f_n N_{n,\mathrm{E}}) \; . \end{aligned}$$

On the other hand, since $R_{\mathrm{E}} = d_n/f_n$ and since we have excluded the exceptional case where $d_i/f_i = d_j/f_j$ for at least one pair $i \neq j$, then $f_i R_{\mathrm{E}} - d_i \neq 0$ for $i = 1,\ldots, n-1$. If $f_i R_{\mathrm{E}} - d_i > 0$ for at least one $i$, then $\lambda_i = 1 + \Delta_t(f_i R_{\mathrm{E}} - d_i) > 1$ and the equilibrium $x_{\mathrm{E}} = (N_{\mathrm{E}}, R_{\mathrm{E}})$ is unstable by Theorem 3.4. If $f_i R_{\mathrm{E}} - d_i < 0$ for $i = 1,\ldots, n-1$, then $0 < \lambda_i = 1 + \Delta_t(f_i R_{\mathrm{E}} - d_i) < 1$ for a small enough $\Delta_t > 0$ and the equilibrium $x_{\mathrm{E}} = (N_{\mathrm{E}}, R_{\mathrm{E}})$ is asymptotically stable by Theorem 3.4.

Thus, among the possible equilibria, the only asymptotic stable equilibrium satisfies the conditions:

$$\begin{cases} R_{\mathrm{E}} = \min\limits_{i=1,\ldots,n} \dfrac{d_i}{f_i} = \dfrac{d_{i_{\mathrm{E}}}}{f_{i_{\mathrm{E}}}} \; , \\[2ex] N_{i,\mathrm{E}} = \begin{cases} \dfrac{S_{\mathrm{E}} - R_{\mathrm{E}} a}{R_{\mathrm{E}} w_{i_{\mathrm{E}}} f_{i_{\mathrm{E}}}} > 0 & \text{if} \quad i = i_{\mathrm{E}} \; , \\ 0 & \text{if} \quad i \neq i_{\mathrm{E}} \; . \end{cases} \end{cases}$$

## A.2 Mathematical proofs of Chap. 4

### Proof of Proposition 4.2.

*Proof.* The proof is by backward induction. At final time $t = T$, it is clear that $\mathbb{V}\mathrm{iab}(T) = \mathbb{A}(T)$.

Assume that the relation (4.11) is true at time $t+1 \leq T$. From the very definition of the viability kernel at time $t$, we can write

$$\mathbb{V}\mathrm{iab}(t) = \left\{ x \in \mathbb{X} \;\middle|\; \begin{array}{l} \exists\,(u(t),\dots,u(T-1)) \text{ and } \exists\,(x(t),\dots,x(T-1),x(T)) \\ \quad \text{such that } x(t) = x, \\ x(s+1) = F\big(s,x(s),u(s)\big),\ u(s) \in \mathbb{B}\big(s,x(s)\big), \\ \forall s = t,\dots,T-1 \\ x(s) \in \mathbb{A}(s),\ \forall s = t,\dots,T-1 \end{array} \right\}$$

$$= \left\{ x \in \mathbb{A}(t) \;\middle|\; \begin{array}{l} \exists\, u \in \mathbb{B}(t,x) \\ \exists\,(u(t+1),\dots,u(T-1)) \text{ and} \\ \exists\,(x(t+1),\dots,x(T-1),x(T)) \\ \quad \text{such that } x(t+1) = F(t,x,u) \\ x(s+1) = F\big(s,x(s),u(s)\big),\ u(s) \in \mathbb{B}\big(s,x(s)\big), \\ \qquad \forall s = t+1,\dots,T-1 \\ x(s) \in \mathbb{A}(s),\ \forall s = t+1,\dots,T-1 \end{array} \right\}$$

$$= \Big\{ x \in \mathbb{A}(t) \mid \exists\, u \in \mathbb{B}(t,x)\ , \quad F(t,x,u) \in \mathbb{V}\mathrm{iab}(t+1) \Big\}$$

and we conclude.

**Proof of Proposition 4.3.**

*Proof.* The unicity of equation (4.13) is clear. For the sake of simplicity, let us denote $W(t,x) := \Psi_{\mathbb{V}\mathrm{iab}(t)}(x)$.

At final time $t = T$, it is clear that $\mathbb{V}\mathrm{iab}(T) = \mathbb{A}(T)$. Expressed with characteristic functions, we have the equivalent of the desired statement:

$$W(T,\cdot) = \Psi_{\mathbb{V}\mathrm{iab}(T)}(\cdot) = \Psi_{\mathbb{A}(T)}(\cdot)\ .$$

At time $t \leq T-1$, let us evaluate the right hand side

$$\inf_{u \in \mathbb{B}(t,x)} \big\{ \Psi_{\mathbb{A}(t)}(x) + W\big(t+1, F(t,x,u)\big) \big\}$$

of (4.13). Whenever $x \in \mathbb{V}\mathrm{iab}(t)$, we know from Proposition 4.2 that there exists $u \in \mathbb{B}(t,x)$ satisfying $F(t,x,u) \in \mathbb{V}\mathrm{iab}(t+1)$. Expressed with characteristic functions, this is equivalent to claiming the existence of $u \in \mathbb{B}(t,x)$ such that $\Psi_{\mathbb{V}\mathrm{iab}(t+1)}\big(F(t,x,u)\big) = 0$. In other words, we have:

$$0 = \min_{u \in \mathbb{B}(t,x)} \big\{ W\big(t+1, F(t,x,u)\big) \big\}\ .$$

Moreover, $x \in \mathbb{A}(t)$ yields $\Psi_{\mathbb{A}(t)}(x) = 0$ and, consequently:

$$0 = \min_{u \in \mathbb{B}(t,x)} \big\{ \Psi_{\mathbb{A}(t)}(x) + W\big(t+1, F(t,x,u)\big) \big\}\ .$$

Thus the assertion holds true since $W(t,x) = \Psi_{\mathbb{V}\mathrm{iab}(t)}(x) = 0$.

Whenever $x \notin \mathbb{V}\mathrm{iab}(t)$, we distinguish two cases.

- If $x \notin \mathbb{A}(t)$, obviously $\Psi_{\mathbb{A}(t)}(x) = +\infty$ and therefore:

$$W(t,x) = +\infty = \inf_{u \in \mathbb{B}(t,x)} \{\Psi_{\mathbb{A}(t)}(x) + W\big(t+1, F(t,x,u)\big)\} .$$

- If $x \in \mathbb{A}(t) \backslash \mathbb{V}\mathrm{iab}(t)$, we know from Proposition 4.2 that, for any $u \in \mathbb{B}(t,x)$, we have $F(t,x,u) \notin \mathbb{V}\mathrm{iab}(t+1)$. Consequently, we obtain

$$W(t,x) = +\infty = \inf_{u \in \mathbb{B}(t,x)} \{W\big(t+1, F(t,x,u)\big)\} ,$$

which concludes the proof.

**Proof of Result 4.10.**

*Proof.* We illustrate how dynamic programming equation (4.11) can be used here. We have $\mathbb{V}\mathrm{iab}(T) = [B^\flat, B^\sharp]$ and:

$$\begin{aligned}
B \in \mathbb{V}\mathrm{iab}(T-1) &\iff \exists e \in [e^\flat, e^\sharp] , \quad B^\flat \le RB(1-e) \le B^\sharp , \\
&\iff \exists e \in [e^\flat, e^\sharp] , \quad \frac{B^\flat}{R(1-e)} \le B \le \frac{B^\sharp}{R(1-e)} \\
&\quad\ \text{where we used the fact that } 0 \le e \le 1 , \\
&\Rightarrow \quad \frac{B^\flat}{R(1-e^\flat)} \le B \le \frac{B^\sharp}{R(1-e^\sharp)} \\
&\quad\ \text{where we used the fact that } 0 \le e^\flat \le e \le e^\sharp \le 1 .
\end{aligned}$$

Denoting $B^\flat(T) = B^\flat$, $B^\flat(T-1) = \frac{B^\flat(T)}{R(1-e^\flat)}$, $B^\sharp(T) = B^\sharp$, $B^\sharp(T-1) = \frac{B^\sharp(T)}{R(1-e^\sharp)}$, we have thus obtained that:

$$\mathbb{V}\mathrm{iab}(T-1) \subset [B^\flat(T-1), B^\sharp(T-1)] .$$

On the other hand, we claim that whenever $B \in [B^\flat(T-1), B^\sharp(T-1)]$, there exists $e \in [e^\flat, e^\sharp]$ such that $B^\flat(T) \le RB(1-e) \le B^\sharp(T)$, thus meaning that:

$$[B^\flat(T-1), B^\sharp(T-1)] \subset \mathbb{V}\mathrm{iab}(T-1) .$$

Indeed, since $B > 0$, we have:

$$B^\flat(T) \le RB(1-e) \le B^\sharp(T) \iff 1 - \frac{B^\sharp(T)}{RB} \le e \le 1 - \frac{B^\flat(T)}{RB} .$$

Hence, our claim is proved as soon as

$$[1 - \frac{B^\sharp(T)}{RB}, 1 - \frac{B^\flat(T)}{RB}] \cap [e^\flat, e^\sharp] \neq \emptyset ,$$

which occurs precisely when $B \in [B^\flat(T-1), B^\sharp(T-1)]$, because

$$B \ge B^\flat(T-1) \Rightarrow 1 - \frac{B^\flat(T)}{RB} \ge 1 - \frac{B^\flat(T)}{RB^\flat(T-1)} = 1 - (1-e^\flat) = e^\flat$$

and, in the same way, for the other bound. For the rest of the proof, we proceed in the same way by induction.

**Proof of Result 4.11.**

*Proof.* To prove this result, we reason backward using the dynamic programming method as in (4.13). First, the viability kernel at final time $T$ is given by:

$$\mathbb{V}\text{iab}(T) = [-\infty, M^\sharp] \times \mathbb{R}_+ .$$

Assume now that at time $t+1$ the viability kernel is:

$$\mathbb{V}\text{iab}(t+1) = [-\infty, M^\sharp(t+1)] \times \mathbb{R}_+ .$$

Using the Bellman equation (4.13), we deduce that:

$$\begin{aligned}
\mathbb{V}\text{iab}(t) &= \{(M,Q) |\ \exists a \in [0,1],\ M + E(Q)(1-a) - \delta(M - M_\infty) \\
&\qquad \leq M^\sharp(t+1),\ Q(1+g) \geq 0\} \\
&= \left\{(M,Q) |\ \ Q \geq 0,\ \inf_{a\in[0,1]} \big(M + E(Q)(1-a) - \delta(M - M_\infty)\big) \leq M^\sharp(t+1)\right\} \\
&= \left\{(M,Q) |\ \ Q \geq 0,\ M - \delta(M - M_\infty) \leq M^\sharp(t+1)\right\} \\
&= \left\{(M,Q) |\ \ Q \geq 0,\ M \leq \frac{M^\sharp(t+1) - \delta M_\infty}{1-\delta}\right\} .
\end{aligned}$$

From definition of $M^\sharp(t)$ in (4.25), we deduce that

$$\frac{M^\sharp(t+1) - \delta M_\infty}{1-\delta} = M^\sharp(t) ,$$

and conclude that $\mathbb{V}\text{iab}(t) = [-\infty, M^\sharp(t)] \times \mathbb{R}_+$.

**Proof of Result 4.12.**

The proof relies on the three Lemmas A.1, A.2 and A.3 that are exposed at the end of the proof. The main proof is detailed below.

*Proof.* First assume that $h_{\text{LIM}} > h_{\text{MSE}}$. Suppose for a moment that $\mathbb{V}\text{iab} \neq \emptyset$ and pick any $B_0 \in \mathbb{V}\text{iab}$. From the very definition of the viability kernel, there exists a path $\big(B(\cdot), h(\cdot)\big)$ with $B(0) = B_0$ that remains in $\mathbb{V}\text{iab}$ in the sense that:

$$h_{\text{LIM}} \leq h(t) \leq B(t) .$$

Since $h_{\text{LIM}} > h_{\text{MSE}}$, Lemma A.1 yields $B_{\text{PA}} = +\infty$. Then, combining Lemmas A.2 and A.3 in this case, we deduce the existence of a time $T$ such that $B(T) \leq B^\flat(T) < h_{\text{LIM}}$. Consequently, a contradiction occurs. Therefore $B_0 \notin \mathbb{V}\text{iab}$ which ends this part of the proof.

Now we assume that $h_{\text{LIM}} \leq h_{\text{MSE}}$. We use the three Lemmas A.1, A.2 and A.3 to obtain the desired assertion. Following Thorem 4.8, we proceed in two steps as follows.

1. We first prove that the set $V = [B_{\text{PA}}, +\infty[$ is a viability domain and a subset of $\mathbb{V}$iab.
2. Second, we prove that $[h_{\text{LIM}}, B_{\text{PA}}[ \subset \mathbb{R}_+ \backslash \mathbb{V}\text{iab}$.

1. From Lemma A.1, we know that $B_{\text{PA}} < +\infty$ and the set $V$ makes sense. Now, consider any $B \in V$, namely $B \geq B_{\text{PA}}$. Pick the lowest admissible catch $h_{\text{LIM}}$. Since $g$ is increasing, we can write:
$$g(B - h_{\text{LIM}}, B) \geq g(B_{\text{PA}} - h^{\flat}) = B_{\text{PA}} .$$
Thus for any $B \in V$, there exists an admissible $h$ such that $g(B-h) \in V$. Consequently, $V$ is a viability domain. Since $B_{\text{PA}} > h_{\text{LIM}}$ we also have $V \subset [h_{\text{LIM}}, +\infty[$. Thus we conclude that $V \subset \mathbb{V}\text{iab}$.
2. Now consider any $B_0$ such that $h_{\text{LIM}} \leq B_0 < B_{\text{PA}}$ and assume for a moment that $B_0 \in \mathbb{V}\text{iab}$. There then exists a solution $B(\cdot)$ starting from $B_0$ that remains in $\mathbb{V}$iab. Combining Lemmas A.3 and A.2, we deduce the existence of a time $T$ such that $B(T) \leq B^{\flat}(T) < h_{\text{LIM}}$. We derive a contradiction with $B_0 \in \mathbb{V}\text{iab}$. Therefore $B_0 \notin \mathbb{V}\text{iab}$ which ends this part of the proof.

The set of viable catches corresponds to:
$$H^{\mathbb{V}\text{iab}}(B) = \left\{ h \geq h_{\text{LIM}},\ g(B-h) \in \mathbb{V}\text{iab} \right\} .$$

Given any initial state $B \in \mathbb{V}\text{iab}$, this set $H^{\mathbb{V}\text{iab}}(B)$ is not empty. Since dynamic $g$ is increasing and $B_{\text{PA}} = g(B_{\text{PA}} - h_{\text{LIM}})$, any $h^*(B) \in H^{\mathbb{V}\text{iab}}(B)$ satisfies:
$$g\big(B - h^*(B)\big) \geq g(B_{\text{PA}} - h_{\text{LIM}}) = B_{\text{PA}} .$$
Since $g$ is increasing, this means equivalently that $B - h^*(B) \geq B_{\text{PA}} - h_{\text{LIM}}$ or the desired result:
$$h_{\text{LIM}} \leq h^*(B) \leq B - B_{\text{PA}} + h_{\text{LIM}} .$$

**Lemma A.1** *We have:*
$$\inf\left\{ B,\ B \geq h_{\text{LIM}},\ g(B - h_{\text{LIM}}) \geq B \right\} = \begin{cases} +\infty & \textit{if} \ \ h_{\text{LIM}} > h_{\text{MSE}} , \\ B_{\text{PA}} < +\infty & \textit{if} \ \ h_{\text{LIM}} \leq h_{\text{MSE}} . \end{cases}$$

*Proof.* (Lemma A.1) Assume that $h_{\text{LIM}} \leq h_{\text{MSE}}$. Consider the set:
$$\mathcal{A} = \{B,\ B \geq h_{\text{LIM}},\ g(B - h_{\text{LIM}}) \geq B\} . \tag{A.1}$$
This set $\mathcal{A}$ is not empty because $B_{\text{MSE}}$ belongs to it. Indeed, $B_{\text{MSE}} \geq h_{\text{MSE}} \geq h_{\text{LIM}}$. Moreover since $h_{\text{LIM}} \leq h_{\text{MSE}}$ and $g$ increasing, we can write that:
$$g(B_{\text{MSE}} - h_{\text{LIM}}) \geq g(B_{\text{MSE}} - h_{\text{MSE}}) = B_{\text{MSE}} .$$
Furthermore $\mathcal{A}$ is bounded from below by $h_{\text{LIM}}$ which implies the existence of the minimum for this set. Such a minimum lies on the lower boundary of the set and satisfies the equality $g(B - h_{\text{LIM}}) = B$. Thus $B_{\text{PA}}$ exists.

**Lemma A.2** *Consider any* $B_0$ *such that* $h_{\text{LIM}} \leq B_0 < B_{\text{PA}}$ *and*

$$M = \max_{B,\, h_{\text{LIM}} \leq B \leq B_0} \Big( g(B - h_{\text{LIM}}) - B \Big) .$$

*Then* $M < 0$.

*Proof.* (Lemma A.2) The maximum is achieved because $g$ is continuous. Assume for a moment that $M \geq 0$. There then exists $B^* \in [h_{\text{LIM}}, B_0]$ such that:

$$g(B^* - h_{\text{LIM}}) - B^* \geq 0 .$$

This situation implies that $B^* \in \mathcal{A}$ where $\mathcal{A}$ is the set defined in (A.1). Thus:

$$B^* \geq \min_{B \in \mathcal{A}} B = B_{\text{PA}} .$$

We derive a contradiction since $B^* \leq B_0 < B_{\text{PA}}$.

**Lemma A.3** *Consider* $B^\flat(t)$ *the solution of* $B^\flat(t+1) = g(B^\flat(t) - h_{\text{LIM}})$ *starting from* $B_0$ *with* $h_{\text{LIM}} \leq B_0 < B_{\text{PA}}$. *Then, for any* $t \geq 0$, *and any admissible solution* $B(\cdot)$, *we have:*

$$B(t) \leq B^\flat(t) \leq B_0 + tM \leq B_0 .$$

*Proof.* (Lemma A.3) Recursive proof. At time $t = 0$, we have:

$$B_0 = B(0) \leq B^\flat(0) \leq B_0 + tM = B_0 \leq B_0 .$$

Assume the relation holds true at time $t$. Then:

$$B^\flat(t+1) = B^\flat(t) + g(B^\flat(t) - h_{\text{LIM}}) - B^\flat(t) \leq B^\flat(t) + M \leq B_0 + (t+1)M \leq B_0 .$$

Moreover, since $g$ is increasing, for any admissible $h(t) \geq h_{\text{LIM}}$ and $B(t) \leq B^\flat(t)$, we claim

$$B(t+1) = g\big(B(t) - h(t)\big) \leq g(B^\flat(t) - h_{\text{LIM}}) = B^\flat(t+1) ,$$

thus concluding the proof.

## A.3 Mathematical proofs of Chap. 5

### Proof of Proposition 5.4

*Proof.* In Proposition 4.4, let us recall (4.16)

$$\mathcal{T}^{ad}(t,x) = \left\{ \big(x(\cdot), u(\cdot)\big) \;\middle|\; \begin{array}{c} x(t) = x, \\ x(t+1) = F\big(t, x(t), u(t)\big),\ t = t, \ldots, T-1 \\ u(t) \in \mathbb{B}^{\text{viab}}\big(t, x(t)\big), \quad t = t, \ldots, T-1 \end{array} \right\},$$

which gives

$$\big(x(\cdot),u(\cdot)\big) \in \mathcal{T}^{ad}(t,x) \iff \begin{cases} x(t) = x \in \mathbb{A}(t) \\ u(t) \in \mathbb{B}^{\text{viab}}(t,x) \\ \big(x(\cdot),u(\cdot)\big) \in \mathcal{T}^{ad}\Big(t+1, F\big(t,x,u(t)\big)\Big) . \end{cases}$$

In the above expression, we may have $\mathcal{T}^{ad}(t,x) = \emptyset$, as well as $\mathbb{B}^{\text{viab}}(t,x) = \emptyset$ (that is, $x \notin \mathbb{V}\text{iab}(t)$).

We have:

$$\begin{aligned} V(t,x) &= \sup_{\big(x(\cdot),u(\cdot)\big)\in\mathcal{T}^{ad}(t,x)} \Big( \sum_{s=t}^{T-1} L\big(s,x(s),u(s)\big) + M\big(T,x(T)\big) \Big) \\ &= \sup_{\begin{cases} x(t) = x \in \mathbb{A}(t) \\ u(t) \in \mathbb{B}(t,x) \end{cases}} \sup_{\big(x(\cdot),u(\cdot)\big)\in\mathcal{T}^{ad}\Big(t+1,F\big(t,x,u(t)\big)\Big)} \\ &\quad \Big( L\big(t,x(t),u(t)\big) + \sum_{s=t+1}^{T-1} L\big(s,x(s),u(s)\big) + M\big(T,x(T)\big) \Big) \\ &= \sup_{u\in\mathbb{B}(t,x), x\in\mathbb{A}(t)} \Big( L(t,x,u) \\ &\quad + \sup_{\big(x(\cdot),u(\cdot)\big)\in\mathcal{T}^{ad}\big(t+1,F(t,x,u)\big)} \sum_{s=t+1}^{T-1} L\big(s,x(s),u(s)\big) + M\big(T,x(T)\big) \Big) \\ &= \sup_{u\in\mathbb{B}(t,x), x\in\mathbb{A}(t)} \Big( L(t,x,u) + V\big(t+1,F(t,x,u)\big) \Big) . \end{aligned}$$

Then we distinguish the cases $x \in \mathbb{V}\text{iab}(t)$ and $x \notin \mathbb{V}\text{iab}(t)$.

## Proof of Proposition 5.5

Just adapt the proof of Proposition 5.4 given above.

## Proof of Proposition 5.20

*Proof.* We proceed similarly to the additive case. Notice that by changing $T$ into $T+1$ and defining

$$L(T,x,u) = M\big(T,x\big) ,$$

the minimax with final payoff may be interpreted as one without payoff on a longer time horizon. We have[1]:

[1] From the relation $\sup_{z\in Z} \min\big(a, b(z)\big) = \min\big(a, \sup_{z\in Z} b(z)\big)$.

$$
\begin{aligned}
V(t,x) &= \sup_{\big(x(\cdot),u(\cdot)\big)\in\mathcal{T}^{ad}(t,x)} \; \min_{s=t,\dots,T-1} \Big(L\big(s,x(s),u(s)\big)\Big) \\
&= \sup_{\begin{cases} x(t)=x\in\mathbb{A}(t) \\ u(t)\in\mathbb{B}(t,x)\end{cases}} \; \sup_{\big(x(\cdot),u(\cdot)\big)\in\mathcal{T}^{ad}\big(t+1,F\big(t,x,u(t)\big)\big)} \\
&\quad \min\left(L(t,x,u), \min_{s=t+1,\dots,T-1}\Big(L\big(s,x(s),u(s)\big)\Big)\right) \\
&= \sup_{u\in\mathbb{B}(t,x),x\in\mathbb{A}(t)} \min\Bigg(L(t,x,u), \\
&\quad \sup_{\big(x(\cdot),u(\cdot)\big)\in\mathcal{T}^{ad}\big(t+1,F\big(t,x,u(t)\big)\big)} \min_{s=t+1,\dots,T-1}\Big(L\big(s,x(s),u(s)\big)\Big)\Bigg) \\
&= \sup_{u\in\mathbb{B}(t,x),x\in\mathbb{A}(t)} \min\Big(L(t,x,u), V\big(t+1,F(t,x,u)\big)\Big) \, .
\end{aligned}
$$

Then we distinguish the cases $x \in \mathbb{V}\mathrm{iab}(t)$ and $x \notin \mathbb{V}\mathrm{iab}(t)$.

## Proof of Proposition 5.13

*Proof.* This proof is by backward induction. At time $T$, according to equation (5.40) for the adjoint state, the equality $q^\star(T-1) = (\frac{\partial M}{\partial x})'(T, x^\star(T))$ holds true. On the other hand, we have $V(T,x) = M(T,x)$, according to induction (5.16). Thus:

$$
q^\star(T-1) = (\frac{\partial M}{\partial x})'(T, x^\star(T)) = (\frac{\partial V}{\partial x})'(T, x^\star(T)) \, .
$$

By assumption, $\mathrm{u}^\star\big(t, x^\star(t)\big)$ in (5.20) is unique. Thus, we may apply the Danskin theorem[2] to (5.16) at point $x = x^\star(t)$:

$$
\begin{aligned}
\frac{\partial V}{\partial x}\big(t, x^\star(t)\big) &= \frac{\partial L}{\partial x}\Big(t, x^\star(t), \mathrm{u}^\star\big(t, x^\star(t)\big)\Big) \\
&\quad + \frac{\partial V}{\partial x}\left(t+1, F\Big(t, x^\star(t), \mathrm{u}^\star\big(t, x^\star(t)\big)\Big)\right)\frac{\partial F}{\partial x}\Big(t, x^\star(t), \mathrm{u}^\star\big(t, x^\star(t)\big)\Big) \, .
\end{aligned}
$$

Using the induction hypothesis

---

[2] We recall here the Danskin theorem (1966) that expresses the derivative of the superior hull of functions in a particular case [3].

**Theorem A.4.** *Let $V$ be a compact of $\mathbb{R}^m$ and $f : \mathbb{R}^n \times V \longrightarrow \mathbb{R}$, jointly continuous, $C^1$ with respect to the first variable. Let us denote $g : \mathbb{R}^n \to \mathbb{R}$, $x \mapsto \max_{y\in V} f(x,y)$ and $\widehat{V}(x) = \{y \in V, f(x,y) = g(x)\}$. Then, $\forall h \in \mathbb{R}^n$, $g$ has a directional derivative in the direction $h$ given by $Dg(x;h) = \max_{y\in\widehat{V}(x)} \langle \frac{\partial f}{\partial x}(x,y), h\rangle$.*

$$q^\star(t) = (\frac{\partial V}{\partial x})'(t+1, x^\star(t+1)) ,$$

we can deduce that

$$\frac{\partial V}{\partial x}(t, x^\star(t)) = \frac{\partial L}{\partial x}\Big(t, x^\star(t), \mathrm{u}^\star\big(t, x^\star(t)\big)\Big) + q^\star(t)'\frac{\partial F}{\partial x}\Big(t, x^\star(t), \mathrm{u}^\star\big(t, x^\star(t)\big)\Big) ,$$

and thus

$$q^\star(t-1) = \frac{\partial L}{\partial x}\Big(t, x^\star(t), \mathrm{u}^\star\big(t, x^\star(t)\big)\Big) + q^\star(t)'\frac{\partial F}{\partial x}\Big(t, x^\star(t), \mathrm{u}^\star\big(t, x^\star(t)\big)\Big) .$$

**Proof of Proposition 5.22**

*Proof.* We proceed in two steps. First, consider the initial time $t_0$ and the initial state $x_0$. Pick some $L^\flat$ such that $x_0 \in \mathbb{V}\mathrm{iab}(t_0, L^\flat)$. From the very definition of the viability kernel, this implies the existence of a feasible path $\big(\widetilde{x}(\cdot), \widetilde{u}(\cdot)\big)$ such that,

$$\begin{cases} \widetilde{x}(t+1) = F\big(t, \widetilde{x}(t), \widetilde{u}(t)\big) , \quad t = t_0, \ldots, T-1 \\ \widetilde{x}(t_0) = x_0 \\ \widetilde{x}(t) \in \mathbb{A}(t) , \quad t = t_0, \ldots, T \\ \widetilde{u}(t) \in \mathbb{B}\big(t, x(t)\big) , \quad t = t_0, \ldots, T-1 \\ L\big(t, \widetilde{x}(t), \widetilde{u}(t)\big) \geq L^\flat , \quad t = t_0, \ldots, T-1 . \end{cases} \tag{A.2}$$

We thus have $\inf_{t=t_0,\ldots,T-1} L\big(t, \widetilde{x}(t), \widetilde{u}(t)\big) \geq L^\flat$ and we deduce that:

$$V(t_0, x_0) = \sup_{\big(\widetilde{x}(\cdot), \widetilde{u}(\cdot)\big) \text{ satisfying } (A.2)} \inf_{t=t_0,\ldots,T-1} L\big(t, \widetilde{x}(t), \widetilde{u}(t)\big) \geq L^\flat .$$

Since the inequality holds for any $L^\flat$, this leads to:

$$V(t_0, x_0) \geq \sup\{L^\flat \in \mathbb{R} \mid x_0 \in \mathbb{V}\mathrm{iab}(t_0, L^\flat)\} .$$

Conversely, from the definition of the value function $V(t_0, x_0)$, for any $n \in \mathbb{N}$, there exists an admissible (satisfying (A.2)) and maximizing sequence $\big(x_n(\cdot), u_n(\cdot)\big)_{n \geq 1}$ in the sense that:

$$V(t_0, x_0) \geq \inf_{t=t_0,\ldots,T-1} L(t, x_n(t), u_n(t)) \geq V(t_0, x_0) - \frac{1}{n} .$$

This situation implies $x_0 \in \mathbb{V}\mathrm{iab}\Big(t_0, V(t_0, x_0) - \frac{1}{n}\Big)$ which leads to

$$V(t_0, x_0) - \frac{1}{n} \leq \sup\{L^\flat \in \mathbb{R} \mid x_0 \in \mathbb{V}\mathrm{iab}(t_0, L^\flat)\}$$

and finally:

$$V(t_0, x_0) \leq \sup\{L^\flat \in \mathbb{R} \mid x_0 \in \mathbb{V}\mathrm{iab}(t_0, L^\flat)\} .$$

Hence the equality holds true.

## A.4 Robust and stochastic dynamic programming equations

We follow [2] for the introduction of the so-called *fear operator*, and for the parallel treatment with the stochastic case.

### Fear and expectation operators

Let $\overline{\mathbb{R}} = \mathbb{R} \cup \{-\infty\} \cup \{+\infty\}$. Consider a general set $\Omega$. The so-called *fear operator* $\mathbb{F}_\Omega$ on $\Omega$ is defined on the set of functions $A : \Omega \to \overline{\mathbb{R}}$ by:

$$\mathbb{F}_\Omega[A] = \mathbb{F}_\omega\left[A(\omega)\right] := \inf_{\omega \in \Omega} A(\omega) . \tag{A.3}$$

When $\Omega = \Omega_1 \times \Omega_2$, we have the formula:

$$\mathbb{F}_\Omega[A] = \mathbb{F}_{(\omega_1,\omega_2)}\left[A(\omega_1,\omega_2)\right] = \mathbb{F}_{\omega_1}\left[\mathbb{F}_{\omega_2}\left[A(\omega_1,\omega_2)\right]\right] . \tag{A.4}$$

Consider a probability space $\Omega$ with $\sigma$-field $\mathcal{F}$ and probability $\mathbb{P}$. The so-called *expectation operator* $\mathbb{E}_{(\Omega,\mathcal{F},\mathbb{P})}$ is defined on the set of measurable and integrable functions $A : \Omega \to \overline{\mathbb{R}}$ by:

$$\mathbb{E}_{(\Omega,\mathcal{F},\mathbb{P})}[A] = \mathbb{E}_\omega\left[A(\omega)\right] = \mathbb{E}_\mathbb{P}\left[A(\omega)\right] = \int_\Omega A(\omega)\, d\mathbb{P}(\omega) . \tag{A.5}$$

When $\Omega = \Omega_1 \times \Omega_2$, $\mathcal{F} = \mathcal{F}_1 \otimes \mathcal{F}_2$ and $\mathbb{P} = \mathbb{P}_1 \otimes \mathbb{P}_2$, we have the Fubini formula:

$$\mathbb{E}_{(\Omega,\mathcal{F},\mathbb{P})}[A] = \mathbb{E}_{(\Omega_1,\mathcal{F}_1,\mathbb{P}_1)}\left[\mathbb{E}_{(\Omega_2,\mathcal{F}_2,\mathbb{P}_2)}\left[A(\omega_1,\omega_2)\right]\right] . \tag{A.6}$$

Let $\mathbb{G}$ denote either $\mathbb{F}$ or $\mathbb{E}$ depending on the context. When $\Omega = \Omega_1 \times \Omega_2$, for an adequate function $A$, we have

$$\mathbb{G}_\Omega[A] = \mathbb{G}_{(\omega_1,\omega_2)}\left[A(\omega_1,\omega_2)\right] = \mathbb{G}_{\omega_1}\left[\mathbb{G}_{\omega_2}\left[A(\omega_1,\omega_2)\right]\right] . \tag{A.7}$$

### Maximization problem

In this proof, we shall stress the dependency of the criterion upon the initial time by considering a *criterion* $\pi$ as a function $\pi : \mathbb{N} \times \mathbb{X}^{T+1} \times \mathbb{U}^T \times \mathbb{W}^{T+1}$. The difference with the definition in Sect. 6.1 is purely notational.

Let us call a criterion $\pi$ in the *Whittle form* [6] whenever it is given by a backward induction of the form:

$$\begin{cases} \pi\big(t, x(\cdot), u(\cdot), w(\cdot)\big) = \psi\Big(t, x(t), u(t), w(t), \pi\big(t+1, x(\cdot), u(\cdot), w(\cdot)\big)\Big) , \\ \qquad\qquad t = t_0, \ldots, T-1 , \\ \pi\big(T, x(\cdot), u(\cdot), w(\cdot)\big) = M(T, x(T), w(T)) . \end{cases} \tag{A.8}$$

The function $\psi : \{t_0, \dots, T-1\} \times \mathbb{X} \times \mathbb{U} \times \mathbb{W} \times \overline{\mathbb{R}} \to \overline{\mathbb{R}}$ is assumed to be $\mathbb{G}$-linear in its last argument in the sense that:

$$\mathbb{G}_{w(t),w(t+1),\dots,w(T)} \left[\psi\left(t, x, u, w(t), A\left(w(t+1), \dots, w(T)\right)\right)\right] =$$

$$\mathbb{G}_{w(t)} \left[\psi\left(t, x, u, w, \mathbb{G}\left[A\left(w(t+1), \dots, w(T)\right)\right]\right)\right] . \tag{A.9}$$

We illustrate what means to be $\mathbb{G}$-linear when $\mathbb{G}$ is either $\mathbb{F}$ or $\mathbb{E}$.

- When $\mathbb{G}$ is the fear operator $\mathbb{F}$, $\psi$ is assumed to be continuously increasing in its last argument[3]. This form is adapted to maximin dynamic programming with $\psi(t, x, u, w, C) = \min(L(t, x, u, w), C)$ and includes both the additive case for which $\psi(t, x, u, w, C) = L(t, x, u, w) + C$ and the multiplicative case for which $\psi(t, x, u, w, C) = L(t, x, u, w) \times C$.
- When $\mathbb{G}$ is the expectation operator $\mathbb{E}$, $\psi(t, x, u, w, C) = g(t, x, u, w) + \beta(t, x, u, w)C$. This form includes the additive and the multiplicative cases.

We consider the dynamics (6.1)

$$x(t+1) = F\big(t, x(t), u(t), w(t)\big) , \quad t = t_0, \dots, T-1 \quad \text{with} \quad x(t_0) = x_0, \tag{A.10}$$

the control constraints (6.5a)

$$u(t) \in \mathbb{B}\big(t, x(t)\big) \subset \mathbb{U}, \tag{A.11}$$

and no state constraints ($\mathbb{A}(t) = \mathbb{X}$ for $t = t_0, \dots, T$). Recall the admissible feedbacks set (6.12)

$$\mathfrak{U}^{ad} = \{\mathfrak{u} \in \mathfrak{U} \mid \mathfrak{u}(t, x) \in \mathbb{B}(t, x) , \quad \forall (t, x)\} , \tag{A.12}$$

and, for any admissible decision strategy $\mathfrak{u} \in \mathfrak{U}^{ad}$, the evaluation (8.4) of the criterion

$$\pi^{\mathfrak{u}}\left(t_0, x_0, w(\cdot)\right) := \pi\big(t_0, x_F[t_0, x_0, \mathfrak{u}, w(\cdot)](\cdot), u_F[t_0, x_0, \mathfrak{u}, w(\cdot)](\cdot), w(\cdot)\big) \tag{A.13}$$

where $t_0 \in \{0, \dots, T-1\}$, $x_0 \in \mathbb{X}$, $w(\cdot) \in \Omega$, with

$$\Omega = \mathbb{S}(t_0) \times \dots \times \mathbb{S}(T) ,$$

and $x_F$, $u_F$ are the solution maps of Sect. 6.2. Define

$$\pi^{\mathfrak{u}}_{\mathbb{G}}(t_0, x_0) := \mathbb{G}_{w(\cdot)} \left[\pi^{\mathfrak{u}}\big(t_0, x_0, w(\cdot)\big)\right] , \tag{A.14}$$

and consider the maximization problem:

$$\pi^{\star}_{\mathbb{G}}(t_0, x_0) := \sup_{\mathfrak{u} \in \mathfrak{U}^{ad}} \pi^{\mathfrak{u}}_{\mathbb{G}}(t_0, x_0) . \tag{A.15}$$

[3] $\psi(t, x, u, w, C^\sharp) \geq \psi(t, x, u, w, C^\flat)$ whenever $-\infty \leq C^\flat \leq C^\sharp \leq +\infty$, and $C_n \to C \Rightarrow \psi(t, x, u, w, C_n) \to \psi(t, x, u, w, C)$.

## Dynamic programming equation

**Definition A.5.** *The* value function $V(t,x)$, *associated with the Whittle criterion (A.8), the dynamics (A.10), the control constraints (A.11) and no state constraints ($\mathbb{A}(t) = \mathbb{X}$ for $t = t_0, \ldots, T$) is defined by the following backward induction, where $t$ runs from $T-1$ down to $t_0$:*

$$\begin{cases} V(T,x) := \mathbb{G}_{w \in \mathbb{S}(T)} \left[ M(T,x,w) \right] , \\ V(t,x) := \sup_{u \in \mathbb{B}(t,x)} \mathbb{G}_{w \in \mathbb{S}(t)} \Big[ \psi\Big(t,x,u,w,V\big(t+1, F(t,x,u,w)\big)\Big) \Big] . \end{cases} \tag{A.16}$$

**Proposition A.6.** *Assume that $\mathbb{A}(t) = \mathbb{X}$ for $t = t_0, \ldots, T$. For any time $t$ and state $x$, assume the existence of the following feedback decision:*

$$\mathfrak{u}^\star(t,x) \in \arg\max_{u \in \mathbb{B}(t,x)} \mathbb{G}_{w \in \mathbb{S}(t)} \Big[ \psi\Big(t,x,u,w,V\big(t+1, F(t,x,u,w)\big)\Big) \Big] . \tag{A.17}$$

*Then $\mathfrak{u}^\star \in \mathfrak{U}^{ad}$ is an optimal feedback of the maximization problem (A.15).*

*Proof.* Let $\mathfrak{u}^\star \in \mathfrak{U}^{ad}$ denote one of the optimal feedback strategies given by dynamic programming (A.17). We perform a backward induction to prove (A.15).

First, the equality at $t = T$ holds true since:

$$\begin{aligned} \pi_{\mathbb{G}}^{\mathfrak{u}^\star}(T,x) &= \mathbb{G}_{w(\cdot) \in \Omega} \Big[ \pi^{\mathfrak{u}^\star}\big(T,x,w(\cdot)\big) \Big] \quad \text{by (A.14)} \\ &= \mathbb{G}_{w(\cdot) \in \mathbb{S}(t_0) \times \cdots \times \mathbb{S}(T)} \Big[ \pi^{\mathfrak{u}^\star}\big(T,x,w(\cdot)\big) \Big] \\ &= \mathbb{G}_{w \in \mathbb{S}(T)} \left[ M(T,x,w) \right] \quad \text{by (A.16)} \\ &= V(T,x) \quad \text{by (A.16).} \end{aligned}$$

Now, suppose that:

$$\pi_{\mathbb{G}}^{\mathfrak{u}^\star}(t+1,x) = \sup_{\mathfrak{u} \in \mathfrak{U}^{ad}} \pi_{\mathbb{G}}^{\mathfrak{u}}(t+1,x) = V(t+1,x) . \tag{A.18}$$

The very definition (A.16) of the value function $V$ combined with (A.19) in Lemma A.7 (proved below) implies that:

$$\begin{aligned} \pi_{\mathbb{G}}^{\mathfrak{u}^\star}(t,x) &= \mathbb{G}_{w \in \mathbb{S}(t)} \Big[ \psi\Big(t,x,w,\mathfrak{u}^\star(t,x), \pi_{\mathbb{G}}^{\mathfrak{u}^\star}\big(t+1, F(t,x,\mathfrak{u}^\star(t,x),w)\big)\Big) \Big] \text{ by (A.19)} \\ &= \mathbb{G}_{w \in \mathbb{S}(t)} \Big[ \psi\Big(t,x,w,\mathfrak{u}^\star(t,x), V\big(t+1, F(t,x,\mathfrak{u}^\star(t,x),w)\big)\Big) \Big] \text{ by (A.18)} \\ &= \max_{u \in \mathbb{B}(t,x)} \mathbb{G}_{w \in \mathbb{S}(t)} \Big[ \psi\Big(t,x,u,w,V\big(t+1, F(t,x,u,w)\big)\Big) \Big] \text{ by (A.17)} \\ &= V(t,x) \quad \text{by (A.16).} \end{aligned}$$

Furthermore, for any $\mathfrak{u} \in \mathfrak{U}^{ad}$, we obtain:

$$
\begin{aligned}
\pi_{\mathbb{G}}^{\mathfrak{u}^\star}(t,x) &= \mathbb{G}_{w\in\mathbb{S}(t)}\Big[\psi\Big(t,x,w,\mathfrak{u}^\star(t,x),\pi_{\mathbb{G}}^{\mathfrak{u}^\star}\big(t+1,F(t,x,\mathfrak{u}^\star(t,x),w)\big)\Big)\Big] \text{ by (A.19)}\\
&\geq \mathbb{G}_{w\in\mathbb{S}(t)}\Big[\psi\Big(t,x,w,\mathfrak{u}(t,x),V\big(t+1,F(t,x,\mathfrak{u}(t,x),w)\big)\Big)\Big] \quad \text{by (A.18)}\\
&\geq \max_{u\in\mathbb{B}(t,x)} \mathbb{G}_{w\in\mathbb{S}(t)}\Big[\psi\Big(t,x,u,w,V\big(t+1,F(t,x,u,w)\big)\Big)\Big]\\
&\qquad \text{since} \quad \mathfrak{u}\in\mathfrak{U}^{ad} \quad \text{hence} \quad \mathfrak{u}(t,x)\in\mathbb{B}(t,x)\\
&= V(t,x) \quad \text{by (A.16)}.
\end{aligned}
$$

Consequently, the desired statement is obtained since

$$
\pi_{\mathbb{G}}^{\mathfrak{u}^\star}(t,x) = V(t,x) \leq \sup_{\mathfrak{u}\in\mathfrak{U}^{ad}} \pi_{\mathbb{G}}^{\mathfrak{u}}(t,x)
$$

yields the equality:

$$
V(t,x) = \pi_{\mathbb{G}}^{\mathfrak{u}^\star}(t,x) = \max_{\mathfrak{u}\in\mathfrak{U}^{ad}} \pi_{\mathbb{G}}^{\mathfrak{u}}(t,x)\,.
$$

**Lemma A.7.** *We have, for* $t = t_0,\dots,T-1$ *and* $\mathfrak{u}\in\mathfrak{U}$*:*

$$
\begin{cases}
\pi_{\mathbb{G}}^{\mathfrak{u}}(T,x) = \mathbb{G}_{w\in\mathbb{S}(T)}\left[M(T,x,w)\right]\\
\pi_{\mathbb{G}}^{\mathfrak{u}}(t,x) = \mathbb{G}_{w\in\mathbb{S}(t)}\left[\psi\left(t,x,w,\mathfrak{u}(t,x),\pi_{\mathbb{G}}^{\mathfrak{u}}\left(t+1,F(t,x,\mathfrak{u}(t,x),w)\right)\right)\right]\,.
\end{cases}
\tag{A.19}
$$

*Proof. (Lemma A.7.)* By (A.8) and (A.13), we have:

$$
\begin{cases}
\pi^{\mathfrak{u}}\big(T,x,w(\cdot)\big) = M\big(T,x,w(T)\big)\\
\pi^{\mathfrak{u}}\big(t,x,w(\cdot)\big) = \psi\Big(t,x,w(t),\mathfrak{u}(t,x),\pi^{\mathfrak{u}}\big(t+1,F(t,x,\mathfrak{u}(t,x),w(t)),w(\cdot)\big)\Big)\,.
\end{cases}
\tag{A.20}
$$

Notice that, according to (6.13), $\pi^{\mathfrak{u}}\big(t,x,w(\cdot)\big)$ depends in fact only upon $\big(w(t),\dots,w(T-1)\big)$ insofar as $w(\cdot)$ is concerned:

$$
\pi^{\mathfrak{u}}\big(t,x,w(\cdot)\big) = \pi^{\mathfrak{u}}\Big(t,x,\big(w(t),\dots,w(T-1)\big)\Big)\,. \tag{A.21}
$$

We have:

$$\begin{aligned}
\pi^{\mathrm{u}}_{\mathbb{G}}(t,x) &= \mathbb{G}_{w(\cdot)\in\Omega}\left[\pi^{\mathrm{u}}(t,x,w(\cdot))\right] \quad \text{by (A.14)} \\
&= \mathbb{G}_{w(\cdot)\in\Omega}\left[\psi\left(t,x,w(t),\pi^{\mathrm{u}}\left(t+1,F(t,x,\mathrm{u}(t,x),w(t)),w(\cdot)\right)\right)\right] \\
&\qquad \text{by (A.20)} \\
&= \mathbb{G}_{w(\cdot)\in\mathbb{S}(t)\times\cdots\times\mathbb{S}(T)}\Big[\psi\big(t,x,w(t),\pi^{\mathrm{u}}(t+1,F(t,x,\mathrm{u}(t,x),w(t)), \\
&\quad w(t+1),\dots,w(T-1))\big)\Big] \text{ by (A.21)} \\
&= \mathbb{G}_{w(t)\in\mathbb{S}(t)}\Big[\mathbb{G}_{(w(t+1),\dots,w(T-1))\in\mathbb{S}(t+1)\times\cdots\times\mathbb{S}(T)}\big[ \\
&\quad \psi\left(t,x,w(t),\pi^{\mathrm{u}}\left(t+1,F(t,x,\mathrm{u}(t,x),w(t)),w(t+1),\dots,w(T-1)\right)\right)\big]\Big] \\
&\quad \text{by (A.7)} \\
&= \mathbb{G}_{w\in\mathbb{S}(t)}\Big[\psi\left(t,x,\mathrm{u}(t,x),w,\mathbb{G}_{(w(t+1),\dots,w(T-1))\in\mathbb{S}(t+1)\times\cdots\times\mathbb{S}(T)}\right. \\
&\quad \left.\times\left[\pi^{\mathrm{u}}\left(t+1,F(t,x,\mathrm{u}(t,x),w(t+1),\dots,w(T-1)\right)\right]\right)\Big] \\
&\quad \text{by (A.9) because } \psi \text{ is } \mathbb{G}\text{-linear in its last argument} \\
&= \mathbb{G}_{w\in\mathbb{S}(t)}\Big[\psi\Big(t,x,w,\mathrm{u}(t,x),\mathbb{G}_{w(\cdot)\in\mathbb{S}(t)\times\cdots\times\mathbb{S}(T)} \\
&\quad \times\Big[\pi^{\mathrm{u}}\Big(t+1,F\big(t,x,\mathrm{u}(t,x),w(\cdot)\big)\Big)\Big]\Big)\Big] \quad \text{by (A.21)} \\
&= \mathbb{G}_{w\in\mathbb{S}(t)}\Big[\psi\left(t,x,w,\mathrm{u}(t,x),\pi^{\mathrm{u}}_{\mathbb{G}}(t+1,F(t,x,\mathrm{u}(t,x),w))\right)\Big] \quad \text{by (A.14).}
\end{aligned}$$

## A.5 Mathematical proofs of Chap. 7

### Proof of Proposition 7.5

*Proof.* Apply the generic Proposition A.6 with the fear operator $\mathbb{G} = \mathbb{F}_\Omega$ and with $\psi(t,x,u,w,C) = \mathbf{1}_{\mathbb{A}(t)}(x) \times C$ and:

$$\pi\big(t_0,x(\cdot),u(\cdot),w(\cdot)\big) = \prod_{t=t_0}^{T} \mathbf{1}_{\mathbb{A}(t)}(x(t)) \,.$$

Notice that:

$$\mathbb{V}\mathrm{iab}_1(t) = \left\{ x_0 \in \mathbb{X} \;\middle|\; \sup_{u(\cdot)} \inf_{w(\cdot)} \pi\big(t_0,x(\cdot),u(\cdot),w(\cdot)\big) = 1 \right\} .$$

### Proof of Proposition 7.10

*Proof.* Apply the generic Proposition A.6 with the expectation operator $\mathbb{G} = \mathbb{E}_{w(\cdot)}$ and with $\psi(t,x,u,w,C) = \mathbf{1}_{\mathbb{A}(t)}(x) \times C$ and:

$$\pi\big(t_0, x(\cdot), u(\cdot), w(\cdot)\big) = \prod_{t=t_0}^{T} \mathbf{1}_{\mathbb{A}(t)}(x(t)) .$$

Notice that:

$$\mathrm{Viab}_\beta(t) = \left\{ x_0 \in \mathbb{X} \;\middle|\; \sup_{u(\cdot)} \mathbb{E}_{w(\cdot)} \big[\pi\big(t_0, x(\cdot), u(\cdot), w(\cdot)\big)\big] \geq \beta \right\} .$$

## A.6 Mathematical proofs of Chap. 8

### Proof of Proposition 8.4

*Proof.* Apply the generic Proposition A.6 with the fear operator $\mathbb{G} = \mathbb{F}_\Omega$ and with $\psi(t, x, u, w, C) = L(t, x, u, w) + C$.

### Proof of Proposition 8.6

*Proof.* Adapt the previous proof of Proposition 8.4 with

$$\widetilde{\psi}(t, x, u, w, C) = \begin{cases} \psi(t, x, u, w, C) & \text{if } x \in \mathbb{A}(t) \\ -\infty & \text{if not,} \end{cases}$$

and

$$\widetilde{M}(T, x, w) = \begin{cases} M(t, x, w) & \text{if } x \in \mathbb{A}(T) \\ -\infty & \text{if not.} \end{cases}$$

Under the additional technical assumptions $\inf_{x,u,w} L\big(t, x, u, w\big) > -\infty$ and $\inf_{x,w} M\big(T, x, w\big) > -\infty$, we prove that

$$x \in \mathrm{Viab}_1(t) \iff V(t, x) > -\infty .$$

### Proof of Proposition 8.8

*Proof.* Apply the generic Proposition A.6 with the fear operator $\mathbb{G} = \mathbb{F}_\Omega$ and with $\psi(t, x, u, w, C) = \min(L(t, x, u, w), C)$.

### Proof of Proposition 8.12

*Proof.* Apply the generic Proposition A.6 with the expectation operator $\mathbb{G} = \mathbb{E}_{w(\cdot)}$ and with $\psi(t, x, u, w, C) = L(t, x, u, w) + C$.

**Proof of Proposition 8.14**

*Proof.* Adapt the previous proof of Proposition 8.12 with

$$\widetilde{\psi}(t,x,u,w,C) = \begin{cases} \psi(t,x,u,w,C) & \text{if } x \in \mathbb{A}(t) \\ -\infty & \text{if not,} \end{cases}$$

and

$$\widetilde{M}(T,x,w) = \begin{cases} M(t,x,w) & \text{if } x \in \mathbb{A}(T) \\ -\infty & \text{if not.} \end{cases}$$

## A.7 Mathematical proofs of Chap. 9

**Proof of Result 9.1.**

*Proof.* We reason backward using the dynamic programming method. We use the notation $\Psi_{\mathbb{A}}(\cdot)$ for the characteristic function of the domain $\mathbb{A}$ defined by:

$$\Psi_{\mathbb{A}}(B) = \begin{cases} 0 & \text{if} \quad B \in \mathbb{A} \\ +\infty & \text{otherwise.} \end{cases}$$

On the one hand, the $V_\beta$ value function at time 1 is given by:

$$\begin{aligned} V_\beta(1,B) &= \inf_{e\in[0,1]} \mathbb{E}_{R(1)}\Big[\rho C(1-e) + \Psi_{\mathbb{A}(2)}\left(R(1)(B(1-e))\right)\Big] \\ &= \inf_{e\in[0,1]}\left(\rho C(1-e) + \mathbb{E}_{R(1)}\left[\Psi_{\frac{\mathbb{A}(2)}{R(1)}}\left(B(1-e)\right)\right]\right). \end{aligned}$$

Using the fact that resource productivity $R(1)$ has a discrete support $\mathbb{S} = \{R^\flat, R^\sharp\}$, and that $\mathbb{P}(R(1) = R^\flat) > 0$ and $\mathbb{P}(R(1) = R^\sharp) > 0$, we write

$$\begin{aligned} \mathbb{E}_R\left[\Psi_{\frac{\mathbb{A}(2)}{R}}(B(1-e))\right] &= \mathbb{P}(R(1) = R^\flat)\Psi_{\frac{\mathbb{A}(2)}{R^\flat}}(B(1-e)) \\ &\quad + \mathbb{P}(R(1) = R^\sharp)\Psi_{\frac{\mathbb{A}(2)}{R^\sharp}}(B(1-e)) \\ &= \Psi_{\frac{\mathbb{A}(2)}{R^\flat}}(B(1-e)) + \Psi_{\frac{\mathbb{A}(2)}{R^\sharp}}(B(1-e)) \\ &= \Psi_{\frac{\mathbb{A}(2)}{R^\flat}\cap\frac{\mathbb{A}(2)}{R^\sharp}}(B(1-e)) \,. \end{aligned}$$

Thus

$$V_\beta(1,B) = \inf_{e\in[0,1]}\left(\rho C(1-e) + \Psi_{\bigcap_{R\in\mathbb{S}}\frac{\mathbb{A}(2)}{R}}\left(B(1-e)\right)\right).$$

The condition $\frac{B^\sharp}{B^\flat} < \frac{R^\sharp}{R^\flat}$ yields that:

$$\bigcap_{R\in\mathbb{S}} \frac{\mathbb{A}(2)}{R} = \bigcap_{R\in\{R^\flat,R^\sharp\}} \frac{[B^\flat, B^\sharp]}{R} = \emptyset \,.$$

Consequently, for any $B$, we obtain $\Psi_{\bigcap_{R\in\mathbb{S}} \frac{\mathbb{A}(2)}{R}}(B) = +\infty$ and thus

$$V_\beta(1, B) = +\infty ,$$

which means that the target cannot be reached no matter what the state $B$ is at time 1. By dynamic programming and backward reasoning, we deduce that $V_\beta(0, B) = +\infty$.

On the other hand, the $V_\gamma$ value function with learning at time 1 is given by:

$$\begin{aligned} V_\gamma(1, B, R) &= \inf_{e\in[0,1]} \Big( \rho C(1-e) + \Psi_{\mathbb{A}(2)}(RB(1-e))) \Big) \\ &= \inf_{\begin{cases} e \in [0,1], \\ RB(1-e) \in \mathbb{A}(2) \end{cases}} \rho C(1-e) . \end{aligned}$$

The constraint set $\{e \in [0, 1],\ RB(1-e) \in \mathbb{A}(2)\}$ is not empty if and only if $B \geq \frac{B^\flat}{R}$. In this case, since $C(1-e) = c(1-e)$ decreases with $e$ and is linear, we can write

$$\inf_{\begin{cases} e \in [0,1], \\ RB(1-e) \in \mathbb{A}(2) \end{cases}} \rho C(1-e) = \rho C(1 - \mathfrak{e}^\star(1, R, B))$$

where $\mathfrak{e}^\star(1, R, B) = 1 - \frac{B^\flat}{RB}$ because $\mathbb{A}(2) = [B^\flat, B^\sharp]$. Thus:

$$V_\gamma(1, B, R) = \rho C(1 - \mathfrak{e}^\star(1, R, B)) + \Psi_{[\frac{B^\flat}{R}, +\infty[}(B) .$$

We now have to distinguish two cases to compute

$$\mathbb{E}_{R(1)}\left[V_\gamma(1, B, R(1))\right] = \upsilon(1, B) .$$

- If $B < \frac{B^\flat}{R^\flat}$, then, since effort cost $C(\cdot)$ is positive:

$$\upsilon(1, B) \geq \Psi_{[\frac{B^\flat}{R^\flat}, +\infty[}(B) = +\infty .$$

- If $B \geq \frac{B^\flat}{R^\flat}$, then, for any $R \in \mathbb{S} = \{R^\flat, R^\sharp\}$, we have $B \geq \frac{B^\flat}{R}$ or equivalently $\Psi_{[\frac{B^\flat}{R}, +\infty[}(B) = 0$ and consequently $\upsilon(1, B) = \mathbb{E}_R\left[\rho C(1 - \mathfrak{e}^\star(1, R, B))\right]$. In that case, using the certainty equivalent $\widehat{R}^{-1} = \mathbb{E}[R^{-1}]$, we write $\upsilon(1, B) = \rho C(1 - \mathfrak{e}^\star(1, \widehat{R}, B))$.

Therefore:

$$\upsilon(1, B) = \rho C(1 - \mathfrak{e}^\star(1, \widehat{R}, B)) + \Psi_{[\frac{B^\flat}{R^\flat}, +\infty[}(B) .$$

Using the dynamic programming method and similar reasonings, we also derive the adaptive precautionary value function at initial time $t = 0$:

$$\begin{aligned}
V_\gamma(0,B) &= \inf_{e\in[0,1]} \mathbb{E}_{R(0)}\left[C(1-e)+v\left(1,R(0)B(1-e)\right)\right] \\
&= \inf_{e\in[0,1]}\left(C(1-e)+\mathbb{E}_{R(0)}\left[v(1,R(0)B(1-e))\right]\right) \\
&= \inf_{e\in[0,1]}\Big(C(1-e)+\mathbb{E}_{R(0)}\left[\rho C(1-\mathfrak{e}^\star(1,\widehat{R},R(0)B(1-e)))\right] \\
&\qquad +\mathbb{E}_{R(0)}\left[\Psi_{[\frac{B^\flat}{R^\flat},+\infty[}(R(0)B(1-e))\right]\Big) \\
&= \inf_{e\in[0,1]}\left(C(1-e)+\rho C(1-\mathfrak{e}^\star(1,\widehat{R},\widehat{R}B(1-e)))+\Psi_{[\frac{B^\flat}{R^{\flat 2}},+\infty[}(B(1-e))\right).
\end{aligned}$$

For any $B\in[\frac{B^\flat}{R^{\flat 2}},+\infty[$, we obtain:

$$V_\gamma(0,B)=\inf_{\begin{cases} e\in[0,1], \\ 1-e\geq \frac{B^\flat}{R^{\flat 2}B}\end{cases}}\left(C(1-e)+\rho C(1-\mathfrak{e}^\star(1,\widehat{R},\widehat{R}B(1-e)))\right).$$

Now we compute the optimal initial feedback $\mathfrak{e}^\star(0,\cdot)$. We combine the first order optimality conditions associated with $V_\gamma(0,B)$ and the linear costs $C(1-e)=c(1-e)$ to obtain:

$$0=c\left(-1+\frac{\rho B^\flat}{\widehat{R}^2B(1-e)^2}\right).$$

The solution is given by $\mathfrak{e}^\star(0,B)=1-\sqrt{\frac{\rho B^\flat}{\widehat{R}^2B}}$ which turns out to be an interior solution if $0<\rho<1$ and $B\in[\frac{B^\flat}{R^{\flat 2}},+\infty[$.

**Proof of Result 9.2.**

*Proof.* To solve the overall optimization problem, we first compute the second period cost minimization with and without information.

Without information, we have:

$$V_\beta(1,\widetilde{M}_1)=\inf_{0\leq a(1)\leq 1}\mathbb{E}_{\theta_0}^{\mathbb{P}_0}\left[\rho C\big(a(1)\big)+\theta(2)D\Big((1-\delta)\widetilde{M}_1+\alpha E_{\mathrm{BAU}}\big(1-a(1)\big)\Big)\right].$$

Similarly, in the case of learning about damage intensity $\theta=\theta_0$, we compute

$$V_\gamma(1,\widetilde{M}_1)=\mathbb{E}_{\theta_0}^{\mathbb{P}_0}\left[\inf_{0\leq a(1)\leq 1}\Big(\rho C\big(a(1)\big)+\theta(2)D\big((1-\delta)\,\widetilde{M}_1+\alpha E_{\mathrm{BAU}}(1-a(1))\big)\Big)\right],$$

In both expressions, appears the quantity

$$f(a_1,\theta)=C(a_1)+\theta D\Big((1-\delta)\widetilde{M}_1+\alpha E_{\mathrm{BAU}}\big(1-a_1\big)\Big),$$

with

$$V_\beta(1,\widetilde{M}_1) = \inf_{0 \le a_1 \le 1} f(a_1,\overline{\theta}) ,$$

where $\overline{\theta} = \mathbb{E}_{\theta_0}^{\mathbb{P}_0}[\theta_0]$, while

$$V_\gamma(1,\widetilde{M}_1) = \mathbb{E}_{\theta_0}^{\mathbb{P}_0}\left[\inf_{0 \le a_1 \le 1} f(a_1,\theta)\right] .$$

Since $C(a) = ca^2$ and $D(\widetilde{M}) = \widetilde{M}^2$ are quadratic, so is $a_1 \mapsto f(a_1,\theta)$, and its minimum is achieved at

$$\overline{a}_1(\theta) = \frac{\theta\alpha E_{\text{BAU}}\big((1-\delta)\widetilde{M}_1 + \alpha E_{\text{BAU}}\big)}{\rho c + \theta\alpha^2 E_{\text{BAU}}^2} ,$$

giving

$$\inf_{a_1 \in \mathbb{R}} f(a_1,\theta) = f(\overline{a}_1(\theta),\theta) = \frac{\theta\rho c}{\rho c + \theta\alpha^2 E_{\text{BAU}}^2}\big((1-\delta)\widetilde{M}_1 + \alpha E_{\text{BAU}}\big)^2 .$$

It can be proved that the function $\theta \mapsto a_1(\theta)$ is increasing. Consequently, using assumption (9.15)

$$0 = \overline{a}_1(0) \le \overline{a}_1(\theta^\flat) \le \overline{a}_1(\theta^\sharp) \le \overline{a}_1\left(\frac{\rho c}{\alpha E_{\text{BAU}}(1-\delta)\big((1-\delta)\widetilde{M}_0 + \alpha E_{\text{BAU}}\big)}\right) .$$

By virtue of inequalities

$$\widetilde{M}_1 \le (1-\delta)\widetilde{M}_0 + \alpha E_{\text{BAU}}(1-a) \le (1-\delta)\widetilde{M}_0 + \alpha E_{\text{BAU}} ,$$

we claim that

$$\overline{a}_1\left(\frac{\rho c}{\alpha E_{\text{BAU}}(1-\delta)\big((1-\delta)\widetilde{M}_0 + \alpha E_{\text{BAU}}\big)}\right) \le 1 .$$

Thus $\overline{a}_1(\theta) \in [0,1]$ for any $\theta \in [\theta^\flat, \theta^\sharp]$ and

$$\inf_{0 \le a_1 \le 1} f(a_1,\theta) = \frac{\theta\rho c}{\rho c + \theta\alpha^2 E_{\text{BAU}}^2}\big((1-\delta)\widetilde{M}_1 + \alpha E_{\text{BAU}}\big)^2 , \quad \forall\theta \in [\theta^\flat, \theta^\sharp] .$$

Thus, since the unknown parameter $\theta_0$ is supposed to follow a probability distribution $\mathbb{P}_0$ having support within $[\theta^\flat, \theta^\sharp]$, we obtain that

$$V_\beta(1,\widetilde{M}_1) = \frac{\mathbb{E}_{\theta_0}^{\mathbb{P}_0}[\theta]\,\rho c}{\rho c + \mathbb{E}_{\theta_0}^{\mathbb{P}_0}[\theta]\,\alpha^2 E_{\text{BAU}}^2}\big((1-\delta)\widetilde{M}_1 + \alpha E_{\text{BAU}}\big)^2 ,$$

and that

$$V_\gamma(1, \widetilde{M}_1) = \mathbb{E}_{\theta_0}^{\mathbb{P}_0}\left[\frac{\theta \rho c}{\rho c + \theta \alpha^2 E_{\text{BAU}}^2}\right]\left((1-\delta)\widetilde{M}_1 + \alpha E_{\text{BAU}}\right)^2 .$$

Hence, we deduce that:

$$V_\gamma(1, \widetilde{M}_1) - V_\beta(1, \widetilde{M}_1) =$$

$$\left(\mathbb{E}_{\theta_0}^{\mathbb{P}_0}\left[\frac{\theta \rho c}{\rho c + \theta \alpha^2 E_{\text{BAU}}^2}\right] - \frac{\mathbb{E}_{\theta_0}^{\mathbb{P}_0}[\theta]\, \rho c}{\rho c + \mathbb{E}_{\theta_0}^{\mathbb{P}_0}[\theta]\, \alpha^2 E_{\text{BAU}}^2}\right)\left((1-\delta)\widetilde{M}_1 + \alpha E_{\text{BAU}}\right)^2 < 0 .$$

This follows from the strict concavity of $\theta \mapsto \dfrac{\theta \rho c}{\rho c + \theta \alpha^2 E_{\text{BAU}}^2}$ and from Jensen inequality.

**Proof of Proposition 9.3.**

**Proposition A.8.** *Let $\mathcal{D} \subset \mathbb{R}$, let $g : \mathcal{D} \to \mathbb{R}$ and $h : \mathcal{D} \to \mathbb{R}$. We denote*

$$\mathcal{D}_g := \arg\max_{u \in \mathcal{D}} g(u) \subset \mathcal{D} \quad \text{and} \quad \mathcal{D}_{g+h} := \arg\max_{u \in \mathcal{D}} (g+h)(u) \subset \mathcal{D} ,$$

*and we assume that $\mathcal{D}_g \neq \emptyset$ and $\mathcal{D}_{g+h} \neq \emptyset$.*

1. *If $h$ is strictly increasing on $]-\infty, \sup \mathcal{D}_g]$, then*
$$\sup \mathcal{D}_g \leq \inf \mathcal{D}_{g+h} .$$
2. *If $h$ is increasing on $]-\infty, \sup \mathcal{D}_g]$, then*
$$\sup \mathcal{D}_g \leq \sup \mathcal{D}_{g+h} .$$
3. *If $h$ is strictly decreasing on $[\inf \mathcal{D}_g, +\infty[$, then*
$$\sup \mathcal{D}_{g+h} \leq \inf \mathcal{D}_g .$$
4. *If $h$ is decreasing on $[\inf \mathcal{D}_g, +\infty[$, then*
$$\inf \mathcal{D}_{g+h} \leq \inf \mathcal{D}_g .$$

*Proof.* We prove the first statement, the others being minor variations.

Let $u_g^\star \in \mathcal{D}_g$. For any $u \in \mathcal{D}$, we have $g(u) \leq g(u_g^\star)$. For any $u \in ]-\infty, u_g^\star[$, we have $h(u) < h(u_g^\star)$ if $h$ is strictly increasing. Thus:

$$u \in ]-\infty, u_g^\star[ \Rightarrow g(u) + h(u) < g(u_g^\star) + h(u_g^\star) .$$

We conclude that $\mathcal{D}_{g+h} \subset [u_g^\star, +\infty[$, so that:

$$\mathcal{D}_{g+h} \subset \bigcap_{u_g^\star \in \mathcal{D}_g} [u_g^\star, +\infty[ = [\sup \mathcal{D}_g, +\infty[ ,$$

thus proving that $\sup \mathcal{D}_g \leq \inf \mathcal{D}_{g+h}$.

# References


[1] K.J. Åström and B. Wittenmark. *Computer Controlled Systems: Theory and Design.* Information and Systems Sciences Series. Prentice-Hall, 1984.

[2] P. Bernhard. A separation theorem for expected value and feared value discrete time control. Technical report, INRIA, Projet Miaou, Sophia Antipolis, Décembre 1995.

[3] F. H. Clarke. *Optimization and Nonsmooth Analysis.* Classics in mathematics. SIAM, Philadelphia, 1990.

[4] E. I. Jury. *Theory and Application of the z-transform Method.* Wiley, New York, 1964.

[5] H. K. Khalil. *Nonlinear Systems.* Prentice-Hall, Englewood Cliffs, second edition, 1995.

[6] P. Whittle. *Optimization over Time: Dynamic Programming and Stochastic Control*, volume 1. John Wiley & Sons, New York, 1982.

# Index

GPSR Compliance

*The European Union's (EU) General Product Safety Regulation (GPSR) is a set of rules that requires consumer products to be safe and our obligations to ensure this.*

*If you have any concerns about our products, you can contact us on ProductSafety@springernature.com*

In case Publisher is established outside the EU, the EU authorized representative is:

Springer Nature Customer Service Center GmbH
Europaplatz 3
69115 Heidelberg, Germany

**Batch number: 10614744**

Printed by Printforce, the Netherlands